Methods in Computational
Molecular Physics

NATO ASI Series

Advanced Science Institutes Series

A series presenting the results of activities sponsored by the NATO Science Committee, which aims at the dissemation of advanced scientific and technological knowledge, with a view to strengthening links between scientific communities.

The series is published by an international board of publishers in conjunction with the NATO Scientific Affairs Division

A	Life Sciences	Plenum Publishing Corporation
B	Physics	London and New York
C	Mathematical and Physical Sciences	D. Reidel Publishing Company Dordrecht, Boston and Lancaster
D	Behavioural and Social Sciences	Martinus Nijhoff Publishers
E	Engineering and Materials Sciences	The Hague, Boston and Lancaster
F	Computer and Systems Sciences	Springer-Verlag
G	Ecological Sciences	Berlin, Heidelberg, New York and Tokyo

Series C: Mathematical and Physical Sciences Vol. 113

Methods in Computational Molecular Physics

edited by

G. H. F. Diercksen

Max-Planck-Institute for Physics and Astrophysics,
Garching, West Germany

and

S. Wilson

University of Oxford, Theoretical Chemistry Department,
Oxford, U.K.

D. Reidel Publishing Company

Dordrecht / Boston / Lancaster

Published in cooperation with NATO Scientific Affairs Division

Proceedings of the NATO Advanced Study Institute on
Methods in Computational Molecular Physics
Bad Windsheim, West Germany,
August 1982

Library of Congress Cataloging in Publication Data

Main entry under title:

Methods in computational molecular physics.

(NATO ASI series. Series C, Mathematical and physical sciences ; no. 113)
"Published in cooperation with NATO Scientific Affairs Division."
Includes index.
1. Molecules—Measurement—Congresses. 2. Molecular structure—Congresses. 3. Mathematical physics—Congresses. I. Diercksen, G. H. F. II. Wilson, S. (Stephen), 1950– III. NATO Advanced Study Institute. IV. North Atlantic Treaty Organization. Scientific Affairs Division. V. Series.
QC173.M4748 1983 539'.6'0151 83–13929

ISBN-13: 978-94-009-7202-5 e-ISBN-13: 978-94-009-7200-1
DOI: 10.1007/978-94-009-7200-1

Published by D. Reidel Publishing Company
P.O. Box 17, 3300 AA Dordrecht, Holland

Sold and distributed in the U.S.A. and Canada
by Kluwer Academic Publishers,
190 Old Derby Street, Hingham, MA 02043, U.S.A.

In all other countries, sold and distributed
by Kluwer Academic Publishers Group,
P.O. Box 322, 3300 AH Dordrecht, Holland

D. Reidel Publishing Company is a member of the Kluwer Academic Publishers Group

CONTENTS

PREFACE

This NATO Advanced Study Institute was concerned with modern ab initio methods for the determination of the electronic structure of molecules. Recent years have seen considerable progress in computer technology and computer science and these developments have had a very significant influence on computational molecular physics. Progress in computer technology has led to increasingly larger and faster systems as well as powerful minicomputers. Simultaneous research in computer science has explored new methods for the optimal use of these resources. To a large extent developments in computer technology, computer science and computational molecular physics have been mutually dependent. The availability of new computational resources, particularly minicomputers and, more recently, vector processors, has stimulated a great deal of research in molecular physics. Well established techniques have been reformulated to make more efficient use of the new computer technology and algorithms which were previously computationally intractable have now been successfully implemented. This research has given a new and exciting insight into molecular structure and molecular processes by enabling smaller systems to be studied in greater detail and larger systems to be studied for the first time. Although significant computational advances have been made over the past few years, perhaps the most exciting developments have resulted from theoretical innovations. The application of the unitary group and the symmetric group to the molecular electronic structure problem has been very fruitful. The direct configuration interaction method continues to be popular. The Green's function and polarization propagator methods have given new physical insight. The many-body perturbation theory has been shown to be an efficient and accurate approach to the electron correlation problem.

In all of the courses given the interplay between theoretical innovations and the development of efficient computational schemes were emphasized.

This NATO Advanced Study Institute was financially sponsored by the NATO Scientific Affairs Division. The organizers wish to express their gratitude for this support.

G.H.F. Diercksen
Dr. S. Wilson

MOLECULAR INTEGRALS FOR GAUSSIAN TYPE FUNCTIONS

V.R. Saunders
Science & Engineering Research Council, Daresbury Laboratory, Daresbury, Warrington WA4 4AD, UK

1. INTRODUCTION

The Gaussian type function (GTF) was introduced into quantum chemistry by Boys [1], who showed that the molecular integrals involving GTFs take a reasonably simple form. The field has been reviewed by Shavitt [2], and the present author [3], the latter in 1975, and it is our present purpose to summarize the considerable progress made since 1975.

The scope of the work will cover the evaluation of all the integrals required to compute the molecular electronic energy using the Schrödinger spin-free and Breit-Pauli spin dependent Hamiltonians, both in the Born-Oppenheimer approximation. The evaluation of the derivatives of the Schrödinger integrals with respect to nuclear displacements is briefly described, as are integrals which arise when one wishes to evaluate one electron properties; for example, dipole, electric field and electric field gradient.

2 BASIS FUNCTIONS

2.1 The 'Simple' GTF.

The unnormalized three dimensional 'simple' GTF is defined by:

$$G[a,A] = \exp[-ar_A{}^2] \qquad (1)$$

where a, the orbital exponent, is real and positive, and where r_A denotes the distance between a fixed point, $\underset{\sim}{A}$, with co-ordinates (A_x,A_y,A_z), and a variable point, normally interpreted as the co-

G. H. F. Diercksen and S. Wilson (eds.), Methods in Computational Molecular Physics, 1–36.

ordinates of an electron, $\underset{\sim}{r} = (x,y,z)$ so that:

$$\underset{\sim}{r}_A = \underset{\sim}{r} - \underset{\sim}{A} = (x_A, y_A, z_A) \tag{2a}$$

$$r_A^2 = \underset{\sim}{r}_A^2 = x_A^2 + y_A^2 + z_A^2 \tag{2b}$$

The most immediately obvious property of G[a,A] is that it is separable in the co-ordinates x, y and z. A further, and extremely important property concerns the 'Gaussian product theorem' [1-3], which states that the product of the two 'simple' GTF, located a centres $\underset{\sim}{A}$ and $\underset{\sim}{B}$, is itself a simple GTF multiplied by a constant factor, K_{AB}, located at a point $\underset{\sim}{P}$ along the line segment AB, with exponent p, where

$$p = a + b \tag{3a}$$

$$\underset{\sim}{P} = (a\underset{\sim}{A} + b\underset{\sim}{B})/p \tag{3b}$$

$$K_{AB} = \exp[-ab\ \overline{AB}^2/p] \tag{3c}$$

$$\overline{AB} = \underset{\sim}{A} - \underset{\sim}{B} \tag{3d}$$

2.2 The Gaussian Lobe Orbitals.

The Gaussian lobe orbitals (GLOs) [4] may be constructed by taking linear combinations of the simple GTF, these GTF being located on different centres. The orbital exponents, linear coefficients and centres of the GTF are normally chosen so that a good approximation to some other category of GTF (for example the Spherical or Cartesian GTF, to be described below) is given. The major disadvantage of the GLOs is that the molecular integrals involving them are expressed as small differences of potentially very large quantities, so that some care is needed if adequate numerical precision is to be obtained. A further disadvantage of the GLOs is that they cannot be made to transform under symmetry operations exactly as the target GTF they are supposed to be simulating, except in the limit that the numerical precision problems become overwhelming, so that special care must be taken when symmetry adapting the basis set. Associated with this symmetry problem is the fact that calculations using GLOs are not strictly invariant to changes in the rotational orientation chosen for the molecule.

The greatest advantage of the GLO is that the formalism for computing the molecular integrals is very simple, but we refer to a large literature [4] for further details.

2.3 The "Trigonometric" GTF.

The trigonometric GTF (TGTF) are defined [5] through:

$$T[a,A',k] = \exp[i\underset{\sim}{k}.\underset{\sim}{r} - a(\underset{\sim}{r} - \underset{\sim}{A}')^2] \tag{4}$$

with T^* the complex conjugate of T. Allison et al.[5] showed that the TGTF could be expressed as the product of a constant factor and a 'simple' GTF, with the latter being of complex origin. Thus let:

$$\underset{\sim}{A}'' = \underset{\sim}{k}/2a \tag{5a}$$

$$\underset{\sim}{A} = \underset{\sim}{A}' + i\underset{\sim}{A}'' \tag{5b}$$

when it may be shown directly that:

$$T[a,A',k] = G[a,A]\exp[a(\underset{\sim}{A}^2 - \underset{\sim}{A}'^2)] \tag{6}$$

It has further been shown that molecular integrals derived for the 'simple' GTF of real origin may be easily generalized to the complex origin case, but we refer to [5] for the details, merely noting that the 'Gaussian product theorem' eqn.(3), also applies to GTF of complex origin.

Perhaps the major potential use of the TGTF (and generalizations thereof) is in the treatment of certain magnetic properties (for example, magnetic shielding constants) in a gauge invariant manner [6] within a finite perturbation theory approach. The availability of formulae for the molecular integrals over the TGTF implies that mixed integrals between GTF and plane wave functions are also available, and this may prove useful in the treatment of certain continuum problems [7], such as charge exchange in heavy particle collisions.

2.4 The Cartesian GTF.

A more general class of function than the 'simple' GTF was first proposed by Boys [1], the cartesian GTF (CGTF), which take the form:

$$G[a,A,i,j,k] = x_A^i\ y_A^j\ z_A^k \exp[-a\ r_A^2] \tag{7}$$

where i,j and k (the 'quantum numbers') are integers $\geqslant 0$. In common with the 'simple' GTF, the CGTF are separable in the co-ordinates x, y and z:

$$G[a,A,i,j,k] = G_x[a,A,i]\ G_y[a,A,j]\ G_z[a,A,k] \tag{8a}$$

$$G_x[a,A,i] = x_A^i \exp[-ax_A^2] \tag{8b}$$

The CGTF may be factored into a product of a polynomial in x_A, y_A and z_A (actually a monomial in this case) and an exponential:

$$G[a,A,i,j,k] = g[A,i,j,k] \exp[-ar_A^2] \tag{9}$$

where the monomial, g, is of degree

$$\mu = i + j + k \tag{10}$$

and obeys the uniquely simple recursion formula:

$$g[A,i+1,j,k] = x_A g[A,i,j,k] \tag{11}$$

with similar results if one wishes to increment j and k.

A normalized CGTF, G_N, may be defined by means of

$$G_N[a,A,i,j,k] = NG[a,A,i,j,k] \tag{12a}$$

$$N = S^{-1/2} \tag{12b}$$

$$S = \int_{-\infty}^{\infty}\int_{-\infty}^{\infty}\int_{-\infty}^{\infty} \exp[-2a(x^2 + y^2 + x^2)]x^{2i}\, y^{2j}\, z^{2k}\, dxdydz \tag{12c}$$

The self overlap integral, S, may be evaluated from (see page 355 of [3]):

$$\int_{-\infty}^{\infty} \exp[-2ax^2]x^{2i}\, dx = \Gamma[i + \tfrac{1}{2}]/(2a)^{i+1/2} \tag{13a}$$

$$\Gamma[i + \tfrac{1}{2}] = (2i - 1)!!\; \pi^{1/2}/2^i \tag{13b}$$

so that the self overlap integral is given by:

$$S = (2i - 1)!!\, (2j - 1)!!\, (2k - 1)!!\, (\pi/2a)^{3/2}/(4a)^{i+j+k} \tag{14}$$

The CGTF is presently the most widely used system of basis function in quantum chemistry, and will receive considerable attention in the present work.

2.5 The Spherical GTF.

The unnormalized complex spherical GTF(SGTF) is defined in a spherical polar co-ordinate system centred at $\underset{\sim}{A}$ by means of:

$$S[a,A,n,\ell,m] = r_A^{2n}\, r_A^{\ell}\, P_{\ell}^{|m|}[\cos\theta] \exp[im\phi] \exp[-ar_A^2] \tag{15}$$

where n and ℓ are integers ≥ 0, $|m| \leq \ell$ and integer, and $P_\ell^{|m|}[\cos\theta]$ is the associated Legendre function. The (unnormalized) complex spherical harmonic in the Schiff [8] phase convention is:

$$Y_{\ell m}[\theta,\phi] = P_\ell^{|m|}[\cos\theta]\ \exp[im\phi] \tag{16}$$

so that the angular part of the SGTF is such a spherical harmonic. The SGTF may be decomposed into a polynomial (in x_A, y_A, z_A) and exponential factor:

$$S[a,A,n,\ell,m] = s[A,n,\ell,m]\ \exp[-ar_A^2] \tag{17}$$

The following recursion formulae for the associated Legendre functions will prove very useful (see page 68 of [9]):

$$P_{\ell+1}^{\ell+1} = (2\ell + 1)\ \sin\theta\ P_\ell^\ell \tag{18a}$$

$$P_{\ell+1}^{|m|} = ((2\ell + 1)\ \cos\theta\ P_\ell^{|m|} - (\ell + |m|)P_{\ell-1}^{|m|})/(\ell - |m| + 1) \tag{18b}$$

where the argument of the associated Legendre function has been suppressed.

Eqns.(15), (17), (18a) give rise to the recursion formulae:

$$s[A,n,\ell+1,\ell+1] = (2\ell + 1)(x_A + iy_A)\ s[A,n,\ell,\ell] \tag{19a}$$

$$s[A,n,\ell+1,-\ell-1] = (2\ell + 1)(x_A - iy_A)\ s[A,n,\ell,-\ell] \tag{19b}$$

where we have also used:

$$r_A\ \sin\theta\ e^{i\phi} = x_A + iy_A \tag{20a}$$

$$r_A\ \sin\theta\ e^{-i\phi} = x_A - iy_A \tag{20b}$$

Eqn.(19b) follows immediately from eqn.(19a) when one notes that s[A,n,ℓ,-m] is the complex conjugate of s[A,n,ℓ,m].

Eqns.(15),(17),(18b) give rise to the recursion formula:

$$s[A,n,\ell+1,m] = ((2\ell+1)z_A s[A,n,\ell,m]-r_A^2(\ell+|m|)s[A,n,\ell-1,m])/(\ell-|m|+1) \tag{21}$$

where we have also used:

$$r_A \cos\theta = z_A \tag{22}$$

Thus to increment ℓ alone, use eqn.(21), whilst to increment ℓ and m (target m positive and equal to ℓ), use eqn.(19a), whilst to increment ℓ and decrement m, (target m negative, $|m| = \ell$), use eqn. (19b). This allows us to reach all possible ℓ, m combinations.

To increment n, use:

$$s[A,n+1,\ell,m] = r_A^2 \, s[A,n,\ell,m] \tag{23}$$

all recursions commencing from s[A,0,0,0] = 1. It is obvious from the recurrence equations, (19), (21), (23), that $s[A,n,\ell,m]$ is a polynomial in x_A, y_A, z_A of degree:

$$\mu = 2n + \ell \tag{24}$$

The self overlap integral of SGTF is required to evaluate the normalization factor, eqn.(12b), and we find

$$\int S^*(a,A,n,\ell,m)\, S(a,A,n,\ell,m)\, d\tau = \int_0^\infty r^{4n+2\ell+2} \exp[-2ar^2]\, dr$$

$$\int_0^\pi \left(P_\ell^{|m|}[\cos\theta]\right)^2 \sin\theta\, d\theta \int_0^{2\pi} d\phi = I_r\, I_\theta\, I_\phi \tag{25}$$

the integral over ϕ is trivial:

$$I_\phi = 2\pi \tag{26}$$

the integral over θ is given on page 66 of [9]:

$$I_\theta = \frac{2}{(2\ell + 1)} \frac{(\ell + |m|)!}{(\ell - |m|)!} \tag{27}$$

whilst the radial integral may be deduced from eqn.(13) when it is recognized that:

$$\int_{-\infty}^{\infty} \exp[-2ax^2]x^{2i}\, dx = 2\int_0^\infty \exp[-2ax^2]x^{2i}\, dx \tag{28}$$

so that

$$I_r = \frac{1}{2}\,\Gamma[2n+\ell+3/2]/(2a)^{2n+\ell+3/2} = (4n+2\ell+1)!!\ (\pi/8a)^{1/2}/(4a)^{2n+\ell+1} \tag{29}$$

The use of the complex SGTF is often not convenient. Thus many SCF and CI codes seek to exploit the simplifications which arise when real basis functions are used. To circumvent this difficulty, we define real SGTF, through:

$$R[a,A,n,\ell,0] = S[a,A,n,\ell,0] \tag{30a}$$

$$R[a,A,n,\ell,m] = (S[a,A,n,\ell,m] + S[a,A,n,\ell,-m])/\sqrt{2} \tag{30b}$$

$$R[a,A,n,\ell,-m] = (S[a,A,n,\ell,m] - S[a,A,n,\ell,-m])/\sqrt{-2} \tag{30c}$$

where m is here to be interpreted as a positive integer. Thus the functions of ϕ associated with positive and negative m are $\cos|m|\phi$ and $\sin|m|\phi$ respectively for the real SGTF.

2.6 The Hermite GTF.

The Hermite GTF (HGTF) were first proposed by Zivkovic and Maksic [10], who considered their use as basis functions. The HGTF were subsequently used by McMurchie and Davidson [11] in the role of useful intermediaries in the evaluation of molecular integrals, rather than as basis functions, and we will stress this role in the present work. The HGTF may be defined through:

$$\Lambda[a,A,i,j,k] = (\partial/\partial A_x)^i\ (\partial/\partial A_y)^j\ (\partial/\partial A_z)^k \exp[-ar_A^2] \tag{31}$$

It is obvious from the definition that the HGTF are separable in x_A, y_A and z_A. Thus

$$\Lambda[a,A,i,j,k] = \Lambda_x[a,A,i]\ \Lambda_y[a,A,j]\ \Lambda_z[a,A,k] \tag{32a}$$

$$\Lambda_x[a,A,i] = (\partial/\partial A_x)^i \exp[-ax_A^2] \tag{32b}$$

and similarly for Λ_y and Λ_z. To establish a conection with the Hermite polynomials, $H_i[\varepsilon]$, we use the standard equation (see page 158 of [9]).

$$H_i[\varepsilon] = (-1)^i \exp[\varepsilon^2](\partial/\partial\varepsilon)^i \exp[-\varepsilon^2] \tag{33}$$

Let

$$\varepsilon = a^{1/2}(x - A_x) = a^{1/2}x_A \tag{34a}$$

so that

$$\partial\varepsilon/\partial A_x = -a^{1/2} \tag{34b}$$

and

$$(-1)^i (\partial/\partial\varepsilon)^i \exp[-\varepsilon^2] = (\partial/\partial A_x)^i \exp(-\varepsilon^2)/a^{i/2} \tag{35}$$

A combination of eqns.(32b), (33), (34), (35) gives the result:

$$\Lambda_x[a,A,i] = a^{i/2}\, H_i[a^{1/2}x_A]\, \exp[-ax_A^2] \tag{36}$$

with similar results for Λ_y and Λ_z. The explicit form for the Hermite polynomial (see page 157 of [9]) is:

$$H_i[\varepsilon] = \sum_{t=0}^{[i/2]} \frac{(-1)^t\, i!}{t!\,(i-2t)!} (2\varepsilon)^{i-2t} \tag{37}$$

where [i/2], the summation limit, denotes the largest integer less than or equal to i/2. Thus $\Lambda_x[a,A,i]$ may be factored into a product of a polynomial in x_A of degree i, and an exponential:

$$\Lambda_x[a,A,i] = \lambda_x[a,A,i]\, \exp[-ax_A^2] \tag{38a}$$

$$\lambda_x[a,A,i] = a^{i/2}\, H_i[a^{1/2}x_A] \tag{38b}$$

and the HGTF itself may be factored into a polynomial in x_A, y_A, z_A of degree

$$\mu = i + j + k \tag{39}$$

and an exponential:

$$\Lambda[a,A,i,j,k] = \lambda[a,A,i,j,k]\, \exp[-ar_A^2] \tag{40}$$

The Hermite polynomials are linearly independent, and it then follows that any polynomial in x_A, y_A, z_A of degree ν may be expanded as a linear combination of the λ-polynomials of degree less than or equal to ν, and we shall utilize this property in the evaluation of molecular integrals.

The recursion formula for the Hermite polynomials (see page 162 of [9]) is:

$$\varepsilon\, H_i[\varepsilon] = \frac{1}{2} H_{i+1}[\varepsilon] + i\, H_{i-1}[\varepsilon] \tag{41}$$

from which may be deduced a recursion formula for the λ_x-polynomials (eqn.(38b)):

$$x_A \lambda_x[a,A,i] = (1/2a)\, \lambda_x[a,A,i+1] + i\lambda_x[a,A,i-1] \tag{42}$$

with similar results for λ_y and λ_z. A similar result may be given for the λ-polynomials (eqn.(40)):

$$x_A\lambda[a,A,i,j,k] = (1/2a)\ \lambda[a,A,i+1,j,k] + i\lambda[a,A,i-1,j,k] \qquad (43)$$

with similar relations for recurrence in the y and z quantum numbers.

2.7 Other Systems of GTF.

A brief discussion of three other systems of GTF is presented. Procedures for integrating over these systems will not be given. It is hoped that our discussion of the CGTF and HGTF systems will demonstrate a general procedure applicable to all polynomial GTF systems.

2.7.1 The Cartesian oscillator GTF. Roberts [12] has defined a basis set consisting of the eigen functions of the isotropic three-dimensional Cartesian oscillator, which may be expressed as a product of three one-dimensional harmonic oscillator functions. The Hamiltonian for the Cartesian oscillator is

$$\hat{H} = -\frac{1}{2}\nabla^2 + 2a^2r^2_A \qquad (44)$$

with eigen functions (unnormalized):

$$CO[a,A,i,j,k] =$$

$$H_i[(2a)^{1/2}x_A]H_j[(2a)^{1/2}y_A]H_k[(2a)^{1/2}z_A]\exp[-ar_A^2] \qquad (45)$$

and eigen values:

$$E_{ijk} = (i + j + k + 3/2)2a \qquad (46)$$

The major advantage of the cartesian oscillator GTF (COGTF) basis is that it is orthogonal, and separable in x, y, z.

2.7.2 The Spherical Oscillator GTF. The isotropic three dimensional oscillator Hamiltonian is spherically symmetric, so that the eigen functions may be written as a product of a spherical harmonic and a radial function. These spherical oscillator GTF (SOGTF) are appropriate linear combinations of degenerate COGTF, the latter being of equal polynomial degree, eqn.(46). Bell [13] has shown that the SOGTF take the explicit form (unnormalized):

$$SO[a,A,n,\ell,m] = L_n^{\ell+1/2}[2ar_A^2]r_A^\ell\ Y_{\ell m}[\theta,\phi]\ \exp[-ar_A^2] \qquad (47)$$

where $L_n^m[\varepsilon]$ denotes an associated Laguerre polynomial, see page 176 of [9]. Maretis [14] has defined procedures for evaluating molecular integrals over the SOGTF. The major advantage of the SOGTF basis is that it is orthogonal and adapted to the three dimensional rotation group.

2.7.3 The Spherical Gradient GTF. The HGTF may be formally related to the CGTF by the substitution $(x_A, y_A, z_A) \to (\partial/\partial A_x, \partial/\partial A_y, \partial/\partial A_z)$ in the polynomial factor. Similarly, the spherical gradient GTF system (SGGTF) may be related to the SGTF by the same substitution. Thus for $r_A{}^2$ use $\partial^2/\partial A_x{}^2 + \partial^2/\partial^2 A_y + \partial^2/\partial^2 A_z$, whilst the solid harmonics, instead of being polynomials in x_A, y_A, z_A are similar polynomials in the gradient operators. Bayman [15] has discussed the properties of the spherical gradient operator in some detail, whilst Fieck [16] has given procedures for evaluating integrals over the SGGTF basis. The SGGTF basis takes the explicit form (unnormalized):

$$SG[a,A,n,\ell,m] = L_n^{\ell+1/2}[ar_A^2]r_A^\ell \, Y_{\ell m}[\theta,\phi] \exp[-ar_A^2] \tag{48}$$

2.8 Overview of Systems of GTF.

The Cartesian, spherical, Hermite, Cartesian oscillator, spherical oscillator and spherical gradient GTF may all prove useful under appropriate circumstances, and all share one property; they may be expressed as a product of an exponential and a homogeneous polynomial of finite degree in x_A, y_A, z_A. Furthermore, recurrence relations to enable the generation of these polynomials to arbitrarily high order can be specified. Thus our primary aim in the present work is to define a procedure for the evaluation of molecular integrals whose viability depends only on the existence of the aforementioned recurrence relations. The CGTF and SGTF systems will be used to illustrate the general technique, whilst the CGTF system will be used to exemplify special techniques applicable to systems which are separable in x, y, z.

The functions discussed above have all involved a single exponential. However, in practice, it is more common to use linear combinations of GTF, usually of common angular factor and centre, the resulting functions being referred to as 'contracted' GTF. Thus the simple exponential of the 'primitive' GTF is replaced by a linear combination of exponential terms, $\sum_\tau b_\tau \exp[-a_\tau r_A{}^2]$, in the 'contracted' GTF, where the b_τ are the contraction coefficients. In general, an integral over such contracted GTF may be written as a linear combination of integrals over uncontracted GTF, so that much of our analysis will refer directly to the uncontracted case. However, some procedures [11,17,18] seek to gain efficiency by exploiting the nature of such contractions, so that it is necessary to explicitly consider the possibility of contraction when discussing such procedures.

Many methods [3,11,18,19] are designed to exploit the 'shell' structure of typical basis sets. Thus one often wishes to evaluate all the integrals involving a group of GTF of common exponent and centre, and differing only in the form of the polynomial factor. Thus, if we use the CGTF as an example, one may choose a shell of all functions whose polynomial factor is of a given degree, μ, giving rise to S,P,D,F etc shells consisting of $(\mu+1)(\mu+2)/2$ functions, or one may choose all functions whose polynomial factor is of degree less than or equal to some maximum, μ, giving rise to S, L, M etc shells [20] consisting of $(\mu+1)(\mu+2)(\mu+3)/6$ functions. In the SGTF case, one commonly uses a shell consisting of all the m components associated with a given ℓ quantum number, giving rise to s,p,d etc shells of dimension $2\ell+1$. In general, the auxiliary functions associated with a given integral are also required for some of the related integrals involving different members of the same shell, so that substantial economies may result from the exploitation of shell structure.

3. OVERLAP DISTRIBUTIONS

The product of two GTF, centred on $\underset{\sim}{A}$ and $\underset{\sim}{B}$ ($\underset{\sim}{A}$ may equal $\underset{\sim}{B}$), which are functions of the co-ordinates of the same electron is referred to as an overlap distribution, and all the integrals with which we are concerned involve at least one such overlap distribution. It has already been noted that the overlap distribution of two simple GTF is itself a 'simple' GTF located at a point $\underset{\sim}{P}$, see eqn.(3). The overlap distribution of two polynomial GTF, of degree μ and ν, and located at $\underset{\sim}{A}$ and $\underset{\sim}{B}$ respectively is therefore a polynomial GTF located at $\underset{\sim}{P}$, of degree $\mu+\nu$ in x_P, y_P z_P. This is because

$$x_A = x_P + P_x - A_x = x_P + \overline{PA}_x \tag{49a}$$

$$x_B = x_P + P_x - B_x = x_P + \overline{PB}_x \tag{49b}$$

with similar results for the y and z co-ordinates. At first sight, it would appear that it would be best to expand the product of the two polynomials as a linear combination of Cartesian monomials located at $\underset{\sim}{P}$, because of the very simple recurrence formula for the CGTF (eqn.(11)). The molecular integrals would then be expressed as linear combinations of integrals over CGTF located at $\underset{\sim}{P}$, and this strategy has been investigated by Taketa et al [21] and studied in depth by the present author [3] and Benard and Barry [22]. The strategy implies that one is able to evaluate integrals over CGTF. However, it turns out to be considerably easier (and more efficient) to evaluate such integrals over the HGTF basis, and thus a better strategy would be to use the HGTF, located at $\tilde{P}$, as expansion functions for the overlap distributions, as first advocated by McMurchie and Davidson [11].

3.1 Expansion of CGTF Overlap Distributions in HGTF.

Because of the separability of the CGTF and HGTF systems, it is sufficient to consider the expansion of the term $x_A^i\, x_B^{i'}$ as a linear combination of the polynomials $\lambda_x[p,P,\tau]$ defined by eqn. (38):

$$x_A^{i}\, x_B^{i'} = \sum_{\tau=0}^{i+i'} e[i,i',\tau]\; \lambda_x[p,P,\tau] \tag{50}$$

where the $e[i,i',\tau]$ are the linear expansion coefficients. Note that $e[0,0,0]$ is equal to unity, because $\lambda_x[p,P,0]$ is also unity, and we wish to derive recurrence relationships to allow us to generate the E-coefficients for arbitrarily high i and i' quantum numbers. We follow closely the procedure of [11]. From eqn.(50) we find

$$x_A^{i+1}\, x_B^{i'} = \sum_{\tau=0}^{i+i'} e_x[i,i',\tau] x_A\; \lambda_x[p,P,\tau] \tag{51}$$

By combining eqn.(42) and (49a) it is found that:

$$x_A \lambda_x[p,P,\tau] = \lambda_x[p,P,\tau+1]/2p + \overline{PA}_x \lambda_x[p,P,\tau] + \tau\; \lambda_x[p,P,\tau-1] \tag{52}$$

so that eqn.(51) may be re-written:

$$x_A^{i+1} x_B^{i'} = \sum_{\tau=0}^{i+i'} e_x[i,i',\tau](\lambda_x[i,i',\tau+1]/2p+\overline{PA}_x \lambda_x[p,P,\tau]+\tau\lambda_x[p,P,\tau-1]) \tag{53}$$

and we deduce the following recurence relation:

$$e_x[i+1,i',t] = e_x[i,i',t-1]/2p+\overline{PA}_x\; e_x[i,i',t]+(t+1)\; e_x[i,i',t+1] \tag{54}$$

with similar results for recursion in i' and the y and z quantum numbers. It should be noted that $e_x[i,i',t]$ is zero in eqn.(54) if t is negative or greater than i+i'.

The full product of two cartesian monomials may be expanded thus:

$$g[A,i,j,k]g[B,i',j',k'] =$$

$$\sum_{\tau}^{i+i'} e_x[i,i',\tau]\, \lambda_x[p,P,\tau] \sum_{\mu}^{j+j'} e_y[j,j',\mu]\, \lambda_y[p,P,\mu]$$

$$\sum_{\nu}^{k+k'} e_z[k,k',\nu]\, \lambda_z[p,P,\nu]$$

$$= \sum_{\tau,\mu,\nu} E[i,j,k,i',j',k',\tau,\mu,\nu]\, \lambda[p,P,\tau,\mu,\nu] \qquad (55)$$

where

$$E[i,j,k,i',j',k',\tau,\mu,\nu] = e_x[i,i',\tau]\, e_y[j,j',\mu]\, e_z[k,k',\nu] \qquad (56)$$

3.2 Expansion of SGTF Overlap Distributions in HGTF.

We first note that Fieck [23] has defined a procedure for expanding products of SGTF on $\underset{\sim}{A}$ and $\underset{\sim}{B}$ as linear combinations of SGTF on $\underset{\sim}{P}$, and has given procedures for evaluating molecular integrals over the latter. We however require an expansion in HGTF on $\underset{\sim}{P}$, of the form

$$s[A,n,\ell,m]\, s[B,n',\ell',m'] =$$

$$\sum_{\tau,\mu,\nu} E[n,\ell,m,n',\ell',m',\tau,\mu,\nu]\, \lambda[p,P,\tau,\mu,\nu] \qquad (57)$$

where $\tau + \mu + \nu \leqslant 2n + 2n' + \ell + \ell'$, and where s and λ are given in eqns.(17) and (40) respectively. We note that s[A,0,0,0] and consequently E[0,0,0,0,0,0,0,0,0,] are unity, and wish to derive recurrence relations to allow the incrementation of the n,ℓ,m parameters of the E-coefficients (and similarly for n',ℓ' and m').

<u>3.2.1 Recursion In n.</u> By combining eqns.(23) and (57), we find

$$s[A,n+1,\ell,m]s[B,n',\ell',m'] =$$

$$\sum_{\tau,\mu,\nu} E[n,\tau,\mu,\nu](x_A^2+y_A^2+z_A^2)\lambda[\tau,\mu,\nu] \qquad (58)$$

where arguments p and P have been suppressed in λ, and ℓ,m,n'ℓ',m' have been suppressed in E, in the interests of brevity. The multiplications by $x_A{}^2$ (and similarly for $y_A{}^2$, $z_A{}^2$) in eqn.(58) are accomplished by a double application of eqn.(52), to yield

$$s[A,n+1,\ell,m]s[B,n',\ell',m'] =$$

$$\sum_{\tau,\mu,\nu} E[n,\tau,\mu,\nu]\Big((\lambda[\tau+2,\mu,\nu] + \lambda[\tau,\mu+2,\nu] + \lambda[\tau,\mu,\nu+2])/(2p)^2$$
$$+ (\overline{PA}_x\lambda[\tau+1,\mu,\nu] + \overline{PA}_y\lambda[\tau,\mu+1,\nu] + \overline{PA}_z\lambda[\tau,\mu,\nu+1])/p$$
$$+ (\overline{PA}^2 + (2\tau + 2\mu + 2\nu + 3)/2p)\lambda[\tau,\mu,\nu] + 2(\overline{PA}_x\tau\lambda[\tau-1,\mu,\nu]$$
$$+ \overline{PA}_y\mu\lambda[\tau,\mu-1,\nu] + \overline{PA}_z\nu\lambda[\tau,\mu,\nu-1]) + \tau(\tau-1)\lambda[\tau-2,\mu,\nu]$$
$$+ \mu(\mu-1)\lambda[\tau,\mu-2,\nu] + \nu(\nu-1)\lambda[\tau,\mu,\nu-2]\Big) \quad (59)$$

whence the recurrence relation:

$$E[n+1,t,u,v] = (E[n,t-2,u,v] + E[n,t,u-2,v] + E[n,t,u,v-2])/(2p)^2$$
$$+ (\overline{PA}_xE[n,t-1,u,v] + \overline{PA}_yE[n,t,u-1,v] + \overline{PA}_zE[n,t,u,v-1])/p$$
$$+ (\overline{PA}^2 + (2t + 2u + 2v + 3)/2p)E[n,t,u,v]$$
$$+ 2(\overline{PA}_x(t+1)E[n,t+1,u,v] + \overline{PA}_y(u+1)E[n,t,u+1,v]$$
$$+ \overline{PA}_z(v+1)E[n,t,u,v+1]) + (t+2)(t+1)E[n,t+2,u,v]$$
$$+ (u+2)(u+1)E[n,t,u+2,v] + (v+2)(v+1)E[n,t,u,v+2] \quad (60)$$

3.2.2. Recursion In ℓ. A combination of eqns.(21) and (57) yields:

$$s[A,n,\ell+1,m]\ s[B,n',\ell',m'] =$$

$$\sum_{\tau,\mu,\nu} \Big(E[\ell,\tau,\mu,\nu](2\ell+1)z_A\lambda[\tau,\mu,\nu]$$
$$- E[\ell-1,\tau,\mu,\nu](\ell+|m|)r_A^2\lambda[\tau,\mu,\nu]\Big)/(\ell-|m|+1) \quad (61)$$

where arguments p,P have again been omitted from λ, and n,m,n',ℓ', m' have been omitted from E. The multiplications by z_A and r_A^2 in the right hand side of eqn.(61) are accomplished using eqn.(43), and using the arguments presented in sections 3.1 and 3.2.1 as a model (for z_A and $r_A{}^2$ respectively) a recursion relation for the E-coefficients may be derived as follows:

$$\begin{aligned}E[\ell+1,t,u,v] = \Big((2\ell+1)(E[\ell,t,u,v-1]/2p + \overline{PA}_z E[\ell,t,u,v]\\
+ (v+1)E[\ell,t,u,v+1]) - (\ell+|m|)((E[\ell-1,t-2,u,v]\\
+ E[\ell-1,t,u-2,v] + E[\ell-1,t,u,v-2])/(2p)^2\\
+ (\overline{PA}_x E[\ell-1,t-1,u,v]) + \overline{PA}_y E[\ell-1,t,u-1,v] + \overline{PA}_z E[\ell-1,t,u,v-1])/p\\
+ (\overline{PA}^2+(2t+2u+2v+3)/2p)E[\ell-1,t,u,v] + 2(\overline{PA}_x(t+1)E[\ell-1,t+1,u,v]\\
+ \overline{PA}_y(u+1)E[\ell-1,t,u+1,v] + \overline{PA}_z(v+1)E[\ell-1,t,u,v+1])\\
+ (t+2)(t+1)E[\ell-1,t+2,u,v] + (u+2)(u+1)E[\ell-1,t,u+2,v]\\
+ (v+2)(v+1)E[\ell-1,t,u,v+2])\Big)\\
/(\ell - |m| + 1)\end{aligned} \quad (62)$$

3.2.3. Recursion in ℓ and m. We first consider the case that ℓ and m are equal (so that m is positive). By combining eqns.(19a) and (57) we find:

$$s[A,n,\ell+1,\ell+1]\, s[B,n',\ell',m'] = \sum_{\tau,\mu,\nu} E[\ell,\ell,\tau,\mu,\nu](x_A + iy_A)\lambda[\tau,\mu,\nu] \quad (63)$$

where arguments p,P have again been omitted from λ, and n,n',ℓ',m' have been omitted from E. The multiplications by x_A and y_A on the right hand side of eqn.(63) are accomplished using eqn.(43), and using the argument presented in section 3.1 as a model a recursion relation for the E-coefficients may be derived:

$$\begin{aligned}E[\ell+1,\ell+1,t,u,v] = E[\ell,\ell,t-1,u,v]/2p + \overline{PA}_x E[\ell,\ell,t,u,v]\\
+ (t+1)E[\ell,\ell,t+1,u,v] + i(E[\ell,\ell,t,u-1,v]/2p\\
+ \overline{PA}_y E[\ell,\ell,t,u,v] + (u+1)E[\ell,\ell,t,u+1,v])\end{aligned} \quad (64a)$$

A combination of eqn.(19b) and (57) gives the recursion relation:

$$\begin{aligned}E[\ell+1,-\ell-1,t,u,v] = E[\ell,-\ell,t-1,u,v]/2p + \overline{PA}_x E[\ell,-\ell,t,u,v]\\
+ (t+1)E[\ell,-\ell,t+1,u,v] - i(E(\ell,-\ell,t,u-1,v)/2p\\
+ \overline{PA}_y E[\ell,-\ell,t,u,v] + (u+1)E[\ell,-\ell,t,u+1,v])\end{aligned} \quad (64b)$$

thus permitting the simultaneous incrementation of ℓ, and decrementation of m where $|m| = \ell$, and where the target m value is negative.

3.2.4. Overview of the Expansion of SGTF Overlap Distributions in HGTF. The recursive procedure defined above may be summarized thus. Commencing at E[0,0,0,0,0,0,0,0,0] = 1, we aim to increment the n,ℓ,m quantum numbers associated with the function on A. To reach a given ℓ,m quantum number, use eqn.(64a) until the desired m quantum number is obtained, followed by eqn.(62) to generate the required ℓ quantum number. Naturally, much advantage is obtained by generating all the positive m components associated with a given ℓ in this process, and notice that it is not necessary to generate the negative m E-coefficients at this stage, because:

$$E[0,\ell,-|m|,0,0,0,t,u,v] = E^{*}[0,\ell,|m|,0,0,0,t,u,v] \tag{65}$$

A similar procedure involving eqn.(62) and (64) may now be used to generate appropriate ℓ' and m' indices, noticing that negative m' quantum numbers should be generated, eqn.(64b). At this stage the n and n' quantum numbers may be incremented using eqn.(60), followed by a transformation to real SGTF using eqn.(30). E-coefficients not explicitly generated may be deduced as complex conjugates of those available:

$$E[n,\ell,-|m|,n',\ell',-|m'|,t,u,v] = E^{*}[n,\ell,|m|,n',\ell',|m'|,t,u,v] \tag{66a}$$

$$E[n,\ell,-|m|,n',\ell',|m'|,t,u,v] = E^{*}[n,\ell,|m|,n',\ell',-|m'|,t,u,v] \tag{66b}$$

3.3 Some General Comments on the Generation of E-coefficients.

One should note that a given set of E-coefficients is normally used in the evaluation of many integrals, particularly in the context of the electron repulsion integrals, so that it may be useful to compute the E-coefficients for all possible overlap distributions at the start of the integral evaluation run, these being subsequently read from backing store as required. By arranging that the electron repulsion integrals are evaluated in reasonably large blocks, the input/output penalty associated with this strategy may be minimized, with a considerable saving in CPU time, particularly for non-separable systems of GTF.

By adopting the convention:

$$E[0,0,0,0,0,0,0,0,0] = K_{AB} \tag{67}$$

in the recursive generation of the E-coefficients (K_{AB} defined in eqn.(3c)), K_{AB} may be incorporated into the integrals without cost. Also normalization factors and contraction coefficients may be absorbed into the E-coefficients, with some benefit and we assume here on that this has been done.

4. ONE-ELECTRON INTEGRALS

4.1 Overlap Integral.

This integral is of the form

$$S_{ij} = \int \phi_i \phi_j d\tau = \sum_{t,u,v} E[i,j,t,u,v] \int \Lambda[p,P,t,u,v]d\tau \qquad (68)$$

where i and j denote a composite of the quantum numbers of the basis functions, and where the overlap distribution has been expanded as a linear combination of HGTF located at $\underset{\sim}{P}$. The integral over the x co-ordinate of a typical term in the summation of eqn. (68) may be written (using eqn.(38)):

$$I_x[p,t] = \int_{-\infty}^{\infty} \exp[-px^2]\ p^{t/2}H_t[p^{1/2}x]dx \qquad (69)$$

After noting that $H_0[\varepsilon] = 1$, and the orthogonality properties of the Hermite polynomials (see page 161 of [9]), it is clear that $I_x[p,t]$ is zero unless $t = 0$. $I_x[p,0]$ may be evaluated using eqn. (13):

$$I_x[p,0] = (\pi/p)^{1/2} \qquad (70)$$

After similar integrations over the y and z co-ordinates, the overlap integral may be reduced to a single term:

$$S_{ij} = E[i,j,0,0,0](\pi/p)^{3/2} \qquad (71)$$

4.2 Dipole Integral.

The x component of the dipole integral is of the form:

$$X_{ij} = \int \phi_i \phi_j x_C d\tau = \sum_{t,u,v} E[i,j,t,u,v] \int x_C\ \Lambda[p,P,t,u,v]d\tau \qquad (72)$$

Noting that $x_C = x_P + \overline{PC}_x$, and using eqn.(43) and (71), we find:

$$X_{ij} = \left(E[i,j,1,0,0] + \overline{PC}_x\ E[i,j,0,0,0]\right)(\pi/p)^{3/2} \qquad (73)$$

with similar results for Y_{ij} and Z_{ij}. The procedure for multipole operators which are quadratic, cubic etc in the co-ordinates is an obvious extension of the above.

4.3 Kinetic Energy Integrals.

These integrals are of the form:

$$T_{ij} = \int \phi_i \hat{T} \phi_j \, d\tau = \int \phi_j \hat{T} \phi_i \, d\tau \tag{74}$$

where

$$\hat{T} = -\frac{1}{2} \nabla^2 = -\frac{1}{2}(\partial^2/\partial x^2 + \partial^2/\partial y^2 + \partial^2/\partial z^2) \tag{75}$$

4.3.1 Kinetic Energy Integrals for CGTF. The following is easily proved from the definition of the CGTF, eqn.(7):

$$\partial G[a,A,i,j,k]/\partial x = iG[a,A,i-1,j,k] - 2aG[a,A,i+1,j,k] \tag{76}$$

so that

$$\partial^2 G[a,A,i,j,k]/\partial x^2 = i(i-1)G[a,A,i-2,j,k]$$

$$- 2a(2i+1)G[a,A,i,j,k] + (2a)^2 G[a,A,i+2,j,k] \tag{77}$$

so that the kinetic energy integral may be reduced to a linear combination of overlap integrals:

$$-\frac{1}{2} \int G[b,B,i',j',k'] \nabla^2 G[a,A,i,j,k] \, d\tau = -\frac{1}{2} \Big(i(i-1)\langle i-2,j,k\rangle$$

$$+ j(j-1)\langle i,j-2,k\rangle + k(k-1)\langle i,j,k-2\rangle - 2a(2i+2j+2k+3)\langle i,j,k\rangle$$

$$+ (2a)^2(\langle i+2,j,k\rangle + \langle i,j+2,k\rangle + \langle i,j,k+2\rangle) \Big) \tag{78}$$

$$\langle i,j,k\rangle = \int G[b,B,i',j',k']G[a,A,i,j,k]d\tau \tag{79}$$

An alternative formulation is based on the identity

$$\int \phi_i \nabla^2 \phi_j d\tau = - \int \underset{\sim}{\nabla}\phi_i \cdot \underset{\sim}{\nabla}\phi_j d\tau \tag{80}$$

given that ϕ_i or ϕ_j belong to the Hilbert space $L^2(R^3)$, so that eqn.(76) may be used to reduce the kinetic energy integral to a linear combination of overlap integrals, but we omit details.

4.3.2 Kinetic Energy Integrals for SGTF. It is convenient to use the kinetic energy operator in spherical polar co-ordinates [24]:

$$\hat{T} = \frac{1}{2}(\hat{L}^2/r^2 - (\partial/\partial r + 1/r)^2) \tag{81}$$

where $\hat{L}^2$ is the square of the angular momentum operator. The spherical harmonics are eigen functions of $\hat{L}^2$, with eigen values $\ell(\ell+1)$. Therefore

$$\hat{L}^2/r^2\, S[a,A,n,\ell,m] = \ell(\ell+1)S[a,A,n-1,\ell,m] \tag{82}$$

The operator $-(\partial/\partial r + 1/r)^2$, the square of the so-called 'radial momentum operator', acts on the SGTF as follows:

$$-(\partial/\partial r + 1/r)^2 S[a,A,n,\ell,m] = -\ (2n+\ell+1)(2n+\ell)S[a,A,n-1,\ell,m] + 2a(4n+2\ell+3)S[a,A,n,\ell,m] - (2a)^2 S[a,A,n+1,\ell,m] \tag{83}$$

A combination of eqn.(81-83) yields the result:

$$\int S[b,B,n',\ell',m']\hat{T}\ S[a,A,n,\ell,m]d\tau = -\ n(2n+2\ell+1)\langle n-1\rangle + a(4n+2\ell+3)\langle n\rangle - 2a^2\langle n+1\rangle \tag{84}$$

$$\langle n\rangle = \int S[b,B,n',\ell',m']\ S[a,A,n,\ell,m]d\tau \tag{85}$$

so that the kinetic energy integral has again been reduced to a linear combination of overlap integrals.

4.4 Nuclear Attraction Integrals.

These integrals are of the form

$$N_{ij} = \int \phi_i \phi_j r_C^{-1} d\tau = \sum_{t,u,v} E[i,j,t,u,v]\ [t,u,v|r_C^{-1}] \tag{86a}$$

$$[t,u,v|r_C^{-1}] = \int \Lambda[p,P,t,u,v] r_C^{-1}\ d\tau \tag{86b}$$

Boys [1] has shown

$$[0,0,0|r_C^{-1}] = (2\pi/p)\ F_o[W] \tag{87a}$$

$$W = p\ \overline{CP}^2 \tag{87b}$$

$$F_m[W] = {}^1\!\!\int_0 \exp[-Ws^2]s^{2m}\ ds \tag{87c}$$

From eqn.(31) and (86b) we can show:

$$[t,u,v|r_C^{-1}] = (2\pi/p)\ R[t,u,v] \tag{88a}$$

$$R[t,u,v] = (\partial/\partial P_x)^t(\partial/\partial P_y)^u(\partial/\partial P_z)^v F_o[W] \tag{88b}$$

We use Leibniz's rule to move the different operators under the integral sign in eqn.(88b):

$$R[t,u,v] = \int_0^1 (\partial/\partial P_x)^t (\partial/\partial P_y)^u (\partial/\partial P_z)^v \exp[-p\, \overline{CP}^2 s^2] ds \quad (89)$$

The quantity under the integral sign in eqn.(89) may be related to the Hermite polynomials using eqn.(33):

$$R[t,u,v] = p^{w/2}$$

$$\int_0^1 \exp[-Ws^2] s^w H_t[p^{1/2}\overline{CP}_x s] H_u[p^{1/2}\overline{CP}_y s] H_v[p^{1/2}\overline{CP}_z s] ds \quad (90a)$$

$$w = t + u + v \quad (90b)$$

Following McMurchie and Davidson [11] we introduce the more general integral:

$$R_j[t,u,v] = p^{w/2}(-2p)^j$$

$$\int_0^1 \exp[-Ws^2] s^{w+2j} H_t[p^{1/2}\overline{CP}_x s] H_u[p^{1/2}\overline{CP}_y s] H_v[p^{1/2}\overline{CP}_z s] ds \quad (91)$$

where, from the recursion relation for the Hermite polynomials, eqn.(41), we find:

$$R_j[t+1,u,v] = -\overline{CP}_x R_{j+1}[t,u,v] + t R_{j+1}[t-1,u,v] \quad (92)$$

with similar results for the incrementation of u and v. Eqn.(92) forms the basis for a recursive scheme for the evaluation of the R[t,u,v] when it is noted that:

$$R[t,u,v] = R_o[t,u,v] \quad (93a)$$

$$R_j[0,0,0] = (-2p)^j F_j[W] \quad (93b)$$

Most frequently one requires a sum of nuclear integrals of the form:

$$\int \phi_i \phi_j \sum_C Z_C r_C^{-1} d\tau =$$

$$(2\pi/p) \sum_{t,u,v} E[i,j,t,u,v] \sum_C Z_C R^C[t,u,v] \quad (94)$$

where Z_C is the charge of nuclear centre C. The summation over C is independent of the labels, i and j, of the GTF. Thus when shells of GTF are involved, it is best to perform the summation over C first in eqn.(94) for all t, u and v, the resultants being useable for all integrals arising from the shell-shell interaction.

4.5 One-Electron Electric Field Integrals.

These integrals are defined by

$$\int \phi_i \phi_j x_C r_C^{-3} d\tau = \sum_{t,u,v} E[i,j,t,u,v]\ [t,u,v|x_C r_C^{-3}] \tag{95}$$

where we consider the x-component of the operator explicitly. Now it is easy to show using Leibniz's rule that:

$$(\partial/\partial C_x)[t,u,v|r_C^{-1}] = [t,u,v|x_C r_C^{-3}] \tag{96}$$

The integral $[t,u,v|r_C^{-1}]$ is invariant to a simultaneous uniform translation of the centres $\underset{\sim}{P}$ and $\underset{\sim}{C}$, so that:

$$(\partial/\partial C_x)[t,u,v|r_C^{-1}] = -\ (\partial/\partial P_x)[t,u,v|r_C^{-1}] \tag{97}$$

and from the definition of the HGTF, eqn.(31):

$$[t,u,v|x_C r_C^{-3}] = -\ [t+1,u,v|r_C^{-1}] \tag{98}$$

so that the electric field integral, eqn.(95), may be expressed as a linear combination of nuclear attraction integrals, whose evaluation is detailed in eqn.(88-93).

Matsuoka [25] has evaluated the electric field integrals over CGTF using the Fourier transform of the electric field operator, producing a method similar to that given by Taketa _et al_ [21] for the nuclear attraction integrals. However, the McMurchie and Davidson scheme [11] presented above is much simpler to implement.

In the evaluation of the derivative of the Schrödinger energy by the 'force method' [26] one requires the derivative of the molecular integrals with respect to nuclear centre positions. It is obvious that the gradient of the nuclear attraction integral with respect to the position vector of the nuclear attractor is an electric field integral, and may be evaluated as described above. For more comments on the evaluation of derivatives of molecular integrals with respect to nuclear displacements, see section 6 below.

4.6 The Diamagnetic Contribution to the Orbital Term in NMR Spin-Spin Coupling.

In the theory of NMR spin-spin coupling [27] the diamagnetic contribution to the orbital term requires the evaluation of the integrals

$$DO_{ij} = \int \phi_i \phi_j x_C r_C^{-3} x_D r_D^{-3} d\tau \tag{99}$$

A Monte-Carlo integration scheme has been proposed by Lee and Schulman [28] which appears to converge only slowly, and is less than satisfactory. A procedure based on using the Gaussian transform for one or both of the electric field operators appearing in eqn.(99) has been suggested, giving rise to a one or two dimensional numerical quadrature over electric field or overlap integrals respectively [29]. Both the methods proposed in [29] appear to provide numerically accurate results, but the procedure based on the quadrature over electric field integrals is more economic. We refer to [29] for details.

4.7 The One-Electron Electric Field Gradient Integrals.

The typical components of the electric field gradient are given by:

$$(\partial^2/\partial C_x{}^2 - \partial^2/\partial C_y{}^2) r_C{}^{-1} = 3(x_C{}^2 - y_C{}^2) r_C{}^{-5} \tag{100a}$$

$$\frac{1}{3}(2\partial^2/\partial C_z{}^2 - \partial^2/\partial C_x{}^2 - \partial^2/\partial C_y{}^2) r_C{}^{-1} = (3z_C{}^2 - r_C{}^2) r_C{}^{-5} \tag{100b}$$

$$(\partial^2/\partial C_x \partial C_y) r_C{}^{-1} = 3x_C y_C r_C{}^{-5} \tag{100c}$$

Notice that $\nabla^2 r_C{}^{-1}$ does not arise in practice and can give rise to divergent integrals, vitiating Leibniz's theorem. It is necessary to use the spherical gradient operators [15] as given in eqn.(100) to guarantee convergent integrals. Using an argument similar to that given for the electric field integrals, based on the translational invariance of the integrals, we can show that:

$$[t,u,v|3(x_C{}^2 - y_C{}^2) r_C{}^{-5}] = [t+2,u,v|r_C{}^{-1}] - [t,u+2,v|r_C{}^{-1}] \tag{101a}$$

$$[t,u,v|(3z_C{}^2 - r_C{}^2) r_C{}^{-5}] =$$
$$(2[t,u,v+2|r_C{}^{-1}] - [t+2,u,v|r_C{}^{-1}] - [t,u+2,v|r_C{}^{-1}])/3 \tag{101b}$$

$$[t,u,v|3x_C y_C r_C{}^{-5}] = [t+1,u+1,v|r_C{}^{-1}] \tag{101c}$$

So that the electric field gradient integrals may be written as linear combinations of the nuclear attraction integrals. The field gradient integrals are required in the treatment of the spin dipolar contribution to NMR spin-spin coupling constants [27]. The nuclear attraction, electric field and electric field gradient integrals represent the energy of interaction of the overlap distribution with point electric monopoles, dipoles and quadrupoles respectively, and may be used to approximate the long range electrostatic interaction energies, for example, in crystals [30] and large molecules.

5. TWO-ELECTRON INTEGRALS

The two-electron repulsion integral involves the interaction of two overlap distributions, centred on $\underset{\sim}{A},\underset{\sim}{B}$ and $\underset{\sim}{C},\underset{\sim}{D}$, and functions of the co-ordinates of electrons 1 and 2 respectively. The interaction is through the coulomb repulsion operator, r_{12}^{-1}, so that:

$$ERI_{ABCD} = \int \int \phi_A(1)\phi_B(1)r_{12}^{-1}\ \phi_C(2)\phi_D(2)d\tau_1 d\tau_2 \qquad (102)$$

Both overlap distributions may be expanded in HGTF, located at centres $\underset{\sim}{P}$ and $\underset{\sim}{Q}$ respectively:

$$ERI_{ABCD} =$$

$$\sum_{t,u,v} E[A,B,t,u,v] \sum_{t',u',v'} E[C,D,t',u',v'][t,u,v|r_{12}^{-1}|t',u',v'] \qquad (103a)$$

$$[t,u,v|r_{12}^{-1}|t',u',v'] =$$

$$(\partial/\partial P_x)^t(\partial/\partial P_y)^u(\partial/\partial P_z)^v(\partial/\partial Q_x)^{t'}(\partial/\partial Q_y)^{u'}(\partial/\partial Q_z)^{v'}[000|r_{12}^{-1}|000] \qquad (103b)$$

5.1 McMurchie/Davidson [11] Algorithm for the Evaluation of the ERI.

The two centre integral over HGTF in eqn.(103b) is invariant to a simultaneous uniform translation of centres $\underset{\sim}{P}$ and $\underset{\sim}{Q}$. Thus:

$$(\partial/\partial P_x)[t,u,v|r_{12}^{-1}|t',u',v'] = -(\partial/\partial Q_x)[t,u,v|r_{12}^{-1}|t',u',v'] \qquad (104)$$

So that, using the definition of the HGTF, eqn.(31):

$$[t+1,u,v|r_{12}^{-1}|t',u',v'] = -[t,u,v|r_{12}^{-1}|t'+1,u',v'] \qquad (105)$$

with similar relationships for the y and z quantum numbers. By successive application of eqn.(105), it can be shown that:

$$[t,u,v|r_{12}^{-1}|t',u',v'] = (-1)^{t'+u'+v'}[t+t',u+u',v+v'|r_{12}^{-1}|0,0,0] \qquad (106)$$

The importance of eqn.(106) should not be underestimated, since by its means a potentially large number of integrals $[t,u,v|r_{12}^{-1}|t',u',v']$ is reduced to a less numerous set of $[t,u,v|r_{12}^{-1}|0,0,0]$. For example if the polynomial degrees of the overlap distributions associated with centres $\underset{\sim}{P}$ and $\underset{\sim}{Q}$ are both μ, then the number of $[t,u,v|r_{12}^{-1}|t',u',v']$ integrals will limit to $\mu^6/36$ as $\mu \to \infty$, whilst the number of $[t,u,v|r_{12}^{-1}|0,0,0]$ integrals will limit to $4\mu^3/3$, an important saving in the high quantum number limit. It is for this reason that the HGTF are to be preferred over the CGTF as an expansion basis, since the latter do not exploit the translation symmetry of the integrals.

Boys [1] evaluated the basic integral as:

$$[0,0,0|r_{12}^{-1}|0,0,0] = 2\pi^{5/2}p^{-1}q^{-1}(p+q)^{-1/2}F_o[W] \tag{107a}$$

$$W = \alpha\,\overline{QP}^2 \tag{107b}$$

$$\alpha = pq/(p+q) \tag{107c}$$

So that, using eqn.(31) and (33) and Leibniz's theorem:

$$[t,u,v|r_{12}^{-1}|0,0,0]$$

$$= 2\pi^{5/2}p^{-1}q^{-1}(p+q)^{-1/2}(\partial/\partial P_x)^t(\partial/\partial P_y)^u(\partial/\partial P_z)^v F_o[W]$$

$$= 2\pi^{5/2}p^{-1}q^{-1}(p+q)^{-1/2}R[t,u,v] \tag{108a}$$

$$R[t,u,v]$$

$$= \alpha^{w/2}\int \exp[-Ws^2]s^w H_t[\alpha^{1/2}\overline{QP}_x s]H_u[\alpha^{1/2}\overline{QP}_y s]H_v[\alpha^{1/2}\overline{QP}_z s]ds \tag{108b}$$

$$w = t + u + v \tag{108c}$$

The R[t,u,v] of eqn.(108) are related to those of eqn.(90) by the substitutions $p \rightarrow \alpha$, $C \rightarrow Q$, so that the recursive scheme for their evaluation, eqn.(89-93), may be applied. By combining eqn. (103,106,108) we obtain the following formula for the ERI over uncontracted GTF:

$$ERI_{ABCD} = 2\pi^{5/2}p^{-1}q^{-1}(p+q)^{-1/2}\sum_{t,u,v} E[A,B,t,u,v]$$

$$\sum_{t',u',v'} E[C,D,t',u',v'](-1)^{t'+u'+v'}\, R[t+t',u+u',v+v'] \tag{109}$$

Eqn.(109) is best evaluated by absorbing the factor $(p+q)^{-1/2}$ into the R[t+t',u+u',v+v'], the factor $2^{1/2}\pi^{5/4}p^{-1}$ into the E[A,B,t,u,v] and the factor $2^{1/2}\pi^{5/4}q^{-1}(-1)^{t'+u'+v'}$ into the E[C,D,t',u',v'] if shell structure is exploited. After these changes of definition of the E-coefficients and R-integrals, which we assume henceforth, the formula for the ERI simplifies to

$$ERI_{ABCD} = \sum_{t,u,v} E[A,B,t,u,v]$$

$$\sum_{t',u',v'} E[C,D,t',u',v']\, R[t+t',u+u',v+v'] \tag{110}$$

McMurchie and Davidson [11] sought to exploit the contracted GTF by implementing eqn.(110) according to fig.1, which for simplicity omits detail concerning the shell organisation of the GTFs.

The cost of the kernels marked ** and * in fig.1 are proportional to $L^4\mu^{10}$ and $L^2\mu^{11}$ respectively, where we assume S,P,D etc

```
I = 0

|- Sum over primitives on A,B
|
|  Set J[t,u,v] zero for all t,u,v
|
|  |- Sum over primitives on C,D
|  |
|  |  Evaluate all R-integrals (scaled by (p+q)^(-1/2))
|  |
|  |  |- Loop over t,u,v                                  *
|  |  |
|  |  |  J[t,u,v] = J[t,u,v]
|  |  |           +   Σ      E[C,D,t',u',v']R[t+t',u+u',v+v']
|  |  |_            t',u',v'
|  |
|  |______
|
| I = I +  Σ    E[A,B,t,u,v]J[t,u,v]                     **
|        t,u,v
|
|__________
```

Figure 1: McMurchie/Davidson [11] Algorithm for the Evaluation of the ERI.

shells of GTF, all four shells being of contraction depth L and polynomial degree μ respectively. One should note that for more asymmetric batches of integrals one should also consider a structure where the loops over the primitives on $\tilde{A},\tilde{B}$ and $\tilde{C},\tilde{D}$ of fig.1 are interchanged, as these two structures are not of identical cost.

McMurchie [31] has devised a scheme for the evaluation of the ERI to exploit the shell structure of GTF in the case the GTF are separable, making use of eqn.(55). McMurchie's [31] scheme has costs proportional to $L^4\mu^8$ and $L^2\mu^{10}$ for S,P,D shells of CGTF, a clear improvement on the McMurchie/Davidson scheme [11]. However, in an implementation on a scalar computer, the McMurchie scheme [31] was slower than the McMurchie/Davidson scheme, because of high overheads due to short loops. However, if the evaluation of the ERIs is vectorized using the 'extrinsic' scheme [32] (that is, evaluate large numbers of batches of integrals simultaneously) then the costs of the short loop overheads may be amortized over long vector lengths, making the McMurchie scheme [31] preferable.

5.2 Rys Quadrature [20] for the Evaluation of the ERI.

Eqn.(110) coupled with the recursive scheme for the evaluation of the R[t,u,v], eqn.(89-93), make it clear that the ERI may be written as a linear combinatiuon of $F_m[W]$:

$$ERI_{ABCD} = \sum_{m=0}^{\lambda} C_m F_m[W] \tag{111}$$

where λ denotes the sum of the polynomial degrees of the four GTF involved. Therefore, using eqn.(87c):

$$ERI_{ABCD} = \int_0^1 S_\lambda[s^2]\exp[-Ws^2]ds \tag{112}$$

where $S_\lambda[s^2]$ is a polynomial in s^2 of degree λ. As first observed in [20], it follows from the theory of orthogonal polynomials that the ERI, eqn.(112), may be evaluated exactly by means of a γ-point Gaussian quadrature [33]:

$$\int_0^1 S_\lambda[s^2]\exp[-Ws^2]ds = \sum_{i=1}^{\gamma} S_\lambda[s_i^2]\Delta_i \tag{113}$$

where the Δ_i are the quadrature weights, and s_i^2 are the roots of the Rys polynomial of order γ, $R_\gamma[s^2,W]$, orthogonal as follows:

$$\int_0^1 R_n[s^2,W]R_m[s^2,W]\exp[-Ws^2]ds = \delta_{nm}J_{nn} \tag{114a}$$

$$R_o[s^2,W] = 1 \tag{114b}$$

the minimum acceptable value of γ (which is of course used in practice) being:

$$\gamma = [\lambda/2] + 1 \tag{115}$$

Notice carefully that the arguments of S_λ and the Rys polynomials is s^2, eqn.(112,114), so that we must seek values of s^2 which generate zeros of the Rys polynomials to compute the s_i^2. A procedure for the determination of the roots and weights has been given in [33] and summarized in [34]. The essence of the Gaussian quadrature is to choose roots and weights so as to make the procedure exact for the monomial integrals, $F_m[W]$, m = 0 to $2\gamma - 1$, so that the 2γ parameters (roots and weights) are determined completely by the 2γ monomial integrals. King and Dupuis [35] show that the Rys roots may be related to the positive roots of the even order Legendre and Hermite polynomials in the limits that W tends to zero and infinity respectively, and have also shown how Tchebychev interpolation may be applied to a table of pre-computed roots and weights, thus much reducing the cost of their determination during integral evaluation.

To complete the algorithm for the evaluation of the ERI, it is necessary to evaluate the $S_\lambda[s_i^2]$ which appear in eqn.(113). Dupuis et al [20] did not specify their procedure in their pioneering work, but it appers from a study of their code, HONDO, that a formalism based on using CGTF as expansion functions for the overlap distributions was used. A scheme based on the HGTF expansion basis [36] turns out to be better, and this we now describe. First consider the evaluation of the R[t,u,v], eqn.(108b), by means of Rys quadrature:

$$R[t,u,v] = \int_0^1 \exp[-Ws^2] G_t[\alpha^{1/2}s, \overline{QP}_x] G_u[\alpha^{1/2}s, \overline{QP}_y] G_v[\alpha^{1/2}s, \overline{QP}_z] ds \qquad (116a)$$

$$G_t[a,b] = a^t H_t[ab] \qquad (116b)$$

so that after quadrature:

$$R[t,u,v] = \sum_{i=1}^{\gamma} \Delta_i G_t[\alpha^{1/2}s_i, \overline{QP}_x] G_u[\alpha^{1/2}s_i, \overline{QP}_y] G_v[\alpha^{1/2}s_i, \overline{QP}_z] \qquad (117)$$

We may devise a recursion scheme for the $G_t[a,b]$ based on the recursion scheme for the Hermite polynomials, eqn.(41):

$$G_{t+1}[a,b] = 2a^2(b\, G_t[a,b] - t\, G_{t-1}[a,b]) \qquad (118)$$

thus enabling the $G_t[\alpha^{1/2}s_i, QP_x]$ to be evaluated, and similarly for the G_u and G_v of eqn.(117). Notice that the factor $\Delta_i/(p+q)^{1/2}$ may be usefully merged with the $G_t[\alpha^{1/2}s_i, QP_x]$ if shell structure is exploited. This we assume henceforth. We now specialize to the case of separable GTF (the McMurchie/Davidson [11] scheme is to be preferred for non-separable GTF), and using eqn.(56,110,117):

$$\mathrm{ERI}_{ABCD} =$$

$$\sum_i \sum_t e_x[A,B,t] \sum_{t'} e_x[C,D,t'] \, G_{t+t'}[\alpha^{1/2}s_i, \overline{QP_x}]$$

$$\sum_u e_y[A,B,u] \sum_{u'} e_y[C,D,u'] \, G_{u+u'}[\alpha^{1/2}s_i, \overline{QP_y}]$$

$$\sum_v e_z[A,B,v] \sum_{v'} e_z[C,D,v'] \, G_{v+v'}[\alpha^{1/2}s_i, \overline{QP_z}] \quad (119)$$

where the factor $2\pi^{5/2}p^{-1}q^{-1}(-1)^{t'+u'+v'}$ has again been appropriately merged with the e-coefficients. We define

$$X_i = \sum_t e_x[A,B,t] \sum_{t'} e_x[C,D,t'] \, G_{t+t'}[\alpha^{1/2}s_i, QP_x] \quad (120)$$

with similar definitives for Y_i and Z_i factors, so that the ERI may be reduced to the form:

$$\mathrm{ERI}_{ABCD} = \sum_i X_i Y_i Z_i \quad (121)$$

The most time consuming element in the Rys quadrature scheme has a cost proportional to $L^4\mu^9$ for S,P,D etc CGTF shells of equal contraction and polynomial degree, clearly an improvement over the McMurchie/Davidson [11] and McMurchie [31] schemes in the limit of high μ, low L. For any particular batch of integrals, either the McMurchie [31] or Rys [20] schemes may prove optimal, and an optimal code would include both algorithms, with the capability to select the appropriate scheme for any given situation. For further discussion of these matters, see [36].

5.3 Special Co-ordinate Systems.

The method to be described [18] requires that the GTF be contracted and that the shell structure is exploited if an economy is to be expected in the evaluation of the ERIs. Let $\phi[A,r]$ denote a contracted GTF located at $\underset{\sim}{A}$, with angular factor (or shell member index) r. $\phi[A,r]$ is a linear combination of primitive GTF of identical angular factor, $\phi[i,A,r]$, so that:

$$\phi[A,r] = \sum_i d[i,A,r] \, \phi[i,A,r] \quad (122)$$

where the d[i,A,r] are the contraction coefficients. The overlap distribution of two contracted GTF takes the form:

$$\phi[A,r]\phi[B,s] = \sum_{ij} d[i,A,r]d[j,B,s]\phi[i,A,r]\phi[j,B,s]$$
$$= \sum_{P,v} E[P,r,s,v]\phi[P,v] \quad (123)$$

where we have converted the sum over ij into one over the centres $\underset{\sim}{P}$ associated with each primitive overlap distribution, and we have used an expansion basis located at $\underset{\sim}{P}$ with angular factors indexed by v. Thus for SP=L shells there are 16 overlap distributions to be expanded into 10 expansion functions for each centre $\underset{\sim}{P}$. A similar expansion may be made for a second overlap distribution:

$$\phi[C,t]\phi[D,u] = \sum_{Q,w} E[Q,t,u,w]\phi[Q,w] \quad (124)$$

so that the ERI may be written:

$$\left[\phi[A,r]\phi[B,s]\middle|\phi[C,t]\phi[D,u]\right] = \sum_{P,v} E[P,r,s,v]J[P,v,t,u] \quad (125a)$$

$$J[P,v,t,u] = \sum_{Q,w} E[Q,t,u,w]\left[\phi[P,v]\middle|\phi[Q,w]\right] \quad (125b)$$

For L shells there would in general be 160 (=4*4*10) J[P,v,t,u] integrals. However, by choosing the special axis system, AXIS-1, shown in fig.2, only 88 such integrals are finite, and there is considerable simplification in the evaluation of the $\left[\phi[P,v]\middle|\phi[Q,q]\right]$. It is important to appreciate that each change of co-ordinate system involves a change (rotation) of the GTF involved in the integrals, and therefore a change in the definition of the integrals themselves.

AXIS-1 has its z axis along the $\overline{CD}$ bond, the x axis being orthogonal thereto and passing through the point $\underset{\sim}{P}$. The y axis is chosen to complete a right handed orthogonal system. The key features of AXIS-1 are that its definition is independent of $\underset{\sim}{Q}$, thus allowing the sum over Q in eqn.(125b) to be performed in a common co-ordinate system, whilst the J[P,v,t,u] integrals whose sum of y quantum numbers is odd are zero.

The J[P,v,t,u] must now be transformed (rotated) to an axis system whose definition does not depend on P (clearly AXIS-1 does not meet this criterion), and for this purpose AXIS-2 is chosen, shown in fig.3. Again the z axis coincides with the CD bond, the x axis is chosen to be orthogonal to both AB and CD (x = AB × CD), so that special care is needed if AB is nearly parallel to CD. The y axis is chosen to complete a right handed orthogonal system.

AXIS-2 is related to AXIS-1 by a translation, which leaves the

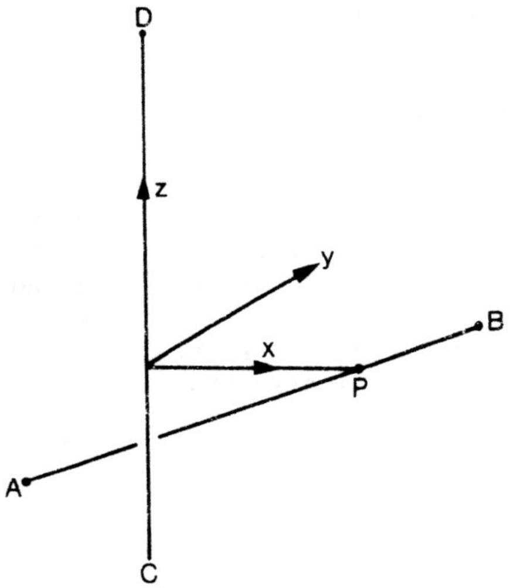

Fig.2: AXIS-1 Co-ordinate System.

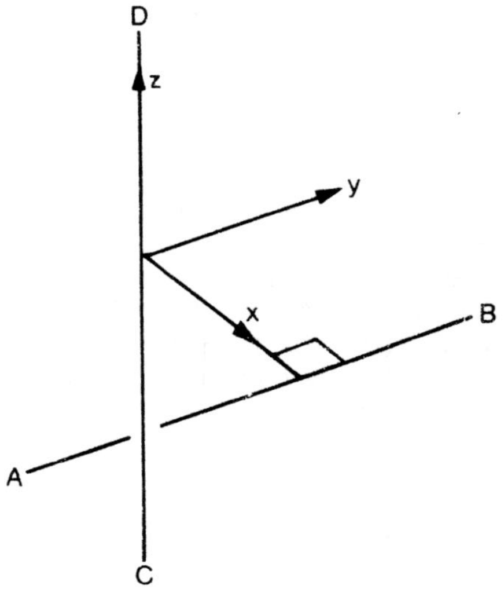

Fig.3: AXIS-2 Co-ordinate System.

J[P,v,t,u] integrals invariant, and so may be ignored and a rotation about the z axis. It is this simple form of rotation which is the key feature of AXIS-2, for it simplifies the rotation of the J[P,v,t,u] integrals. After this rotation there will in general be 160 J[P,v,t,u] integrals (for L shells). The sum over P in eqn. (125a) is carried out in AXIS-2, to produce 256 integrals (for L shells). After the sum over P is complete, these integrals must be rotated to AXIS-3, the global co-ordinate system. The special co-ordinate system method has been employed in the code GAUSSIAN70 using CGTF as the expansion basis. The method could probably be marginally improved if the HGTF were used as the expansion basis; this would require knowledge of the rotational properties of the HGTF, but fortunately these are known [37].

Another interesting feature is that the general formulae for the ERI have ben specialized for all possible cases up to L shells [18]. This strategy eliminates short loop overheads (not so significant if an 'extrinsic' vectorized version [32] is coded), and also such absurdities as multiplication by unity or addition of zero. Benard and Barry [22] have extended the principle to F shells in the code ASTERIX.

5.4 Two-Electron Electric Field Integrals.

The x component of the two-electron electric field operator is:

$$x_{12}r_{12}^{-3} = \partial/\partial x_2 r_{12}^{-1} = -\ \partial/\partial x_1 r_{12}^{-1} \tag{126}$$

Therefore, letting $\phi_P(1)$ denote a basis function of the co-ordinates of electron 1, and centred at $\underset{\sim}{P}$:

$$\int \phi_P(1) \int x_{12}r_{12}^{-3}\phi_Q(2)d\tau_2 d\tau_1 = -\int \phi_P(1)\partial/\partial x_1 \int r_{12}^{-1}\phi_Q(2)d\tau_2 d\tau_1 \tag{127}$$

where Leibniz's theorem has been used to move the differential operator outside the integral over the co-ordinates of electron 2. Now the integral $\int r_{12}^{-1}\phi_Q(2)d\tau_2$ is invariant to a uniform translation of the centre $\underset{\sim}{Q}$ and the co-ordinate of electron 1, so that:

$$-\ \partial/\partial x_1 \int r_{12}^{-1}\phi_Q(2)d\tau_2 = \partial/\partial Q_x \int r_{12}^{-1}\phi_Q(2)d\tau_2 \tag{128}$$

This is the same trick as that used in the one-electron electric field operator, section 4.5. Therefore:

$$[\phi_P|x_{12}r_{12}^{-3}|\phi_Q] = \partial/\partial Q_x[\phi_P|r_{12}^{-1}|\phi_Q] \tag{129}$$

where Leibniz's theorem has again been used. Thus if we now specialize to the case that ϕ_P and ϕ_Q are HGTF, and using eqn.(31):

$$[t,u,v|x_{12}r_{12}^{-3}|t',u',v'] = [t,u,v|r_{12}^{-1}|t'+1,u',v']$$

with similar results for the y and z components of the operator, so that integrals over the two-electron electric field operator may be related to electron repulsion integrals.

5.5 Two-Electron Electric Field Gradient Operator.

The gradient formulae for the two-electron electric field gradient operators may be deduced by analogy with the one-electron case, eqn.(100), and again it is necessary to use spherical gradient [15] operators. For example:

$$3x_{12}y_{12}r_{12}^{-5} = (\partial^2/\partial x_2\partial y_2)r_{12}^{-1} = (\partial^2/\partial x_1\partial y_1)r_{12}^{-1} \tag{130}$$

Using tricks similar to those employed for the two-electron electric field operators, we find, for example:

$$[t,u,v|3x_{12}y_{12}r_{12}^{-5}|t',u',v'] = [t,u,v|r_{12}^{-1}|t'+1,u'+1,v'] \tag{131}$$

so that the electric field gradient integrals may be reduced to electron repulsion integrals. For details of the other components of the tensor, see [11]. The two-electron electric field gradient operator appears in the spin-spin interaction of the Breit-Pauli Hamiltonian [38].

5.6 The $F_m[W]$ Integrals.

The evaluation of the nuclear attraction, electron repulsion and related integrals requires the determination of the $F_m[W]$ integrals, eqn.(87c). Satisfactory procedures have been outlined in [2,3], power series expansion being used for small values of W, and a repeated fraction formulation being useful at larger values. However, in practice these integrals are usually produced using interpolation techniques [11,18,39], where the range of W is broken into many sub-intervals, in the interests of efficiency. In the Rys quadrature method [20] one uses interpolation of pre-computed roots and weights [35], by-passing the evaluation of the $F_m[W]$ in so doing.

6. GRADIENT INTEGRALS

Integrals arise which involve derivatives of the basis functions with respect to electron or nuclear co-ordinates, the latter most frequently in the context of the 'force' method [26] for evaluating the gradient of molecular energies. However, the two categories of gradient integral are not independent, because the GTF involve only factors such as $x - A_x$, so that:

$$(\partial/\partial x)\phi_A = -(\partial/\partial A_x)\phi_A \tag{132}$$

where ϕ_A is a GTF located at $\underset{\sim}{A}$.

6.1 Operators Involving Gradients of Electron Co-ordinates.

We have already met one such operator in section 4.3, the kinetic energy. Other examples are (ignoring factors of $\sqrt{-1}$ and spin dependent terms):

$$\hat{M}_x = \partial/\partial x \qquad \text{(x component of momentum)} \tag{133a}$$

$$\hat{L}_{zA} = x_A\partial/\partial y - y_A\partial/\partial x \qquad \text{(z component of angular momentum)} \tag{133b}$$

$$\hat{K}_{zA} = r_A^{-3}\hat{L}_{zA} \qquad \text{(z component of spin orbit)} \tag{133c}$$

Now for certain systems of GTF, the gradient of a given function is easily expressed as a linear combination of the GTF, the most simple case being for HGTF, using eqn.(31,132):

$$\partial/\partial x\ \Lambda[a,A,i,j,k] = -\Lambda[a,A,i+1,j,k] \tag{134}$$

A simple relationship also exists for the CGTF, eqn.(76). This means that integrals of the form:

$$O_{AB} = \int \phi_A \hat{O}_C \phi_B \, d\tau \qquad (135)$$

may be converted to gradient free integrals, where the gradient dependent overlap distribution may be expanded in HGTF using the known result for $(\partial/\partial x)\phi_B$ and the techniques outlined in section 3. In this way the momentum, angular momentum and spin orbit integrals may be reduced to linear combinations of overlap (section 4.1), dipole (section 4.2) and electric field (section 4.3) integrals respectively.

In the case that $(\partial/\partial x)\phi_B$ is not easily converted into a linear combination of GTF of the same system, it is necessary to find the E-coefficients of the HGTF expansion of $\phi_A(\partial/\partial x)\phi_B$ by recursive techniques similar to and based on the procedures described in section 3, but we will not present details.

We should mention that two-electron operators involving gradient factors arise, such as the two-electron spin orbit, which may be reduced to a linear combination of electric field integrals (section 5.4) [11]. The evaluation of spin orbit integrals has also been considered in [38,40].

6.2 Gradients with Respect to Nuclear Co-ordinates.

Using and the techniques of section 6.1, gradients with respect to nuclear co-ordinates may be evaluated, the GTF being assumed to be located at and 'following' [26] such centres. Because of the translational invariance of the integrals, I, we have:

$$\sum_A (\partial/\partial A_x) I = 0 \qquad (136)$$

where the sum over A involves all the centres in I. This means that not all the gradients are independent, a feature which may be exploited to reduce computational expense. Eqn.(136) also implies that if a HGTF basis is chosen, many integrals are inter-related by simple formulae, resulting in potential savings, and this is perhaps the most pressing reason for considering the HGTF for use as a molecular basis set. We should note that other relationships (for example eqn.(80)) are known [41,42] between molecular integrals involving gradients, which may prove useful in providing cross-checks or in speeding up the calculations. For further details on the subject of gradient integrals, consult [43].

7. CONCLUSION

It is hoped that the present work will convince the reader that accurate, rapid procedures exist for the evaluation of the Schrödinger and Breit-Pauli integrals over many different systems of GTF. Expectation values of many kinds of one-electron operator may be determined. These happy circumstances serve to explain the current pre-eminent position of the GTF as a basis function in quantum chemistry.

REFERENCES

[1] Boys, S.F.,: 1950, Proc. Roy. Soc., A200 p542.
[2] Shavitt, I.,: 1963, Methods in Computational Physics, 2 p1.
[3] Saunders, V.R.,: 1975, in 'Computational Techniques in Quantum Chemistry and Molecular Physics', edited by Diercksen, G.H.F.D. Sutcliffe, B.T. and Veillard, A., (Reidel) p347.
[4] Preuss, H.: 1956, Z. Naturforschung, 11a p823;
Witten, J.L.,: 1963, J. Chem. Phys., 39 p349;
Sambe, H.,: 1965, J. Chem. Phys., 42 p1732;
Witten, J.L.,: 1966, J. Chem. Phys., 44 p359;
Harrison, J.F.,: 1967, J. Chem. Phys., 46 p1115;
Powell, R.E.,: 1968, J. Chem. Educ., 45 p45;
Petke, J.D., Witten, J.L. and Douglas, A.W.,: 1969, J. Chem. Phys., 51 p256;
Shillady, D.D. and Richardson, F.S.,: 1970, Chem. Phys. Lett., 6 p359;
Chu, S.Y. and Frost, R.A.,: 1971, J. Chem. Phys., 54 p760;
Driessler, F. and Ahlrichs, R.,: 1973, Chem. Phys. Lett., 23 p571;
Maggiora, G.M. and Christoffersen, R.E.,: 1976, J. Am. Chem. Soc., 98 p8325;
Schmelzer, A.,: 1977, Int. J. Quant. Chem., 11 p561;
Le Rouzo, H. and Silvi, B.,: 1978, Int. J. Quant. Chem., 13 p297;
Le Rouzo, H. and Silvi, B.,: 1978, Int. J. Quant. Chem., 13 p311;
Le Rouzo, H. and Silvi, B.,: 1978, Int. J. Quant. Chem., 13 p325;
Le Rouzo, H.,: 1978, Int. J. Quant. Chem., 14 p121;
Le Rouzo, H.,: 1978, Int. J. Quant. Chem., 15 p57;
Le Rouzo, H.,: 1978, Int. J. Quant. Chem., 19 p11;
[5] Allison, D.J., Handy, N.C. and Boys, S.F.,: 1973, Mol. Phys., 26 p715.
[6] Ditchfield, R.,: 1972, J. Chem. Phys., 56 p5688.
[7] Bransden, B.H.,: 1972, Reports on Progress in Physics, 35 p949
[8] Schiff, L.,: 1949, 'Quantum Mechanics', p72 (McGraw-Hill).
[9] Bell, W.W.,: 1968, 'Special Functions for Scientists and Engineers', (Van Nostrand).
[10] Zivkovic, T. and Maksic, Z.B.,: 1968, J. Chem. Phys., 49 p3083;
Golebiewski, A. and Mrozek, J.,: 1973, Int. J. Quant. Chem., 7 p623;

Golebiewski, A. and Mrozek, J.,: 1973, Int. J. Quant. Chem., 7 p1021;
Van Der Velde, G.A.,: 1974, Ph.D. Thesis (Groningen).
[11] McMurchie, L.E. and Davidson, E.R.,: 1978, J. Comp. Phys., 26 p218.
[12] Roberts, P.J.,: 1965, J. Chem. Phys., 42 p879.
[13] Bell, S.,: 1969, J. Phys.B: At. Mol. Phys., 2 p1001.
[14] Maretis, D.M.,: 1979, J. Chem. Phys., 71 p917
[15] Bayman, B.F.,: 1978, J. Math. Phys., 19 p2558.
[16] Fieck, G.,: 1980, Theor. Chim. Acta (Berlin), 54 p323.
[17] Ahlrichs, R.,: 1974, Theor. Chim. Acta (Berlin), 33 p157.
[18] Pople, J.A. and Hehre, W.J.,: 1978, J. Comp. Phys., 27 p161.
[19] Raffenetti, R.C.,: 1973, J. Chem. Phys., 58 p4452.
[20] Dupuis, M., Rys, J. and King, H.F.,: 1976, J. Chem. Phys., 65 p111.
[21] Taketa, H., Huzinaga, S. and O-ohata, K.,: 1966, J. Phys. Soc. (Japan), 11 p2313.
[22] Benard, M. and Barry, M.,: 1979, Computers and Chemistry, 3 p121.
[23] Fieck, G.,: 1979, J. Phys. B: At. Mol. Phys. 12 p1063.
[24] Messiah, A.,: 1961, 'Quantum Mechanics - Volume 1', translated by G.M. Temmer, (North-Holland) p344.
[25] Matsuoka, O.,: 1971, Int. J. Quant. Chem., 5 p1.
[26] Pulay, P.,: 1969, Mol. Phys., 17 p197;
Pulay, P.,: 1977, 'Modern Theoretical Chemistry - Volume 4', edited by H.F. Schaefer, (Plenum) p153
[27] Ramsay, N.F.,: 1953, Phys. Rev., 91 p303
[28] Lee, W.S. and Schulman, J.M.,: 1979, J. Chem. Phys., 70 p1530.
[29] Matsuoka, O. and Aoyama, T.,: 1980, J. Chem. Phys., 73 p5718.
[30] Pisani, C. and Dovesi, R.,: 1980, Int. J. Quant. Chem., 17 p501.
[31] McMurchie, L.E.,: 1977, Ph.D. Thesis (University of Seattle), p47
[32] Saunders, V.R. and Guest, M.F.,: 1982, Comp. Phys. Comm. 26 p389.
[33] Stroud, A.H. and Secrest, D.,: 1966, 'Gaussian Quadrature Formulas' (Prentice-Hall).
[34] Delaney, J.J., Saunders, V.R. and Hillier, I.H.,: 1981, J. Phys. B: At. Mol. Phys., 14 p819.
[35] King, H.F. and Dupuis, M.,: 1976, J. Comp. Phys., 21 p144.
[36] Saunders, V.R.,: 1978, "Proceedings of the 4th Seminar on Computational Methods in Quantum Chemistry, Orenas' (Max-Planck Institut, Garching, Munchen).
[37] Zivkovic, T.,: 1971, J. Chem. Phys., 55 p1672.
[38] Langhoff, S.R. and Kern, C.W.,: 1977, 'Modern Theoretical Chemistry - Volume 4', edited by H.F. Schaefer, (Plenum) p381.
[39] Davidson, E.R.,: 1977, 'Algorithms for Chemical Computations', edited by R.E. Christoffersen, (American Chemical Society) p21.

[40] Geller, M. and Griffith, R.W.,: 1964, J. Chem. Phys., 40 p2309;
Hinkley, R.K., Walker, T.E.H. and Richards, W.G.,: 1970, J. Chem. Phys., 52 p5975;
Abegg, P.W. and Ha, T.K.,: 1974, Mol. Phys., 27 p763;
Langhoff, S.R.,: 1974, J. Chem. Phys., 61 p1708;
Breulet, J.,: 1981, J. Comp. Chem., 2 p244.
[41] Matcha, R.L. and Kern, C.W.,: 1970, Phys. Rev. Lett., 25 p981.
[42] Abegg, P.W.,: 1975, Mol. Phys., 30 p579.
[43] Almlof, J. and Helgakar, T.,: 1981, Chem. Phys. Lett., 83 p125;
Schlegel, H.B.,: 1982, J.Chem. Phys., 77, p3676.

ON THE EVALUATION OF EXPONENTIAL (SLATER) TYPE INTEGRALS

E. Otto Steinborn

Institut für Physikalische und
Theoretische Chemie, Universität Regensburg
D-8400 Regensburg, Fed. Rep. Germany

1. INTRODUCTION

Molecular electronic structure calculations are usually based upon the LCAO-MO approach in which the molecular orbitals (MO's) are represented by linear combinations of atomic orbitals (AO's). For the quantum mechanical calculation of energies and properties of molecules one has to evaluate the matrix elements of the Hamiltonian. With the AO's being centered at the various atomic nuclei in the molecule under consideration, these matrix elements become the infamous multicenter molecular integrals: The use of AO's which resemble physically meaningful one-electron wave functions for the one-nucleus one-electron Coulomb problem (i.e. the hydrogen atom) like exponentially decaying Slater-type orbitals (STO's) leads to multicenter integrals which are notoriously difficult to evaluate for poly-atomic molecules.

As this problem had been a stumbling block to molecular *ab initio* calculations for a long time, Boys (1950)

G. H. F. Diercksen and S. Wilson (eds.), Methods in Computational Molecular Physics, 37–69.

suggested the use of Gaussian-type orbitals (GTO's) as basis functions because the multi-center molecular integrals can be evaluated easily for GTO's. Nowadays, with advanced computational techniques, the vast majority of molecular quantum-mechanical calculations is carried out with GTO basis sets. However, the usage of GTO's is also not without problems in large scale calculations, mainly due to the immensely great number of GTO's to be used. Therefore, interest in applying STO's has recently again risen, especially for investigations in which the behavior of the wave function at very short or long distances from a nucleus is essential, but also for cutting down the number of basis functions to be used in a calculation.

In this article we deal with techniques for the evaluation of the matrix elements of the (nonrelativistic) molecular Hamiltonian, i.e. the multicenter molecular integrals, over Slater-type orbitals (STO's) or, more general, exponential-type functions (ETF's).

Each molecular integral (MI) represents a well defined problem which in principle can be dealt with without referring to other MI's. Most of the MI's have been treated with different methods by different authors. As is well known, not all of these evaluations can be used with sufficient efficiency in molecular calculations. It may be true, though, that in order to finally evaluate efficiently all MI's which are needed in a molecular calculation it will be necessary to rely on different methods for different MI's.

Instead of discussing each MI individually, it seems preferable to discuss some of the methods which have been developed for the evaluation of MI's. Of course, for the evaluation of a certain MI, various mathematical tools may be needed. Nevertheless, the most important features of the evaluation procedure of a given MI may be assigned to one of the methods which we are going to discuss in the following.

In section 2 we give a classification of the ETO's under consideration, in section 3 we discuss the classification of molecular integrals (MI's) which occur. The following sections are devoted to the discussion of the most important methods for the evaluation of MI's for ETO's. In section 4 we discuss the use of addition theorems for the evaluation of MI's, in section 5 we consider various aspects of the Fourier transform method. Miscellaneous methods are discussed in section 6, which is followed by a summary and conclusion and a list of references.

2. CLASSIFICATION OF ORBITALS

2.1. General definitions and remarks

For the definitions and for the well-known formulas and theorems for the special functions of mathematical physics which we use in this context, we refer to the literature. For the more frequently references quoted we use the following abbreviations:

MOS: W.Magnus, F.Oberhettinger, R.P.Soni, Formulas and Theorems for the Special Functions of Mathematical Physics, Springer Verlag, New York 1966.

GR: I.S.Gradshteyn, I.M.Ryzhik, Table of Integrals, Series and Products, Academic Press, New York 1980.

AS: M.Abramowitz, I.S.Stegun, Handbook of Mathematical Functions, Dover, New York 1972.

EH: A.Erdély, W.Magnus, F.Oberhettinger, and F.G.Tricomi, Higher Transcendental Functions, Mc Graw-Hill, New York 1953. Vols. I (EH I), II (EH II), III (EH III).

ET: A.Erdély, W.Magnus, F.Oberhettinger, and F.G.Tricomi, Tables of Integral Transforms, McGraw-Hill, New York 1954, Vols. I (ET I), II (ET II).

Our considerations are based upon the LCAO-MO approximation. For simplicity, we concentrate on the spatial dependencies of the orbitals and disregard spin and relativistic effects. In the LCAO-MO approach an MO ψ_n is constructed by a linear combination of various AO's ϕ_i centered at nuclei with positions $\vec{A}_\mu$, i.e. we want to use the approximation

$$\psi_n(\vec{r}_1) = \sum_{i,\mu} c^n_{i,\mu} \, \phi_i(\vec{r}_1 - \vec{A}_\mu) \tag{2.1-1}$$

Each AO from the set of functions ϕ is characterized by i and $\vec{A}_\mu$. Different indices i of ϕ characterize different functions of the same set ϕ. For instance, under i we may summarize certain parameters or pseudo-quantum numbers like $n, \ell, m, \alpha, \zeta$, specifying various functional shapes of the AO's as defined by the following Eqs. (2.1-3,4). In addition, each AO ϕ_i is defined with respect to a certain center in coordinate space, in general the position of the nucleus of the atom the respective AO belongs to. This means, the AO ϕ_i is a function of the local vector

$$\vec{r}_{A1} = \vec{r}_1 - \vec{A} \tag{2.1-2}$$

which points from center A (the origin of $\vec{r}_{A1}$) to the field point P, i.e. the "position" of the electron no. 1. Different centers may be called A, B, C, D, ..., or A_μ with $\mu = 1,2,3,4,...$, respectively. The positions of these various points (centers) in the coordinate space are defined by the local vectors $\vec{A}$, $\vec{B}$, $\vec{C}$, $\vec{D}$, ..., or $\vec{A}_\mu$ with $\mu = 1,2,3,4,...$, respectively. All these local vectors, just as $\vec{r}_1$, originate in O. In general, the orbital centers coincide with the nuclei of the atoms in the molecule. If, sometimes, the chosen centers do not coincide with all the nuclei, one talks about "floating orbitals".

In this article, we assume the AO's ϕ_{aA} to be given by exponential-type basis functions (ETF's) which shall have the general form

$$G^m_{n,\ell}(\vec{r}_A;\alpha,\zeta) = N^{(g)}(n,\ell,\alpha,\zeta)\cdot e^{-\zeta r_A}\cdot g_{n,\ell}(\alpha r_A)\cdot \mathcal{Y}^m_\ell(\alpha\vec{r}_A) \tag{2.1-3}$$

with

$$\mathcal{Y}^m_\ell(\alpha\vec{r}_A) = (\alpha r_A)^\ell \; Y^m_\ell(\theta_A,\phi_A). \tag{2.1-4}$$

In Eq. (2.1-3), the factor $N^{(g)}$ is usually (but not always) the normalization constant for the function G, depending on the "quantum numbers" n and ℓ and the parameters α and ζ. Of course, the $N^{(g)}$'s are different for different g's. The function $g_{n,\ell}$ is usually given by a polynomial in $(\alpha|\vec{r}_A|)$. In most cases, the function $g_{n,\ell}$ can be represented by one of the standard polynomials as dealt with in the theory of special functions. The definitions of these polynomials shall not be repeated here. In Eqs. (2.1-3), (2.1-4), the $\mathcal{Y}_\ell^m(\alpha\vec{r}_A)$ is a regular solid spherical harmonic. For the spherical surface harmonics Y_ℓ^m we use the convention of Condon and Shortley (1953), p.51. Of course, r_A, θ_A, ϕ_A are the spherical coordinates of the field point P specified by $\vec{r}_A$ in an atomic coordinate system with its origin in A. In a molecule with several centers (nuclei), we assume the atomic coordinate systems to differ by parallel displacements only, i.e. to have their x-axes parallel with respect to each other, etc. For the evaluation of the multicenter integrals, rotations and/or translations of these coordinate systems may be necessary, requiring heavy use of transformation theory.

2.2. Slater-type orbitals

Slater-type orbitals (STO's) are defined by (Slater, 1930)

$$\chi_{n,\ell}^m(\alpha\vec{r}) = N^{(\chi)}(n,\alpha)\; e^{-\alpha r}\; (\alpha r)^{n-1}\; Y_\ell^m(\theta,\phi); \tag{2.2-1a}$$

$$N^{(\chi)}(n,\alpha) = \alpha^{-n+1}\; \left[(2\alpha)^{2n+1}/\Gamma(2n+1)\right]^{1/2} \tag{2.2-1b}$$

$$\alpha > 0,\; n - 1 \geq \ell \geq 0. \tag{2.2-1c}$$

The Slater 1s, $2p_o$, $2p_1$, $2p_{-1}$ orbitals are identical with hydrogen-like orbitals of appropriate Z values. The parameter α is an orbital exponent. The parameter n is an effective principal quantum number which need not be integer, but will be assumed to be integer for mathematical reasons and greater convenience.

2.3. Laguerre-type orbitals

For Laguerre-type orbitals (LTO's), the function $g_{n,\ell}(\alpha r)$ in Eq. (2.1-3) is given by associated Laguerre polynomials. Various notations are common for Laguerre functions. Our notation for $L_n^{(k)}$, following that of MOS and EH II, differs from the notation used by other authors according to (Kaijser and Smith, 1977, p.48)

$$(k+n)!\ L_n^{(k)}(x) = (-1)^k\ {}^{P}L_{k+n}^{(k)}(x) = {}^{M}L_n^{(k)}(x) \qquad (2.3\text{-}1)$$

Here, the left superscript P specifies the notation for L used by Shull and Löwdin (1955), or Podolsky and Pauling (1929), respectively. The left superscript M specifies the notation used by Messiah (1967). In order to avoid confusion, it is recommendable to use confluent hypergeometric functions only, according to

$$L_n^{(\beta)}(x) = \frac{(\beta+1)_n}{n!}\ {}_1F_1(-n;\beta+1;x). \qquad (2.3\text{-}2)$$

We distinguish the following LTO's:

(a) The hydrogen-like functions $H_{n,\ell}^m$ which solve Schrödinger's equation for one electron in the central field of a nucleus with charge Z, n being the principal quantum number (Schrödinger, 1926; Pilar, 1968, p.155, p.248):

$$H_{n,\ell}^m(r;Z/n) =$$

$$= N^{(H)}(n,\ell,Z/n)\cdot e^{-Zr/n}\cdot L_{n-\ell-1}^{(2\ell+1)}(2Zr/n)\cdot(2Zr/n)^{\ell}\cdot Y_\ell^m(\theta,\phi); \qquad (2.3\text{-}3a)$$

$$N^{(H)}(n,\ell,Z/n) = \left[\left(\frac{2Z}{n}\right)^3 \frac{(n-\ell-1)!}{2n(n+\ell)!}\right]^{1/2} \qquad (2.3\text{-}3b)$$

(b) The modified hydrogen-like functions $h_{n,\ell}^{m}$ are obtained from the hydrogen-like functions by putting $Z/n = \alpha$, i.e. by omitting the principal quantum number n in the argument (Schrödinger, 1926, Pilar, 1968, p.248):

$$h_{n,\ell}^{m}(\vec{r};\alpha) =$$

$$= N^{(h)}(n,\ell,\alpha)\cdot e^{-\alpha r}\; L_{n-\ell-1}^{(2\ell+1)}(2\alpha r)\cdot(2\alpha r)^{\ell}\cdot Y_{\ell}^{m}(\theta,\phi) \qquad (2.3\text{-}4a)$$

$$N^{(h)}(n,\ell,\alpha) = \left[(2\alpha)^{3}\,\frac{(n-\ell-1)!}{2n(n+\ell)!}\right]^{1/2} \qquad (2.3\text{-}4b)$$

(c) The Λ functions are ETF's based upon $L_{n-\ell-1}^{(2\ell+2)}$ (Filter and Steinborn, 1980) defined by

$$\Lambda_{n,\ell}^{m}(\alpha\vec{r}) =$$

$$= N^{(\Lambda)}(n,\ell,\alpha)\cdot e^{-\alpha r}\cdot L_{n-\ell-1}^{(2\ell+2)}(2\alpha r)\cdot(2\alpha r)^{\ell}\cdot Y_{\ell}^{m}(\theta,\phi); \qquad (2.3\text{-}5a)$$

$$N^{(\Lambda)}(n,\ell,\alpha) = \left[(2\alpha)^{3}\,\frac{(n-\ell-1)!}{(n+\ell+1)!}\right]^{1/2} \qquad (2.3\text{-}5b)$$

2.4. B functions

The so-called B functions (Filter and Steinborn, 1978b)

$$B_{n,\ell}^{m}(\alpha\vec{r}) = \left[2^{n+\ell}\,(n+\ell)!\right]^{-1}\,\hat{k}_{n-1/2}(\alpha r)\cdot(\alpha r)^{\ell}\,Y_{\ell}^{m}(\theta,\phi) \qquad (2.4\text{-}1)$$

are closely related to reduced Bessel functions (RBF's)

$$\hat{k}_{\nu}(z) = (2/\pi)^{1/2}\,z^{\nu}\,K_{\nu}(z) \qquad (2.4\text{-}2)$$

which are constructed from modified Bessel functions $K_\nu(z)$ of the second kind (MOS, p.66).

In the notation of Eq. (2.1-3), with $\alpha = \zeta$, the $g_{n,\ell}(\alpha r)$ is the Bessel polynomial $\Theta_n(z)$ (Grosswald, 1978, p.18; Weniger, 1982, p.39):

$$\Theta_n(z) = e^z \, \hat{k}_{n+1/2}(z) = 2^{-n} \sum_{\rho=0}^{n} \frac{(n+\rho)!}{\rho!(n-\rho)!} (2z)^{n-\rho} = \qquad (2.4\text{-}3a,b)$$

$$= z^n \, {}_2F_0(-n, n+1 \,; -\frac{1}{2z}) = 2^{-n} \, (n+1)_n \cdot {}_1F_1(-n;-2n;2z) \qquad (2.4\text{-}3c,d)$$

3. CLASSIFICATION OF MOLECULAR INTEGRALS

Quantum mechanical calculations require in general the diagonalization of the matrix of the Hamiltonian operator or the application of the Ritz variation principle. In each case it is necessary to calculate the matrix elements of the Hamiltonian or its expectation values first, which lead to the molecular integrals.

We restrict our considerations to non-relativistic Hamiltonians within the realm of the Born-Oppenheimer approximation:

$$\hat{H} = \sum_{i=1}^{N} \{-\frac{1}{2}\Delta_i - \sum_{\mu=1}^{K} \frac{Z_\mu}{|\vec{r}_i - \vec{A}_\mu|} + \sum_{\substack{j=1 \\ i<j}}^{N} \frac{1}{r_{ij}}\} + \sum_{\substack{\mu,\nu=1 \\ \mu<\nu}}^{K} \frac{Z_\mu Z_\nu}{A_{\mu\nu}} \qquad (3\text{-}1)$$

The indices i,j refer to the numbers of the N electrons, the indices μ,ν refer to the numbers of the K nuclei with charges Z_μ, Z_ν at positions $\vec{A}_\mu$, $\vec{A}_\nu$ with $A_{\mu\nu} = |\vec{A}_\mu - A_\nu|$.

In the MO scheme, the molecular wave function $\Psi(1,2,\ldots,N)$, depending on the coordinates of the N electrons, is constructed by antisymmetrized products of N spin orbitals. The spatial part ψ_n of these one-electron functions is given by Eq. (2.1-1). Therefore, the matrix elements of the Hamiltonian in this representation are given by sums of molecular integrals which are either one-electron integrals or two-electron integrals.

Only three types of one-electron integrals occur, according to the three different one-electron operators, i.e. the identity, the Laplacian, and the Coulomb operator $1/|\vec{r}-\vec{C}|$. We distinguish:

Overlap integrals

$$S(A,B) = \int \chi_a^*(\vec{r}-\vec{A})\, \chi_b(\vec{r}-\vec{B})\, d\vec{r} \tag{3-2}$$

The value of this 2-center overlap integral depends on the amount of "overlap" of the two AO's χ_a and χ_b. S_{AB} becomes a 1-center overlap integral for A = B.

Kinetic energy integrals

$$T(A,B) = \int \chi_a^*(\vec{r}-\vec{A})\left(-\frac{1}{2}\,\frac{\partial}{\partial\vec{r}}\cdot\frac{\partial}{\partial\vec{r}}\right) \chi_b(\vec{r}-\vec{B})\, d\vec{r} \tag{3-3}$$

There are 2-center ($A \neq B$) and 1-center ($A = B$) kinetic energy integrals.

Nuclear attraction integrals

$$I_{NA}(A,B;C) = \int \chi_a^*(\vec{r}-\vec{A})\, \frac{1}{|\vec{r}-\vec{C}|}\, \chi_b(\vec{r}-\vec{B})\, d\vec{r} \tag{3-4}$$

In the most general (3-center) case is $A \neq B \neq C$. If two or all of the three centers A,B,C coincide, one obtains 2- or 1-center integrals.

The two-electron integrals are classified with respect to the products of AO's $\chi_a^*(\vec{r}_{A1})\chi_b(\vec{r}_{B1})$ and $\chi_c^*(\vec{r}_{C2})\chi_d(\vec{r}_{D2})$ occurring in the integrands. They are referred to as charge distributions, and may be one-center or two-center charge distributions depending upon whether or not centers A and B, or C and D, coincide. In the Hamiltonian Eq. (3-1) the only two-electron operator is the electronic Coulomb repulsion term $1/r_{12}$. Therefore, the various two electron integrals originate from different constellations of the centers of AO's.

The most general two-electron integral originating from the Hamiltonian Eq. (3-1) in the LCAO-MO method is the 4-center electron repulsion integral

$$\iint \chi_a^*(\vec{r}_1-\vec{A})\ \chi_b(\vec{r}_1-\vec{B})\ \frac{1}{|\vec{r}_1-\vec{r}_2|}\ \chi_c^*(\vec{r}_2-\vec{C})\ \chi_d(\vec{r}_2-\vec{D})\ d\vec{r}_1\ d\vec{r}_2 \qquad (3\text{-}5)$$

with $A \neq B \neq C \neq D$.

The Coulomb integral is an electron repulsion integral whose charge distributions are both 1-center. There are 2-center ($A = B \neq C = D$) and 1-center ($A = B = C = D$) Coulomb integrals.

A hybrid integral is an electron repulsion integral with one charge distribution being 2-center, the other being 1-center. $A = B \neq C \neq D$ specifies a 3-center, $A = B = C \neq D$ a 2-center hybrid integral.

An exchange integral as a counterpart of the Coulomb integral is the 2-center integral defined by Eq. (3-5) with $A = C \neq B = D$. Sometimes any electron repulsion integral as defined by Eq. (3-5) whose charge distributions are both 2-center, is called a (general) exchange integral, including the 3-center case ($A \neq B = C \neq D$) as well as the 4-center case ($A \neq B \neq C \neq D$).

The nomenclature for Coulomb and exchange integrals used here, which is the usual one used in quantum chemistry, does not agree with the nomenclature used in atomic physics.

4. ADDITION THEOREMS

4.1. Properties of addition theorems

An addition theorem is an expansion of a given function $f(\vec{r}-\vec{R})$ in terms of functions $U_\mu(\vec{r})$ and $V_\nu(\vec{R})$ according to

$$f(\vec{r}-\vec{R}) = \sum_{\mu,\nu} C_{\mu,\nu} \, U_\mu(\vec{r}) \, V_\nu(\vec{R}) \qquad (4.1\text{-}1)$$

where the function U_μ and V_ν belong to well-known sets of functions U and V, respectively, and the coefficients $C_{\mu\nu}$ are given by closed form expressions. In this expansion Eq. (4.1-1), the variables $\vec{r}$ and $\vec{R}$ are completely separated. Therefore, addition theorems can be used for simplifying and evaluating molecular integrals because they make it possible to separate the integration variables in the integrand of a MI.

It is difficult and in many cases it has not yet been possible to find an addition theorem for a given function f. Unfortunately, in most cases only addition theorems of the less favorable form

$$f(\vec{r}-\vec{R}) = \sum_{\mu,\nu} C'_{\mu,\nu}\, U_\mu(\vec{r}_<)\, V_\nu(\vec{r}_>) \tag{4.1-2}$$

can be found. Here $\vec{r}_<$ and $\vec{r}_>$ refer to the greater and lesser, respectively, of $\vec{r}$ and $\vec{R}$, depending on the locations in space where $\vec{r}$ and $\vec{R}$ are pointing to.

The functional form of an addition theorem depends on the coordinates chosen. In spherical coordinates, one may expand Eq. (4.1-2) further in terms of spherical harmonics Y_ℓ^m (which absorb all angular dependencies) and other known functions u_μ^ℓ and $v_\nu^{\ell'}$ according to

$$f(\vec{r}-\vec{R}) = \sum_{\mu,\nu} \sum_{\ell,m} \sum_{\ell',m'} c_{\mu\nu}^{\ell\ell'mm'} \cdot$$
$$\cdot\; u_\mu^\ell(r_<)\, v_\nu^{\ell'}(r_>)\, Y_\ell^{m*}(\hat{r}_<)\, Y_{\ell'}^{m'}(\hat{r}_>) \tag{4.1-3}$$

A well-known example for an addition theorem of the kind as given by Eq. (4.1-3) is the Laplace expansion of the Coulomb potential of an electric unit point charge which is shifted by $\vec{R}$ from the origin O:

$$\frac{1}{|\vec{r}-\vec{R}|} = \sum_{\ell=0}^{\infty} \sum_{m=-\ell}^{\ell} [4\pi/(2\ell+1)]\, r_<^{\ell} r_>^{-\ell-1}\, Y_\ell^{m*}(\hat{r}_<)\, Y_\ell^m(\hat{r}_>) \tag{4.1-4a}$$

$$r_< = \min(r,R), \qquad r_> = \max(r,R) \tag{4.1-4b}$$

This is an expansion in terms of regular and irregular solid spherical harmonics, $r^\ell Y_\ell^m(\hat{r})$ and $r^{-\ell-1} Y_\ell^m(\hat{r})$, respectively, which are defined with respect to the origin O. The solid angle (θ,ϕ) is specified by $\hat{r}$.

An addition theorem for a given function f(r) will exhibit a discontinuity according to Eq. (4.1-4b), if the corresponding power series

$$f(r) = \sum_{n=o}^{\infty} a_n \, r^{\lambda+n} \qquad (4.1-5)$$

exists and contains not only powers of r with $\lambda+n$ being even (Filter 1978, p.56). Otherwise the addition theorem will be symmetric (without discontinuity). Examples for that are the addition theorems for the Gaussian $e^{-\alpha r^2}$ function and $(\alpha r)^{-\nu} J_\nu(\alpha r)$ (Steinborn and Filter, 1975).

In an ellipsoidal coordinate system, the Coulombic repulsion energy of two unit point charges can be expressed by the expansion introduced by Neumann (1878):

$$\frac{1}{r_{12}} = \frac{1}{|\vec{r}_1-\vec{r}_2|} = \frac{4}{R} \sum_{\ell=o}^{\infty} \sum_{m=-\ell}^{\infty} (-1)^m \cdot \frac{(\ell-|m|)!}{(\ell+|m|)!} P_\ell^{|m|}(\xi_<) \, Q_\ell^{|m|}(\xi_>) \cdot \mathcal{P}_\ell^{|m|}(\eta_1) \mathcal{P}_\ell^{|m|}(\eta_2) \, e^{im(\phi_1-\phi_2)} \qquad (4.1-6)$$

The ellipsoidal coordinate system (ξ,η,ϕ) is based on foci at A and B separated by $R = |\vec{A}-\vec{B}|$; then $\xi = (r_a+r_b)/R$, $\eta = (r_a-r_b)/R$ with r_a and r_b representing the distances from the coordinate point to $\vec{A}$ and $\vec{B}$, respectively. The azimuthal angle ϕ is measured from a reference halfplane bounded by a line through A and B. In Eq. (4.1-6), the coordinates of the two electrons are (ξ_1,η_1,ϕ_1) and (ξ_2,η_2,ϕ_2), and $\xi_<$ and $\xi_>$ refer to the lesser and greater, respectively, of ξ_1 and ξ_2 . P_ℓ^m and Q_ℓ^m are the associated Legendre functions of the first and the second kind, respectively; the $\mathcal{P}_\ell^m$ are the normalized P_ℓ^m.

4.2. Derivation of addition theorems

If the given function f can be expanded into a three-dimensional Taylor series, the application of the translation operator

$$\mathcal{T}_{\vec{R}} = \exp\left(-\vec{R}\cdot\frac{\partial}{\partial\vec{r}}\right) \tag{4.2-1}$$

onto f gives

$$\mathcal{T}_{\vec{R}}\, f(\vec{r}) = \sum_{\nu=0}^{\infty} \frac{1}{\nu!}\left(-\vec{R}\cdot\frac{\partial}{\partial\vec{r}}\right)^{\nu} f(\vec{r}) = f(\vec{r}-\vec{R}) = F(\vec{r}) \tag{4.2-2}$$

Usually the differentiation cannot be accomplished in closed form in order to derive Eq. (4.1-1) with $C_{\mu\nu}$ being given explicitly, as it is essential. For instance, three- dimensional functions which transform like an irreducible tensor, have characteristic addition theorems with expansion coefficients containing Clebsch-Gordan coefficients (Steinborn and Filter, 1979).

One possibility for the derivation of addition theorems is the application of Fourier transformations as discussed in section 5. For another approach, it seems most helpful to exploit special properties of the functions f. Some functions f are, for instance, generating functions of orthogonal polynomials leading to expansions which are, in fact, rearranged Taylor series. As is well known, the Coulomb potential is the generating function of the Legendre polynomials. On the other hand, Bessel functions are generating functions of the Gegenbauer polynomials (MOS sections 3.9, 5.3). With the help of these relationships addition theorems for several functions could be derived (Steinborn and Filter, 1975), for instance for r^{λ}, log (kr), the Gaussian function e^{-kr^2}, for $r^{-\nu}\, Z_{\nu}(r)$ with Z_{ν} representing a Bessel or modified Bessel function, and the Slater-s-type function $r^{n-1}\, e^{-\alpha r}$. A special case of the function $r^{-\nu}\, Z_{\nu}(r)$ is the reduced Bessel function $\hat{k}_{\nu}(r)$. The addition theorems for the corresponding anisotropic functions could be derived by applying the addition theorems for the solid spherical harmonics (Steinborn and Ruedenberg, 1973).

Due to space considerations, only some references from the extensive literature on addition theorems can be given. Sack (1964 a,b) used properties of differential equations for the derivation of addition theorems for, among others, functions $r^N Y_L^M(\theta,\phi)$. Löwdin (1956) introduced his α function technique for the expansion of a wave function on one center in spherical harmonics about another centers. Nozawa (1966) developed addition theorems for Helmholtz harmonics. Sharma (1968, 1976) gave a closed form expression for Löwdin's α function and derived an addition theorem for STO's.

So far, the methods described can be applied to orbitals, which are elements of Hilbert spaces, as well as to operators which do not belong to the corresponding Hilbert spaces. For orbitals we can use expansions which converge with respect to the norm of the Hilbert space. If $\{\phi_k\}_{k=1}^{\infty}$ is a complete orthonormal set in $L^2(\mathbb{R}^3)$, then any function $f \in L^2$ can be expanded in the sense of the norm according to

$$f(\vec{r}-\vec{R}) = \sum_{k} C_k(\vec{R})\ \phi_k(\vec{r}) \tag{4.2-3a}$$

$$C_k(\vec{R}) = \int \phi_k^*(\vec{r})\ f(\vec{r}-\vec{R})\ d\vec{r} \tag{4.2-3b}$$

Expanding the scalar products $C_k(\vec{R})$ with respect to the function $\phi_\ell(\vec{R})$, we obtain

$$f(\vec{r}-\vec{R}) = \sum_{k,\ell} c_{k,\ell}\ \phi_k(\vec{r})\ \phi_\ell(\vec{R}) \tag{4.2-4}$$

The only complete orthonormal set (CONS) in the Hilbert space $L^2(\mathbb{R}^3)$ which consists of functions that decline exponentially for large distances and are proportional to r^ℓ in the vicinity of the origin are the Λ functions defined by Eq. (2.3-5). Because of this reason, they have been used by various authors (Hylleraas, 1929; Löwdin and Shull, 1956; Howell and Shull, 1959). We used the Λ functions as a CONS to derive new types of addition theorems which converge with respect to the norm of the Hilbert space $L^2(\mathbb{R}^3)$ for Λ functions and B functions (Filter and Steinborn, 1980, p.2733, Eqs. (5.11), (5.12). It is essential to use a CONS for expansions with respect to the norm because otherwise the existence of an expansion like Eq. (4.2-3) is not guaranteed.

Smeyers (1966) and Guseinov (1978, 1980) used a similar approach to express an STO in terms of shifted STO's. However, because STO's do not form a CONS, such series expansions need not exist.

4.3. One-center expansions for integrals

Addition theorems are in fact one-center expansions if one variable - like $\vec{R}$ in Eq. (4.1-1) - is fixed, i.e.

expansions of a given function $f(\vec{r})$ about another origin $\vec{R}$. Such expansions have been widely used for MI evaluations. This approach was probably first suggested by Coolidge (1932) and developed by Landshoff (1936) and Löwdin (1947). Barnett and Coulson (1951) developed a computational scheme and MI program for STO's by expanding an s-tpye STO in terms of Legendre functions and so-called zeta functions containing the radial dependencies. Non-s-type STO's were expressed with the help of the addition theorem of solid spherical harmonics. Zeta functions with higher indices had to be computed by time-consuming two-dimensional recurrence formulas (Barnett, 1963). However, they can also be represented in closed form (Steinborn and Filter, 1975).

The α function technique by Löwdin (1956) is closely related to Barnett-Coulson's approach. One-center expansions for multi-center MI's over STO's were also extensively applied by Harris and Michels (1965, 1966, 1967) who shifted a complex STO along the z-axis and computed the expansion coefficients V also by recurrence relations. Similar expansions for STO's were derived by Silverstone (1967b) who used the Fourier-transform convolution theorem and contour integration.

The application of addition theorems which possess a discontinuity, i.e. a two-range form, leads to indefinite integrals of special functions which in most cases are unknown. This problem is avoided if norm-convergent one-range addition theorems like Eq. (4.2-4) are used. The norm-convergent addition theorem for the Λ function has been used for the evaluation of the three-center nuclear attraction integral (Trivedi and Steinborn, 1982a) which produces satisfactory accuracy but still is not fast enough.

These norm-convergent addition theorems are not only useful for the evaluation of MI's, but they can also be used for economizing calculations with single-center expansions of wave functions (Kranz and Steinborn, 1982).

5. FOURIER-TRANSFORM METHODS

5.1. Properties of Fourier transformations

A given function $f(\vec{r})$ and its Fourier transform $\overline{f}(\vec{p})$ are connected by the relationships (Judd, 1975, p.51)

$$\overline{f}(\vec{p}) = (2\pi)^{-3/2} \int e^{-i\vec{p}\cdot\vec{r}} f(\vec{r})\, d\vec{r} \qquad (5.1\text{-}1)$$

$$f(\vec{r}) = (2\pi)^{-3/2} \int e^{i\vec{r}\cdot\vec{p}}\, \overline{f}(\vec{p})\, d\vec{p} \qquad (5.1\text{-}2)$$

which can be verified with the help of

$$\int e^{i\vec{r}\cdot\vec{p}} d\vec{p} = (2\pi)^3\, \delta(\vec{r}) \qquad (5.1\text{-}3)$$

The Fourier and inverse Fourier transforms exist for orbitals which are square integrable and therefore elements of $L^2(R^3)$. The Fourier transform of the Coulomb operator $1/r$ exist only in the sense of generalized functions, being given by (Gel'fand and Shilov, 1964, p.194)

$$(2\pi)^{-3/2} \int e^{-i\vec{p}\cdot\vec{r}}\, \frac{1}{r}\, d\vec{r} = \left(\frac{2}{\pi}\right)^{1/2} \frac{1}{p^2} \qquad (5.1\text{-}4)$$

The Laplacian in coordinate space corresponds to $-p^2$ in momentum space. One often needs the Rayleigh expansion of a plane wave in terms of spherical Bessel functions j_ℓ and spherical harmonics Y_ℓ^m (Weissbluth, 1978, p.11):

$$e^{\pm i\vec{k}\cdot\vec{r}} = 4\pi \sum_{\ell=o}^{\infty} \sum_{m=-\ell}^{\ell} (\pm i)^\ell \, j_\ell(kr)\cdot Y_\ell^{m^*}(\vec{k}/k)\, Y_\ell^m(\vec{r}/r) \tag{5.1-5}$$

If the representation of $f(\vec{r})$ as given by Eq. (5.1-2) is introduced into Eq. (4.2-2) and the Rayleigh expansion Eq. (5.1-5) applied twice, one obtains (Ruedenberg, 1967; Silverstone 1967b)

$$f(\vec{r}-\vec{R}) = (32\pi)^{1/2} \sum_{\ell,\ell'} \sum_{m,m'} i^{\ell-\ell'}\, Y_\ell^m(\vec{r}/r)\, Y_{\ell'}^{m'}(\vec{R}/R)\cdot$$
$$\cdot\int Y_\ell^{m^*}(\vec{p}/p)\, Y_{\ell'}^{m'^*}(\vec{p}/p)\, j_\ell(rp)\, j_{\ell'}(Rp)\, \overline{f}(\vec{p})\, d\vec{p}. \tag{5.1-6}$$

This is a formal solution of the translation problem as discussed in section 4.1., based upon the translation properties of the plane waves $e^{i\vec{p}\cdot\vec{r}}$ according to

$$\mathcal{T}_{\vec{R}}\, e^{i\vec{p}\cdot\vec{r}} = e^{-\vec{R}\cdot\partial/\partial\vec{r}}\, e^{i\vec{p}\cdot\vec{r}} = e^{i\vec{p}\cdot(\vec{r}-\vec{R})} \tag{5.1-7}$$

Whether in Eq. (5.1-6) the remaining integral can be expressed in closed form or not depends largely on the functional form of the Fourier transform $\overline{f}(\vec{p})$.

5.2. Fourier transforms of exponential-type basis sets

A Fourier transform of an STO is

$$(2\pi)^{-3/2}\int e^{-i\vec{p}\cdot\vec{r}}\, \chi_{n,\ell}^m(\alpha\vec{r})\, d\vec{r} = \frac{\alpha^{2n-\ell-1}\,(n+\ell+1)!}{(2\pi)^{1/2}(1/2)_{\ell+1}\left[\alpha^2+p^2\right]^{n+\ell+1}}\cdot$$
$$(-i/2)^\ell\, \mathcal{Y}_\ell^m(\vec{p})\; {}_2F_1\left(\frac{\ell-n}{2}, \frac{\ell-n+1}{2};\; \ell+3/2;\; -p^2/\alpha^2\right) \tag{5.2-1}$$

This formula is obtained by inserting $\chi_{n,\ell}^{m}(\alpha\vec{r})$ for $f(\vec{r})$ into Eq. (5.1-1), applying Rayleigh's expansion Eq. (5.1-5), and representing the remaining radial integral $\int_0^\infty r^{n-1}\, j_\ell(pr)\, e^{-\alpha r} dr$ by a hypergeometric series ${}_2F_1$ (Watson, 1966, p. 385, Eq. (3)), which terminates after a finite number of terms due to $n \geq \ell + 1$ with $n, \ell \in \mathbb{N}$, because either $n-\ell$ or $n-\ell-1$ is an even integer.

Other Fourier transforms of an STO may be obtained by taking advantage of transformation properties of the ${}_2F_1$ (MOS, p. 47). In addition, several other Fourier transforms of STO's are given in literature (Geller, 1962, p. 2626; Geller, 1963, p.85; Silverstone, 1966, p.4338; Henneker and Cade, 1968, p.576). All of these representations are of similar complexity as Eq. (5.2-1).

Fourier transforms of hydrogenic functions are as complicated as those of STO's. Instead of the hypergeometric function like that in Eq. (5.2-1), one has then a Gegenbauer polynomial (Podolsky and Pauling, 1929, p.114) which is just a special terminating hypergeometric polynomial ${}_2F_1$ (MOS, p.220). We may conclude that functions which are simple in coordinate space need not be simple in momentum space.

For the Fourier transform of a B function, one obtains from Eq. (5.1-1), using Rayleigh's expansion Eq. (5.1-5),

$$(2\pi)^{-3/2} \int_0^\infty e^{-i\vec{p}\cdot\vec{r}}\, B_{n,\ell}^{m}(\alpha\vec{r})\, d\vec{r} = (2/\pi)^{1/2}\, (-i\alpha)^{\ell} \cdot$$
$$\cdot 2^{-n-\ell}\, [(n+\ell)!]^{-1}\, Y_\ell^m(\vec{p}/p) \int_0^\infty r^{\ell+2}\, j_\ell(pr)\, \hat{k}_{n-1/2}(\alpha r) dr \tag{5.2-2}$$

The radial integral in Eq. (5.2-2) can be evaluated using (MOS, p. 100)

$$\int_0^\infty t^{\mu+\nu+1} K_\mu(\alpha t)\, J_\nu(\beta t)dt = \frac{2^{\mu+\nu}\, \alpha^\mu\, \beta^\mu\, \Gamma(\mu+\nu+1)}{(\alpha^2+\beta^2)^{\mu+\nu+1}} \tag{5.2-3}$$

This finally leads to (Weniger, 1982, p.160)

$$(2\pi)^{-3/2} \int e^{-i\vec{p}\cdot\vec{r}}\, B^m_{n,\ell}(\alpha\vec{r})\, d\vec{r} = \frac{(2/\pi)^{1/2}\, \alpha^{2n+\ell-1}}{[\alpha^2+p^2]^{n+\ell+1}} \cdot (-i)^\ell \mathcal{Y}^m_\ell(\vec{p}) \tag{5.2-4}$$

B functions seem to have the simplest Fourier transforms of all exponentially declining functions. Therefore, they seem to be better suited than STO's for all applications where analytical manipulations in momentum space are important.

5.3. Overlap integrals

Prosser and Blanchard (1962) could show that with the help of Eqs. (5.1-2) and (5.1-3) an overlap integral can be transformed into a one-center momentum space integral:

$$\int f^*(\vec{r})\, g(\vec{r}-\vec{R})\, d\vec{r} = \int e^{-i\vec{R}\cdot\vec{p}} \cdot \overline{f}^*(\vec{p})\, \overline{g}(\vec{p})d\vec{p} \tag{5.3-1}$$

This representation of the overlap integral makes it possible to obtain a separation of the integration variables without using an addition theorem. This is especially advantageous if f and g are spherical tensors and if Rayleigh's expansion Eq. (5.1-5) is used. Unfortunately, the remaining radial integrals contain spherical Bessel functions which require a lot of manipulative skill. For overlap integrals over STO's this method has been applied by many authors (Geller,

1962; Silverstone, 1966, 1967a; Harris and Michels, 1967; Todd, Kay and Silverstone, 1970; Avery and Cook, 1974; Edwards, Gottlieb and Doddrell, 1979). However, these authors were not able to derive compact analytical results. On the other hand, it could be shown that the overlap integrals of B functions can be represented quite conveniently in terms of B functions again: (Filter and Steinborn, 1978b)

$$\int d\vec{r}\; B_{N_1,L_1}^{M_1 *}(\alpha\vec{r})\; B_{N_2,L_2}^{M_2}[\alpha(\vec{r}-\vec{R})] =$$

$$= 4\pi\alpha^{-3}(-1)^{L_2} \sum_{\ell} \langle L_2M_2|L_1M_1|\ell m\rangle \cdot$$

$$\cdot \sum_{t}(-1)^{t}\binom{\Delta\ell}{t}\; B_{N_1+N_2+L_1+L_2-\ell-t+1,\ell}^{m}(\alpha\vec{R}), \quad (5.3\text{-}2a)$$

$$\Delta\ell = (L_1+L_2-\ell)/2 \quad (5.3\text{-}2b)$$

The overlap integral over B functions with different scaling parameters is of similar complexity (Filter and Steinborn, 1978a, p.80, Eqs. (3.1), (3.2); Filter and Steinborn 1978b, p.4, Eqs. (4.6), (4.7). These convolution theorems for B functions were originally derived using the addition theorem of the modified Helmholtz harmonics (Filter and Steinborn, 1978a, p.80, Eq. (3.7)). The convolution theorems can also be derived with the help of the Fourier transform method according to Eq. (5.3-1) (Weniger, 1982). The simplicity of these theorems follows from the simplicity of the Fourier transform of B functions Eq. (5.2-4). A comparison of the Fourier transforms of B functions and STO's, i.e. Eqs. (5.2-4) and (5.2-1), shows that similarly compact formulas probably cannot be derived for STO's.

The convolution theorem for B functions were used for the derivation of the norm-convergent addition theorems for Λ and B functions (Filter and Steinborn, 1980, Eqs. (5.11), (5.12),and for compact analytical representations of Coulomb integrals (see next section).

5.4. Interelectronic repulsion integrals

The electron repulsion integrals are of the general type

$$I(\Omega_1,\Omega_2) = \iint \Omega_1^*(\vec{r}_1) \frac{1}{|\vec{r}_1-\vec{r}_2|} \Omega_2(\vec{r}_2)\, d\vec{r}_1 d\vec{r}_2 \qquad (5.4\text{-}1)$$

Geller and Griffith (1964) could show that with the help of Eq. (5.1-2) and Eq. (5.1-3) the above integral can be represented as an inverse Fourier transform

$$I(\Omega_1,\Omega_2) = 4\pi \int \overline{\Omega}_1^*(\vec{p}) \frac{1}{p^2} \overline{\Omega}_2(\vec{p})\, d\vec{p} \qquad (5.4\text{-}2)$$

Hence, a six-dimensional integral in coordinate space with non-separated integration variables $\vec{r}_1$ and $\vec{r}_2$ can be transformed into a three-dimensional momentum space integral in which the angular integration can be performed immediately if Ω_1 and Ω_2 are spherical tensors.

If we have two one-center charge distributions Ω_1 and Ω_2 which are located at different centers, we obtain Coulomb integrals

$$\iint \Omega_1^*(\vec{r}_1) \frac{1}{|\vec{r}_1-\vec{r}_2-\vec{R}|} \Omega_2(\vec{r}_2) d\vec{r}_1 d\vec{r}_2 =$$

$$= 4\pi \int e^{-i\vec{R}\cdot\vec{p}}\, \overline{\Omega}_1^*(\vec{p}) \frac{1}{p^2} \overline{\Omega}_2(\vec{p}) d\vec{p} \qquad (5.4\text{-}3)$$

These integrals resemble very much the convolution integrals Eq. (5.3-2). This cannot be seen if the problem is treated in coordinate space only. The representation of the Coulomb integral as an inverse Fourier integral, Eq. (5.4-3), which was first noticed by Geller and Griffith (1964), has subsequently been applied quite frequently for the computation of Coulomb integrals of STO's (Geller, 1964; Silverstone, 1966; Harris and Michels, 1967; Harris, 1969; Todd, Kay and Silverstone, 1970).

The simplicity of the convolution theorems for B functions made it possible to find extremely compact representations for the Coulomb integrals over B functions, for instance

$$\iint B_{n_1,\ell_1}^{m_1 *}(\alpha\vec{r}_1)\,\frac{1}{|\vec{R}+\vec{r}_2-\vec{r}_1|}\,B_{n_2,\ell_2}^{m_2}(\alpha\vec{r}_2)\,d\vec{r}_1 d\vec{r}_2 =$$

$$= (4\pi)^2\,\alpha^{-5}(-1)^{\ell_2}\sum_{\ell} \langle \ell_2 m_2|\ell_1 m_1|\ell m\rangle \,\cdot$$

$$\cdot\sum_t (-1)^t \binom{\Delta\ell}{t}\Big[(2\ell-1)!!\,(\alpha R)^{-\ell-1}\,Y_\ell^m(\vec{R}/R) - \sum_q B_{q-\ell,\ell}^m(\alpha\vec{R})\Big]. \tag{5.4-4}$$

Coulomb integrals with different scaling parameters have a similar complexity (Filter and Steinborn, 1978b).

The Fourier transform method can also be applied to two-center charge distributions. Therefore, also the most general four-center integral can be treated via the Fourier transform method. However, then one needs the Fourier transform of the two-center charge distribution

$$(2\pi)^{-3/2}\int e^{-i\vec{p}\cdot\vec{r}}\,f^*(\vec{r})\,g(\vec{r}-\vec{R})\,d\vec{r} \tag{5.4-5}$$

There is an extensive literature on two-center Fourier transforms of STO's (Bonham, Peacher and Cox, 1964; Laurenzi 1969; Silverstone and Todd, 1971; Monkhorst and Harris, 1972; Graovac et al., 1973, 1980; Guidotti, Arrighini and Marinelli, 1979; Junker, 1980, 1981; Trivedi and Steinborn, 1982b). In all cases, though, one had to resort to a numerical integration. Nevertheless, it seems that this approach, despite its drawbacks, may numerically be more efficient than certain other expansion techniques.

6. MISCELLANEOUS METHODS

In this section we discuss some procedures which are not directly related to the methods considered so far, but which proved helpful for the evaluation of STO-MI's, or at least for certain types of these integrals.

The work on many-center MI's in ellipsoidal coordinates turned out to be very fruitful at least for diatomic molecules for which the ellipsoidal coordinate system is the natural choice. Ruedenberg (1951) introduced the ellipsoidal-coordinate expansion method for exchange integrals. By using the Neumann expansion Eq. (4.1-6) and various procedures to get a well-organized program, for instance, by using intermediate quantities like densities repeatedly, the computational evaluation of the diatomic MI's has been achieved, among others, by Harris and Michels (1967), Silver and Ruedenberg (1968), Mehler and Ruedenberg (1969), and Hagstrom (1975). Efforts to generalize effectively these approaches to MI's with more than two centers proved not to be very successful so far.

In a so-called brute-force method Wahl (1967) et al. used numerical integration procedures extensively and systematically.

Attempts have been made to utilize the fact that the Poisson equation relates a Coulomb integral to an integral over overlap integrals (O-Ohata and Ruedenberg, 1966), but this scheme is still rather complicated. Harris (1969) used properties both of Fourier transforms and single-center expansions of STO's to obtain formulas for Coulomb integrals as finite linear combinations of overlap and nuclear attraction integrals. The method by Silver and Ruedenberg (1968) to calculate Coulomb integrals in terms of auxiliary functions related to hypergeometric functions provides adequate numerical stability.

An interesting approach to the STO-MI problem is the application of the Gaussian-transform method by Shavitt and Karplus (1962) and Shavitt (1963). This is based on the integral transformation of a 1s-STO

$$e^{-\alpha r} = (\alpha/2)\, \pi^{-1/2} \int_0^\infty s^{-3/2}\, e^{-(\alpha^2/4s)-sr^2}\, ds \qquad (6\text{-}1)$$

Introduced into the many-center MI after some transformation of variables and manipulations, one obtains integrals between Gaussians and multidimensional integrals over dummy variables which have to be carried out numerically. As in the case of MI's for Gaussian-type orbitals (GTO's), formulas for integrals involving higher orbitals are obtained by the application of differential operators to the basic 1s formula.

7. CONCLUSION

For the evaluation of a given MI, it may be necessary to use procedures at the same time which so far have been assigned to different methods. Therefore, the distinction of the various methods as given in this article is somewhat arbitrary. However, it may help to order the tremendous amount of work put into the problem of STO-MI evaluations by numerous authors until recently.

An excellent review of the work done before 1967 has been given by Huzinaga (1967). In the present article, only a selection of papers, with emphasis on more recent work, could be discussed, and due to the lack of space, many potentially important papers could not be mentioned.

Although a unified approach to the whole MI problem is desirable, it seems necessary to rely on several different methods in order to finally evaluate the MI's over STO's, as for each MI a specially suited method may prove to be the most practical one. Therefore, the efforts of the various authors attacking this problem by different methods will not be in vain, and despite all difficulties an optimistic view seems most reasonable although the final solution of the MI problem is not yet at hand. In several groups programs for STO-MI's - sometimes for certain quantum numbers only - are apparently in use. However, as long as these programs are not generally available and competitive with existing GTO-MI programs, at least for some special projects, it is fair to say that we are still in need of a satisfactory solution of the STO-MI problem. Of course, finally only the performance of a whole MI computer program package will be essential, and this requires much concern for the over-all organization of such a program. On the other hand, for instance by repeated use of often needed quantities, one may be able to produce a good program for all MI's needed in a calculation even if <u>some</u> MI's may still need an inconveniently long time for being calculated. This is especially true as it is not the time needed for the computation of <u>one</u> integral which counts, but the time needed for all integrals. This fact finally may well be in favor of the STO's for large basis set calculations, because of the cut down on the size of basis sets needed for a molecular calculation in comparison with the use of GTO basis sets, due to the better physical behavior of STO's in the vicinity of and in large distances from the origin.

ACKNOWLEDGMENTS

The author is very much indebted to Dr. G.H.F. Diercksen and Dr. S. Wilson, the directors of the NATO Advanced Study Institute on *Methods in Computational Molecular Physics* 1982, for inviting him to participate as a lecturer in this institute. The author thanks Dr. E.J.Weniger for many fruitful discussions. Meetings with Dr. David M. Silver made possible by NATO Research Grant No.1861 were also very helpful and stimulating.

LITERATURE

AVERY, J., COOK, M., (1974), Theoret.Chim.Acta 35, 99

BARNETT, M.P., COULSON, C.A. (1951) Phil.Trans.Roy. Soc. A 243, 221

BARNETT, M.P. (1963) in Methods in Computational Physics 2, 95, Academic Press, New York

BONHAM, R.A., PEACHER, J.L., COX, H.L. (1964), J.Chem. Phys. 40, 3083

BOYS, S.F. (1950), Proc.Roy.Soc. A 200, 542

CONDON, E.U., SHORTLEY, G.H. (1970), The Theory of Atomic Spectra, At the University Press, Cambridge

COOLIDGE, A.S. (1932), Phys.Rev. 42, 189

EDWARDS, S.A., GOTTLIEB, H.P.W., DODRELL, D.M. (1979), Mol.Phys. 38, 1147

FILTER, E. (1978), Dissertation, Universität Regensburg

FILTER, E., STEINBORN, E.O. (1978a), J.Math.Phys. 19, 79

FILTER, E., STEINBORN, E.O. (1978b), Phys.Rev. A 18, 1

FILTER, E., STEINBORN, E.O. (1980), J.Math.Phys. 21, 2725

GELLER, M. (1962), J.Chem.Phys. 36, 2424

GELLER, M. (1963), J.Chem.Phys. 39, 84

GELLER, M. (1964), J.Chem.Phys. 41, 4006

GELLER, M., GRIFFITH, R.W. (1964), J.Chem.Phys. 40, 2309

GEL'FAND, I.M., SHILOV, G.E. (1964), Generalized Functions I, Academic Press, New York

GRAOVAC, A., MONKHORST, H.J., ZIVKOVIC, T. (1973), Int.J.Quantum Chem. 7, 233

GRAOVAC, A., KOVACEVIC, K., MAKSIC, Z.B., VESELI, A. (1980), Int.J.Quantum Chem. 17, 747

GROSSWALD, E. (1978), Bessel Polynomials, Springer-Verlag, Berlin

GUIDOTTI, C., ARRIGHINI, G.P., MARINELLI, F. (1979), Theoret.Chim.Acta 53, 165

GUSEINOV, I.I. (1978), J.Chem.Phys. 69, 4990

GUSEINOV, I.I. (1980), Phys.Rev. A 22, 369

HAGSTROM, S. (1975), QCPE, No. 252

HARRIS, F.E., MICHELS, H.H. (1965), J.Chem.Phys. 43, S 165

HARRIS, F.E., MICHELS, H.H. (1966), J.Chem.Phys. 45, 116

HARRIS, F.E., MICHELS, H.H. (1967), Adv.Chem.Phys. 13, 205

HARRIS, F.E. (1969), J.Chem.Phys. 4770

HENNECKER, W.H., CADE, P.E. (1968), Chem.Phys.Lett. 2, 575

HOWELL, K.M., SHULL, H. (1959), J.Chem.Phys. 30, 627

HUZINAGA, S. (1967), Progr. Theor.Phys.Suppl. 40, 52

HYLLERAAS, E.A. (1929), Z.Phys. 54, 347

JUDD, B.R. (1975), Angular Momentum Theory for Diatomic Molecules, Academic Press, New York

JUNKER, B.R. (1980), J.Phys. B 13, 1049

JUNKER, B.R., (1981), Comput.Phys. Commun. 23, 377

KAIJSER, P., SMITH, V.H. (1977), Adv.Quantum Chem. 10, 37

KRANZ, H.H., STEINBORN, E.O. (1982), Phys.Rev. A 25, 66

LANDSHOFF, R. (1936), Z.Physik 102, 201

LAURENZI, B.J. (1969), Int.J.Quantum Chem. 3, 489

LÖWDIN, P.O. (1947), Ark.Mat.Fys.Astr. 35A, No. 9

LÖWDIN, P.O. (1956), Adv.Phys. 5, 1

LÖWDIN, P.O., SHULL, H. (1956), Phys.Rev. 101, 1730

MEHLER, E.L., RUEDENBERG, K. (1969), J.Chem.Phys. 50, 2575

MESSIAH, A. (1967), Quantum Mechanics, North-Holland, Amsterdam

MONKHORST, H.J., HARRIS, F.E. (1972), Int.J.Quantum Chem. 6, 601

NEUMANN, F.E. (1878), Vorlesungen über die Theorie des Potentials und der Kugelfunktionen, Teubner, Leipzig

NOZAWA, R. (1966), J.Math.Phys. 7, 1841

O-OHATA, K., RUEDENBERG, K. (1966), J.Math.Phys. 7, 547

PILAR, F.L. (1968), Elementary Quantum Chemistry, McGraw-Hill, New York

PODOLSKY, B., PAULING, L. (1929), Phys. Rev. 34, 109

PROSSER, E.P., BLANCHARD, C.H. (1962), J.Chem.Phys. 35, 1112

RUEDENBERG, K. (1951), J.Chem.Phys. 19, 1459

RUEDENBERG, K. (1967), Theoret.Chim.Acta 7, 359

SACK, R.A. (1964a), J.Math.Phys. 5, 245

SACK, R.A. (1964b), J.Math.Phys. 5, 252

SCHRÖDINGER, E. (1926), Ann.Physik 79, 361

SHARMA, R.R. (1968), J.Math.Phys. 9, 505

SHARMA, R.R. (1976), Phys.Rev. A13, 517

SHAVITT, I., KARPLUS, M. (1962), J.Chem.Phys. 36, 550

SHAVITT, I. (1963), in Methods in Computational Physics 2, 1, Academic Press, New York

SHULL, H., LÖWDIN, P.O. (1955), J.Chem.Phys. 23, 1362

SILVER, D.M., RUEDENBERG, K. (1968), J.Chem.Phys. 49, 4306

SILVERSTONE, H.J. (1966), J.Chem.Phys. 45, 4337

SILVERSTONE, H.J. (1967a), J.Chem.Phys. 46, 4368

SILVERSTONE, H.J. (1967b), J.Chem.Phys. 47, 537

SILVERSTONE, H.J., TODD, H.D. (1971), Int.J.Quantum Chem. 4, 371

SLATER, J.C. (1930), Phys.Rev. 36, 57

SMEYERS, Y.G. (1966), Theoret.Chim.Acta 4, 452

STEINBORN, E.O., FILTER, E. (1975), Theoret.Chim.Acta 38, 273

STEINBORN, E.O., FILTER, E. (1979), Theoret.Chim.Acta 52, 189

STEINBORN, E.O., RUEDENBERG, K. (1973). Adv.Quantum Chem. 7, 1

TODD, H.D., KAY, K.G., SILVERSTONE, H.J. (1970), J.Chem.Phys. 53, 3951

TRIVEDI, H.P., STEINBORN, E.O. (1982a), Phys.Rev. A25, 113

TRIVEDI, H.P., STEINBORN, E.O. (1982b), Phys.Rev. A, Submitted for Publication

WAHL, A.C., LAND, R.H. (1967), Int.J.Quantum Chem. 1S, 375

WATSON, G.N. (1966), A Treatise on the Theory of Bessel Functions, At the University Press, Cambridge

WEISSBLUTH, M.(1978), Atoms and Molecules, Academic Press, New York

WENIGER, E.J. (1982), Dissertation, Universität Regensburg

BASIS SETS

S. Wilson

Theoretical Chemistry Department,
University of Oxford, 1 South Parks Road, Oxford

The problem of constructing basis sets for use in molecular electronic structure calculations is discussed. Recent progress is reviewed paying particular attention to the use of even-tempered basis sets, universal basis sets and systematic sequences of basis sets. The importance of basis functions containing higher harmonics in calculations which take account of electron correlation effects is briefly examined.

1. INTRODUCTION

Most applications of the methods of quantum mechanics to molecular electronic structure begin with the choice of a one-electron basis set in terms of which the many-electron wavefunction is parameterized. It is this choice of basis set which ultimately determines the accuracy of a given calculation. No amount of configuration interaction or extension of perturbation series to higher order will compensate for an inadequate basis set. In atomic studies, spherical symmetry permits a separation of the problem into an angular part, which can be treated analytically, and a radial part, which involves a one-dimensional numerical integration (see, for example, Froese-Fischer (1)). For diatomic molecules, progress can be made using numerical integration techniques, if a suitable coordinate system is taken (see, for example, Christiansen and McCullough (2)). However, for arbitrary polyatomic molecules it is, in general, impossible to factorize the molecular field and therefore, the numerical approach becomes intractable making the use of a basis set almost obligatory.

In this article, we review some aspects of the problem of constructing basis sets for use in molecular calculations. This prob-

G. H. F. Diercksen and S. Wilson (eds.), Methods in Computational Molecular Physics, 71–93.

lem is common to all of the different methods described in this volume for accurate studies of molecular systems:- configuration interaction, coupled electron pair approximations, many-body perturbation theory, etc. Particular attention will be paid to the relation of the basis sets to the accuracy achieved in the study of correlation effects. In the calculation of molecular properties, particularly correlation energies, there are two truncation errors: that arising from the use of a restricted finite basis set and that resulting from the truncation of the expansion for the correlation energy, the correlated wavefunction or other expectation value. It is important, particularly using methods such as the coupled electron pair approximation or many-body perturbation theory, which are not variationally bound, that apparently good results are not obtained by a fortuitous cancellation of these two errors.

In section 2, a brief description of the magnitude of the basis set truncation error is given. An outline of the relation between the properties of the model problem obtained when a finite basis set is employed and those of the exact problem is present in section 3. Section 4 is devoted to one-centre and multi-centre expansions and discusses the relation between the two.

Different types of basis functions, which have been suggested for molecular calculations, are briefly reviewed in section 5. In section 6, we attempt to survey the current state of the art of choosing a basis set. Even-tempered basis sets are discussed in section 7 and universal basis sets in section 8. In section 9, the use of systematic sequences of basis sets is surveyed and, finally, the convergence of harmonic expansions is addressed in section 10. Section 11 contains a brief overview of some further aspects of the construction of molecular basis sets.

2. BASIS SET TRUNCATION ERRORS

Ideally, when calculating molecular properties, it is preferable to calculate both upper and lower bounds to expectation values. However, the determination of lower bounds is particularly difficult and, in a recent survey of atomic calculations, Bunge (3) summarises the situation as follows: "A few significant calculations have apparently exhausted the present possibilities of current theories... In general today's rigorous error bounds are too large to be of much use..." The situation with respect to the calculation of rigorous error bounds for molecules is even less satisfactory than for atoms.

The vast majority of contemporary molecular calculations adopt a pragmatic approach; no error bounds are determined and accuracy of a calculation in assessed by comparison with quantities derived from experiment. Although it can be difficult to separate the error associated with basis set truncation from that associated with truncation of expansions for correlation effects, this procedure enables the quality of the basis set to be established and an empirical range of validity of the basis set to be obtained which can then be used to make educated guesses of the accuracy of a computed molecular property. The art of selecting a basis set is based on previous experience in treating similar systems using basis sets of comparable quality and will be discussed further in section 6.

There can be little doubt that the error attributable to the truncation of the basis set is the largest source of error in most accurate calculations of electron correlation energies and correlation effects. This is illustrated in Table 1 where the results of calculations on the ground states of the FH, N_2 and CO molecules at their equilibrium geometry demonstrate that the use of a larger basis set is often more important than the evaluation of higher-order correlation effects. Calculated correlation energies for negative ions and for Rydberg states are particularly sensitive to the quality of the basis set.

Table 1

Relative importance of basis set truncation and higher order correlation effects[a]

Molecule	Correlation Energy[b]			
	Basis set A[c]		Basis set B[c]	Empirical Correlation energy
	E(3)	E(4)	E(3)	
FH	-0.2207	-0.2276	-0.3135	-0.381
N_2	-0.3193	-0.3486	-0.4447	-0.538
CO	-0.3028	-0.3312	-0.4294	-0.525

(a) based on the work of Krishnan, Binkley, Seeger and Pople (4). Wilson and Silver (5), Wilson and Guest (8). Atomic units are employed.

(b) calculated by diagrammatic many-body perturbation theory, E(i) denotes the correlation energy through ith order, E(3) includes only configurations which are doubly excited with respect to the reference determinant while E(4) also includes singly, triply and quadruply excited configurations.

(c) basis set A, which is that of Krishnan, Binkley, Seeger and Pople (4), is smaller than basis set B, which is that of Wilson and Silver (5).

Basis set truncation errors are often particularly significant in calculations of properties other than the energy. In Table 2, we display the variation of the dipole moment of the water molecule within the matrix Hartree-Fock approximation for increasing size of basis set. Some properties depend on the quality of the basis set in particular regions of space. Accurate polarisabilities can be computed by employing basis sets capable of describing deformation of the electronic charge cloud on applying an electric field which should, therefore, include diffuse basis functions. On the other hand, when calculating properties such as spin-orbit coupling constants it is necessary to include basis functions capable of providing an accurate description of the wavefunction close to the nucleus. The magnetic vector potential can be defined in an infinite number of ways since it can be subjected to a gauge-transformation, such as a change of origin. Since physical observables should clearly be gauge-invariant, calculated magnetic properties should not depend on the choice of gauge and this is indeed the case if complete basis sets are employed. However, in practice, a finite and incomplete basis set has to be used and calculated magnetic properties are not gauge-invariant. The extent to which calculated magnetic properties are gauge-invariant is an indication of the quality of the calculation and, in particular, the quality of the basis set.

Table 2

Dipole moment of the ground state of the water molecule within the matrix Hartree-Fock approximation using different basis sets[(a)]

Basis set	Dipole moment	Total energy
[5s4p/3s]	1.0672	-76.0207
[5s4p/3s1p]	0.9012	-76.0471
[5s4p1d/3s1p]	0.8576	-76.0559
[5s4p2d/3s2p]	0.7994	-76.0604

(a) The dipole moment and energy are given in atomic units. Basis sets consist of contracted Gaussian functions.

3. THE ALGEBRAIC APPROXIMATION

The determination of the electronic structure of atoms and molecules involves the determination of an appropriate eigenvalue and associated eigenfunctions of a semi-bounded, self-adjoint hamiltonian operator $\hat{\mathcal{H}}$, in Hilbert space, h. In most applications of quantum mechanics to molecules, the N-electron wavefunction is expressed in terms of the Nth rank direct product space V^N generated by a finite dimensional single particle space, V^1, that is

$$V^N = V^1 \otimes V^1 \otimes \ldots\ldots \otimes V^1 \qquad (1)$$

The one-electron functions, $\phi \in V^1$, are parametrized by expansion in terms of a finite basis set $\{\chi_\mu\}$

$$\phi_i = \sum_\mu \chi_\mu C_{i\mu} \qquad (2)$$

where the $C_{i\mu}$'s are expansion coefficients. Then, for example, it is well known that the integro-differential Hartree-Fock equations become a set of algebraic equations for the coefficients. This algebraic approximation is fundamental to most molecular calculations and results in the restriction of the domain of the operator $\hat{\mathcal{H}}$ to a finite dimensional subspace, S, of the Hilbert space, h. The subspace S, can be further subdivided according to the total spin quantum number, S, and the dimension of the total subspace corresponding to each S value is given by Weyl's formula (7).

$$\Gamma_{m,N,S} = \frac{2S+1}{m+1} \binom{m+1}{\frac{1}{2}N-S} \binom{m+1}{m-\frac{1}{2}N-S} \qquad (3)$$

where m is the number of basis functions. $\Gamma_{m,N,S}$ increases rapidly with m, as is demonstrated in Table 3. Klahn and Bingel (8) have investigated the problem of establishing the convergence of an expansion for the wavefunction expression, in terms of all the N-electron configuration functions for a given S and for a given basis set of size m, to the exact result as the number of basis functions tends to infinity.

Table 3

Dependence of Weyl's number on the basis set size

	m	$\Gamma_{m,N,S}$
N=4, S=0	10	825
	20	13300
	30	67425
	40	213200
N=8, S=0	20	5799465
	30	172459665
	40	1851013060

4. ONE-CENTRE AND MULTI-CENTRE EXPANSIONS

In principle, any molecular wavefunction can be expanded in terms of a complete set of functions centred at any convenient point in space. In practice, the one-centre method, otherwise known as the central field approximation or united atom method, usually provides a poor representation of off-centre nuclei and most applications have been limited to hydrides with a single heavy atom, which is used as an expansion centre (9). Although a number of advantages of the one-centre method may be given,(for example, the ease of evaluating the integrals over the basis functions which can lead to a very efficient correlation energy calculation; the control over the convergence of the calculation and the degree of linear dependence; the ease with which energy derivatives with respect to nuclear co-ordinates can be determined) the method is not widely used.

An interesting development of the one-centre method has been proposed by Ladik and Čížek (10) for calculations of correlation energies for large molecules. They suggest that the expression for the total energy of a molecule

$$E = E_{\text{reference}} + E_{\text{correlation}}$$

$$= \langle\Phi|\mathscr{H}|\Phi\rangle + \langle\Phi|V|\chi\rangle, \tag{4}$$

where Φ is the Hartree-Fock function, χ is the correlation correction to the wavefunction and V the fluctuation potential, be modified to give

$$E = \langle\Phi_{I}|\mathscr{H}|\Phi_{I}\rangle + \langle\Phi_{II}|V|\chi_{II}\rangle, \tag{5}$$

where the subscripts I and II denote that multicentre and one-centre basis sets are used, respectively. A multicentre expansion is employed in the Hartree-Fock calculation whilst a one-centre expansion is used in the calculation of correlation effects. Such an approach may prove useful in the treatment of valence correlation effects but atom-centred functions will certainly be required to obtain accurate descriptions of inner-shell correlation.

In the vast majority of contemporary molecular calculations, the difficulties associated with the description of off-centre nuclei in the one-centre expansion approach, are avoided by using basis functions located on a number of centres. Accurate calculations for molecules containing more than one non-hydrogenic nucleus can then be performed. The use of such basis sets does, however, give rise to new problems. As is evident from the articles by Saunders (11) and by Steinborn (12) in the present volume, the evaluation of multicentre, two-electron integrals over basis functions is more complicated than in the one-centre case. Furthermore, as the number of functions in a multicentred basis set tends to infinity, it becomes overcomplete; that is one of the functions could be removed and the basis set would still be complete (8).

The use of multicentre basis sets often leads to overestimates of interaction energies between molecules. This is mainly attributable to the superposition error or counterpoise error (13). This error is particularly significant when small basis sets are used. For a system AB, the description of the subsystem A is improved by the basis functions centred on B and vice versa. The magnitude of the superposition error can be estimated by performing a calculation on A(B) using the basis set employed in the study of AB. McLean and Chandler (14) have recently re-emphasized that lack of control of the superposition error may completely vitiate the calculations of van der Waals energies of interaction or other small energy differences. In a detailed study of the He...H_2 system in the region of the van der Waals minimum, Meyer, Hariharan and Kutzelnigg (15) have demonstrated that the minimum in a previously reported treatment of the system was an artefact resulting from the basis set superposition error. Kutzelnigg (16) has made a formal analysis of wavefunctions employed in calculations of interaction between molecules, paying particular attention to the relation of basis sets used for the "supermolecule" and the fragments and their respective completeness condition.

5. BASIS FUNCTIONS

When implementing the algebraic approximation, there is a considerable degree of freedom in the choice of basis functions since any set of functions which form a complete set can be used. The set of exponential functions $e^{-\zeta_i x}$, where the ζ_i are different orbital exponents, can be shown, for example, to be complete in $L^2(R+)$ if and only if the condition

$$\sum_{i=1}^{\infty} \frac{\zeta_i}{1 + \zeta_1^{2}} = \infty \tag{6}$$

is fulfilled (8,17,18,19). The choice of functional form for the basis functions is mainly influenced by two factors (a) the rate of convergence of orbitals when expanded in the basis set; (b) the ease with which integrals over the basis functions can be evaluated.

The most frequently used basis functions in molecular calculations are exponential-type functions

$$\chi_{n,\ell,m}(r,\theta,\phi) = N_n\, r^{n-1}\, e^{-\zeta r}\, Y_{\ell,m}(\theta,\phi) \tag{7}$$

$$N_n = (2\zeta)^{n+\frac{1}{2}} / \sqrt{(2n)!}$$

where $Y_{\ell,m}(\theta,\phi)$ is a spherical harmonic, and Gaussian-type function

$$\chi_{n,\ell,m}(r,\theta,\phi) = N_n\, r^{n-1}\, e^{-\zeta r^2}\, Y_{\ell,m}(\theta,\phi) \tag{8}$$

$$N_n = \left[\frac{2^{2n}(n-1)!}{(2n-1)} \Big/ \sqrt{\frac{(2S)^{2n+1}}{\pi}}\right]^{\frac{1}{2}}$$

Exponential-type functions usually provide the more rapidly convergent expansion for orbitals and lead to good representations of wavefunctions in regions of space close to the nucleus and in the long range region. Unfortunately, multicentre integrals over exponential type functions are difficult to evaluate and, moreover, the accuracy with which they are evaluated deteriorates for non-linear polyatomic systems. Gaussian-type functions on the other hand, are not well suited to the description of eigenfunction of hamiltonians involving Coulomb potentials and, therefore, relatively large numbers of basis functions are required. The disadvantage is, however, often more than offset by the ease, particularly for the multicentre case, with which the integrals can be evaluated. Furthermore, the efficiency of Gaussian-type basis set can be increased by using contracted rather than primitive functions. Contracted Gaussian-type functions are linear combinations of primitive Gaussian-type functions, the coefficients of which are held fixed during the calculation and not treated as variational parameters. The problem of devising suitable contraction schemes for molecular computation is best summarised by quoting van Duijneveldt (20).

"In order to reduce the number of integrals that must be stored and handled in setting up the HF-matrix, and in order to reduce the size of this matrix, it is useful to contract the primitive basis of GTOs to some extent. Molecular calculations show that contraction of the innermost s- and p- GTOs can be carried through without any loss of accuracy. However, in contracting GTOs in the valence region, one should be very careful. In many cases, the gain in flexibility by <u>not</u> contracting the outer GTOs, more than offsets the increase in computation time (which for an efficient program should be small anyway). An additional advantage of loosely contracted basis sets is that the free-atom exponents need no scaling when used in a molecular environment". Čàrsky and Urban (21) present a detailed comparison of the advantages and disadvantages of exponential-type and Gaussian-type functions in molecular calculations. They conclude that "Gaussian and Slater basis sets of the same size and quality give comparable energies and other molecular properties". If a basis set is sufficiently large and flexible, different choices of basis functions are not reflected in calculated molecular properties.

Some researchers have explored the use of alternative types of basis functions. For example, ellipsoidal functions are sometimes used in studies of diatomic molecules (22). Basis sets of bicentre orbitals, i.e. solutions of the Schrödinger equation for the motion of an electron in the field of two fixed nuclei, have been investigated by Staryk and Kapyshev (23) and by Mogilevsky, Ponomerev and Vakajlovic (24). Silverstone, Carroll and Silver (25) have explored the possibility of using piecewise polynomial basis functions. Hall (26) has discussed various possible basis functions including hydrogenic functions (which must include the continuum functions in order to obtain a complete set), Dirichlet functions and atomic self-consistent-field functions. The most complete review of the different choice of basis functions appears to be that of Burden and Wilson (27). There has been some interest in the use of mixed basis sets in which basis functions of differing types are employed (28).

6. SELECTION OF BASIS SETS

A great deal of knowledge and experience has been accumulated over the years on the selection of basis sets capable of providing a certain accuracy within a given model, (see, for example, ref. (21,29,30,31)). Within molecular orbital theory, the energy corresponding to a minimum basis set of either exponential-type or Gaussian-type functions can be significantly improved by passing to a "double-zeta" basis set which has two functions for each function in the minimum basis set. Double zeta basis sets can be most significantly improved by adding polarization functions, that is, d functions in the case of atoms of the first and second-row. These "double zeta plus polarization" basis sets are widely used. Basis sets which are larger than this, usually referred to as extended basis sets, lead to slow convergence to the Hartree-Fock limit.

It is not proposed to document the large number of different basis sets, particularly Gaussian-type basis sets, which have been proposed for use in molecular calculations. For such details, the reader is referred to recent reviews by Dunning and Hay (29), by Čársky and Urban (21) and by Ahlrichs and Taylor (30). The orbital exponents are often determined by investigation of the variation of the energy, and contraction coefficients are determined from atomic self-consistent-field calculations. Linear combinations of Gaussian-type functions fitted to exponential-type functions are also popular (32). Additional felxibility is sometimes achieved without increasing the number of basis functions by allowing them to 'float' (21, p.38-43 and references therein).

The choice of basis set for a particular application is, on the whole, determined by three factors:- (i) The molecular property of interest. For example, double-zeta basis sets within the self-consistent-field approach will often yield bond lengths, for molecules containing first-row atoms, within 0.01Å of experimental values (21) (a degree of agreement which often deteriorates on adding polarisation functions). On the other hand, in calculations of polarizabilities, extended basis sets including diffuse functions are required, although this can be circumvented by employing electric-field-variant basis sets (33) which depend explicitly on the strength of the external electric field; (ii) The accuracy required. For example, in electron correlation energy calculations, large basis sets are required if a significant fraction of the correlation energy is to be recovered; (iii) Computational demands. These are closely related to the number of basis functions employed. In table 4, the approximate dependence of the various stages of a typical molecular calculation are presented. It should be noted that for extended systems, the dependence of each of these stages on m is reduced (34), for example, the integral evaluation stage then goes as m^2 and the four-index orbital transformation as m^3.

Table 4

Approximate dependence of the various stages of a molecular calculation on the number of basis functions, m

Integral evaluation	m^4
Self-consistent-field iterations	
Fock matrix construction	m^4
Fock matrix diagonalization	m^3
Orbital transformation	
Two-index	m^3
Four-index	m^5
Correlation energy evaluation	
Diagrammatic perturbation theory	
Second-order	m^4
Third-order	m^6

7. EVEN-TEMPERED BASIS SETS

The determination of orbital exponents for Gaussian-type functions or for exponential-type functions by invoking the variation theorem involves a time-consuming, non-linear optimization. The work of Cade and Huo (35) demonstrated, at an early stage, that it is usually much more profitable to improve a basis set by adding extra functions than to embark on the painstaking optimization of the exponents. The large basis sets which are ultimately required for calculations of high accuracy can be efficiently generated by means of the idea of an even-tempered basis set. An even-tempered basis set consists of pure exponential or pure Gaussian functions multiplied by a real solid spherical harmonic. Even-tempered basis sets, therefore, consist of 1s, 2p, 3d.... functions. Furthermore, the orbital exponents are chosen so that their logarithm is a linear function of k, the orbital number. Thus

$$\zeta_k = \alpha\beta^k \qquad k = 1,2,.... \qquad \text{(a)}$$

The use or orbital exponents which form a geometric progression was originally advocated by Reeves (36) and the idea was revived and extensively employed by Ruedenberg and his coworkers (37,38, 39,40).

A number of advantages accrue to the use of even-tempered basis sets:-

(i) Only two parameters, α and β, have to be determined for each group of atomic functions belonging to the same symmetry species. This leads to a tractable approach to the determination of optimal exponents, subject to the constraint that they form a geometric series.

(ii) The further restriction of using the same exponents for all values of ℓ, so that there are only two non-linear parameters per atom, does not produce a very large difference in the calculated energies.

(iii) The question of the proper mixing of functions of different principle quantum number is completely avoided.

(iv) The basis set approaches a complete set in the limit $\alpha \to 0$, $\beta \to 1$, as $k_{max} \to \infty$.

(v) The basis set cannot become linearly dependent if $\beta > 1$.

(vi) Practical linear dependence in the basis set can be controlled.

(vii) The metric matrix, for both exponential and Gaussian basis sets, has the structure

$$\langle i|j\rangle = \langle i+1|j+1\rangle$$

suggesting that even-tempered basis sets have a unique 'space covering' property.

(viii) Restriction of the basis functions to fewer analytic forms leads to simpler and thus more efficient integral evaluation procedures.

8. UNIVERSAL BASIS SETS

In order to achieve high accuracy in molecular calculations, particularly when taking account of electron correlation effects, it is necessary to use moderately large basis sets. Modern vector processing and parallel processing computers are providing the possibility of using large basis sets routinely (41). Since the flexibility of a basis set generally increases with its size and, therefore, the optimization of orbital exponents provides little benefit, the use of a single moderately large basis set which can be transferred from system to system without regard to the nuclear charge or molecular environment is attractive. Such basis sets have been termed universal basis sets (42),(48) and several advantages accrue to their use:-

(a) For a given set of nuclear positions the integrals over basis functions can be evaluated once and then used in all subsequent studies without regard to the identity of the constituent atoms. Alternatively, intermediate quantities which arise in the evaluation of two-electron integrals may be precomputed and the integrals themselves evaluated rapidly from these, perhaps even determining the integrals as they are required in a calculation and thus avoiding, to some extent, the need to handle large data files. For diatomic molecules, a universal basis set of ellipsoidal basis functions could be employed. Since the integrals in this case depend on the internuclear distance only as a multiplicative factor (22), they can be evaluated once and used for different nuclear geometries as well as different nuclear charges.

(b) Almost by definition, a universal basis set provides a uniform description of a series of atoms and molecules.

(c) Since a universal basis set is not optimized with respect to the total energy or any other property, it is expected to provide a uniform description of a range of properties.

(d) The universal basis set is large and, therefore, accurate.

(e) It often possesses a higher degree of symmetry than the

molecule under study.

A universal basis set can be very usefully chosen to be also an even-tempered basis set which leads to the additional advantages listed in section 7. The use of universal even-tempered basis sets in correlation calculations using diagrammatic perturbation theory has been shown to lead to the recovery of significant portions of the correlation energy (5,47). This is illustrated in Table 5 where some typical results are presented. In this Table, a comparison of diagrammatic perturbation theory/universal basis set results with calculations employing a variety of techniques for handling correlation and basis sets designed specifically for each system is made.

9. SYSTEMATIC SEQUENCES OF BASIS SETS

In 1963, Schwarz (59) wrote:- "The first essential in talking of convergence rates is to have an orderly plan of procedure. That is, one must choose a set of basis functions to be used and then gradually add more and more terms to the variational calculation in some systematic manner. The old habit of picking the 'best' (chosen by art) choice of a fixed number of terms is to be discarded if one wants to see how the problem converges".

Ruedenberg and his coworkers (60,67) and the present author (62) have reiterated the view that the convergence of calculations with respect to the size of basis set is a most important problem.

Ruedenberg et al (60) have devised empirical schemes for systematically extending even-tempered basis sets. As the number of functions, m, in such a basis set tends to infinity, the parameters α and β must satisfy the limits

$$\lim_{m \to \infty} \alpha = 0 \qquad (10)$$

$$\lim_{m \to \infty} \beta = 1 \qquad (11)$$

$$\lim_{m \to \infty} \beta^m = \infty \qquad (12)$$

in order that the basis set be complete in this limit. Thus Ruedenberg et al suggest that α and β be determined by the functions

$$\ell n(\ell n\beta) = b\ \ell n(m) + b' \qquad -1 < b < 0 \qquad (13)$$

and

$$\ell n(\alpha) = a\,\ell n(\beta-1) + a' \qquad a > 0 \tag{14}$$

In Table 6, the results of some matrix Hartree-Fock calculations on some first-row atoms are presented in which basis sets of even-tempered Gaussian functions are generated by the above empirical functions and by optimization with respect to variation of the energy. It can be seen that there is little loss in accuracy of the calculated energies by using the logarithmic relations to determine the parameters. Schmidt and Ruedenberg (60) demonstrated that the Hartree extrapolation technique (63,64) can be applied to a series of energies obtained by using a systematic sequence of even-tempered basis sets to obtain an empirical upper bound to the basis set limit:

$$E_\infty(m_3)=(E(m_3)E(m_1)-E(m_2)^2)/(E(m_3)-2E(m_2)+E(m_1)) \tag{15}$$

where m_1,m_2,m_3 denote the dimension of basis sets of increasing size. They also proposed methods for the determination of empirical lower bounds

$$\hat{E}_\infty(m) = E_\infty(m) - [E(m)-E_\infty(m)] \tag{16}$$

"best" estimates of the basis set limit

$$E_{av}(m) = \tfrac{1}{2}(E_\infty(m) + \hat{E}_\infty(m)) \tag{17}$$

and a measure of its accuracy

$$D(m) = \tfrac{1}{2}(E_\infty(m) - \hat{E}_\infty(m)). \tag{18}$$

The use of these techniques in calculations within the matrix Hartree-Fock model and in calculations which take account of correlation effects is illustrated in Table 7. The greater sensitivity of the correlation energy to the size of the basis set is apparent from this Table.

The use of a universal systematic sequence of even-tempered basis sets has been proposed (65). This approach has been shown to be useful not only in calculations of the total energy (65,66, 67) and its components (68), but also in the calculation of properties such as spin-orbit coupling constants (69,70).

10. CONVERGENCE OF THE HARMONIC EXPANSION

The basis sets employed in molecular calculations are often restricted to functions of s, p and d symmetry. In orbital models, only functions corresponding to the first few values of the angular quantum numbers, ℓ, contribute significantly to the energy and other expectation values. In calculations which take account of electron correlation effects, however, the higher harmonics can be important.

Table 5

Comparison of correlation energies obtained by using universal basis sets with those obtained by using basis sets specifically designed for a given system.[(a)]

Method	E_{total}	$\%e_{exp}$
LiH (Empirical correlation energy, e_{exp} = - 0.083)		
Bender and Davidson (48,49) CI	-8.0606	88.3%
Meyer and Rosmus (50), PNO CI	-8.0647	93.3%
CEPA	-8.0660	94.8%
Wilson and Silver (51) DPT, E[2/1]	-8.0643	92.8%
$\tilde{E}$[2/1]	-8.0652	93.9%
Diagrammatic perturbation E[2/1]	-8.0653	94.0%
theory/Universal basis sets (47) $\tilde{E}$[2/1]	-8.0661	94.9%
Li_2 (Empirical correlation energy, e_{exp} = - 0.126)		
Werner and Reinsch (52) MCSCF-CI	-14.9649	74.3%
Diagrammatic perturbation E[2/1]	-14.9842	89.6%
theory/Universal basis set (53) $\tilde{E}$[2/1]	-14.9845	89.9%
FH (Empirical correlation energy, e_{exp} = - 0.381		
Bender and Davidson (48,49) CI	-100.3564	75.1%
Meyer and Rosmus (50) PNO-CI	-100.3274	67.5%
CEPA	-100.3392	70.6%
Wilson and Silver (51) E[2/1]	-100.3727	79.4%
$\tilde{E}$[2/1]	-100.3707	78.9%
Diagrammatic perturbation E[2/1]	-100.3837	82.3%
theory/Universal basis set (47) $\tilde{E}$[2/1]	-100.3770	80.6%
N_2 (Empirical correlation energy, e_{exp} = - 0.538)		
Langhoff and Davidson (54) CI	-109.2832	58.4%
Wilson and Silver DPT (55) E[2/1]	-109.4180	79.5%
Diagrammatic perturbation theory/Universal basis set (5) E[2/1]	-109.443	83.7%
CO (Empirical correlation energy, e_{exp} = - 0.525)		
Siu and Davidson (56) CI	-113.1456	69.4%
Bartlett et al DPT (57) E[2/1]	-113.1952	77.4%
Diagrammatic perturbation theory/Universal basis set (5) E[2/1]	-113.2286	82.7%
BF (Empirical correlation energy, e_{exp} = - 0.531)		
Bender and Davidson (49) MCSCF	-124.235	
Wilson et al. DPT (58) E[2/1]	-124.5028	65.2%
Diagrammatic perturbation theory/Universal basis set (5) E[2/1]	-124.5782	77.1%

(a) All energies are in atomic units. The following abbreviations are used:- CI:configuration interaction; PNO-CI:pair natural orbital-configuration interaction; CEPA:coupled electron pair approximation; DPT:diagrammatic perturbation theory; MCSCF:multiconfiguration self-consistent field; E[2/1] denotes the [2/1] Padé approximant to the perturbation series based on the Hartree-Fock model zero-order hamiltonian and $\tilde{E}[2/1]$ denotes the [2/1] Padé approximant to the shifted denominator expansion.

Table 6

Matrix Hartree-Fock energies for the ground states of the beryllium and the oxygen atoms using even-tempered basis sets optimized with respect to the energy and basis set generated by means of equations (13) and (14)[a]

Basis set	Optimal energy	Energy obtaining using equations
	Beryllium	
6S	-14.53520087	-14.53489699
8S	-14.56652191	-14.56644167
10S	-14.57173762	-14.57172696
12S	-14.57270690	-14.57270546
14S	-14.57295145	-14.57295146
16S	-14.57300189	-14.57300182
20S		-14.57302132
26S		-14.57302307
28S		-14.57302313
	Oxygen	
6S/3P	-74.570775	-74.569652
8S/4P	-74.758151	-74.757800
10S/5P	-74.796929	-74.796859
12S/6P	-74.806038	-74.806024
14S/7p	-74.808421	-74.808418
20S/10P		-74.809356
22S/11P		-74.809381

(a) taken from the work of Schmidt and Ruedenberg (60).

A considerable amount of data is available on the convergence of the harmonic expansion for atoms. Using second-order perturbation theory, Schwartz (59,71) shows that

$$E_{2,\ell} = -\frac{45}{256}(\ell+\tfrac{1}{2})^{-4}\{1-\frac{19/8}{(\ell+\tfrac{1}{2})^2}+O(\ell^{-4})\} \qquad (19)$$

where $E_{2,\ell}$ is the component of the second- order energy corresponding to the quantum number ℓ. Carroll et al. (72) have established similar expressions for the convergence of the correlation

Table 7

Empirical upper bounds and lower bounds to the basis set limit in calculations for the ground state of the neon atom within the matrix Hartree-Fock model and including correlation effects(a)

Basis set	E(SCF)	E[2/1]
(a) Empirical upper bounds		
10S/5P	-128.542358	-128.729415
12S/6P	-128.546031	-128.733333
14S/7P	-128.546726	-128.733972
16S/8P	-128.546935	-128.734451
18S/9P	-128.547071	-128.734688
(b) Empirical lower bounds		
10S/5P	-128.565387	-128.759054
12S/6P	-128.552702	-128.742228
14S/7P	-128.548707	-128.736629
16S/8P	-128.547568	-128.735433
18S/9P	-128.547325	-128.735128
(c) Average value		
10S/5P	-128.553873	-128.744235
12S/6P	-128.549366	-128.737781
14S/7P	-128.547716	-128.735301
16S/8P	-128.547252	-128.734942
18S/9P	-128.547198	-128.734908
(d) Deviation		
10S/5P	0.011514	0.014819
12S/6P	0.003336	0.004447
14S/7P	0.000990	0.001329
16S/8P	0.000317	0.000491
18S/9P	0.000127	0.000220

(a) taken from the work of Wilson (62).
(b) [2/1] Padé approximant to the diagrammatic perturbation theory expansion for the correlation energy.

energy of $1s^2$He. They obtained rough estimates, shown in Table 8, of the maximum value of ℓ required to obtain the energy of $1s^2$He to a given accuracy. In Table 9, the convergence of the harmonic expansion for some electron pairs in the ground state of the neon atom are displayed (73). It can be seen that the convergence for the $2p^2\,{}^1D$ electron pair in this system is particularly unfavourable.

The importance of higher terms in the harmonic expansion for molecular systems is illustrated in Table 10 (74) where calculations are presented for the CO molecule using a basis set of s,

Table 8

Rough estimate of the maximum value of ℓ required to obtain the energy of $1s^2$He to a given accuracy[a]

Accuracy in atomic units	maximum value of ℓ
10^{-4}	6
10^{-5}	14
10^{-6}	32
10^{-7}	69
10^{-10}	690
10^{-12}	3204

(a) taken from the work of Carroll, Silverstone and Metzger (73).

p, d functions and a basis set of s,p,d, f functions. Matrix Hartree-Fock energies and correlation energies were found as a function of internuclear distance. The inclusion of f basis function leads to a much greater energy lowering in the correlated calculation but it should be noted that this lowering is not as sensitive to changes in the internuclear distance as the energy lowering in the matrix Hartree-Fock calculations.

Table 9

Convergence of the harmonic expansion for some electron pairs energies in the ground state of the neon atom[a]

ℓ	Electron pair energy[b]		
	$(1s2s)\,^1S$ Ne	$(1s2s)\,^3S$ Ne	$(2p^2)\,^1D$ Ne
2	422	73	29950
3	116	12	7265
4	43	3	5900
5	20	1	2240
6	10		1030
7	6		545
8	4		306
9	2		185

(a) taken from the work of Jankowski and Malinowski (74) who used second order Rayleigh-Schrödinger perturbation theory.

(b) energies are in microhartree with signs reversed.

Table 10

Energy lowering obtained on adding f functions to basis set in calculations in the ground state of the CO molecule.

R(Å)	δ_{MHF} (b)	$\delta_{correlation}$ (c)
0.9283	5.5	24.6
0.9783	4.7	24.8
1.0283	4.0	25.0
1.0783	3.4	25.1
1.1283	2.9	25.1
1.1783	2.5	25.0
1.2283	2.2	24.9
1.2783	2.0	24.7
1.3283	1.8	24.4

(a) taken from the work of Wilson (74).

(b) Energy lowering obtained in a matrix Hartree-Fock calculation in millihartree.

(c) Energy lowering obtained in a calculation of the correlation energy using diagrammatic perturbation theory in millihartree.

11 FURTHER ASPECTS

In this section, we discuss, very briefly, some other aspects of the problem of constructing basis sets for use in molecular electronic structure studies.

In a series of papers by Eggarter and Eggarter (75-78) and by Jankowski and his co-workers (73,79-84), accurate calculations of correlation energies using low-order Rayleigh-Schrödinger perturbation theory have been reported for a variety of atomic systems. Such calculations have clearly demonstrated the need to include basis functions corresponding to higher angular quantum numbers if the error due to basis set truncation is to be significantly reduced. Jankowski and Malinowski (73), for example, showed that the second-order energy for the ground state of the neon atom provides a very accurate approximation to the total correlation energy (99.7%) if functions corresponding to higher ℓ quantum numbers are included in the basis set. These atomic calculations point to the need to extend basis set used in molecular calculations to include functions of f,g,h,i... symmetry in order to improve their accuracy.

Optimization of orbital exponents in molecular calculations at every nuclear configuration considered, is computationally prohibitive. Poirier, Daudel, Mezey and Csizmadia (85) have explored the possibility of allowing molecular basis sets to vary with nuclear geometry by assuming that the exponents have a quadratic dependence on a single parameter, Q, which is treated as a variational para-[2] meter. Poirer et al (85) report studies of a number of molecules containing carbon atoms. It should be remembered, however, that if a sufficiently large and flexible basis set is employed, variation of the exponents with nuclear geometry will not be necessary.

Krishnan, Binkley, Seeger and Pople (4) have obtained basis sets of contracted Gaussian-type functions for use in calculations of correlation energies by choosing the exponents to yield the lowest atomic ground state energy within second-order perturbation thoery. Strictly this procedure is incorrect from a variational point of view (as the authors acknowledge).

Finally, we mention the work of McDowell (86-88) on the incomplete basis set problem. He suggests the use of a diagrammatic perturbation theoretic approach, which he reformulates within a differential equation format.

12 FINAL COMMENT

Once the algebraic approximation has been invoked, the vast majority of the computation involved in the study of molecular electronic structure from first principles, can be cast in the form of simple matrix operations. The parallel computer (89) is well suited to such operations (the CRAY-1 computer will perform matrix multiplication at a rate or $\sim$135 MFlops (41,90)) and will therefore, lead to the possibility of using much larger basis sets. This, on the one hand, will allow calculations to be made on larger molecules and, on the other hand, will allow small molecules to be handled with greater accuracy. Provided that the large basis sets, which are obviously going to be required to perform accurate calculations for small molecules, can be designed so as to avoid numerical problems, it is likely that the next few years will see us able to calculate some 95-99% of the correlation energy of a molecule such as CO (cf., e.g., (74)) routinely.

The author is an S.E.R.C. Advanced Fellow

REFERENCES

1. Froese Fischer, C. 1977: *The Hartree-Fock Method for Atoms*, Wiley, New York.
2. Christiansen, P.A., and McCullough Jr., E.A.: 1977, J. Chem Phys. 67, 1877.
3. Bunge, C.F.: 1980, Physica Scripta 21, 328.
4. Krishnan, R, Binkley, J.S., Seeger, R., and Pople, J.A.: 1980, J. Chem. Phys. 72, 650.
5. Wilson, S. and Silver, D.M.: 1980, J. Chem. Phys. 72, 2159.

6. Wilson, S, and Guest, M.F.: 1981, J. Phys. B: At. Mol. Phys. 14 1709.
7. See, for example, J. Paldus, 1976, Theoretical Chemistry, Advances and Perspectives 2,131.
8. Klahn, B, and Bingel, W.: 1977, Intern. J. Quantum Chem. 11 934; Klahn, B.: 1981, Adv. Quantum Chem. 13, 155
9. See, for example, Bishop, D.M., 1967 Adv. Quantum Chem. 3,25.
10. Ladik, J. and Cizek, J.: 1980, J. Chem. Phys. 73, 2357.
11. Saunders, V.R.: This volume.
12. Steinborn, O.: This volume.
13. Boys, S.F. and Bernadi, F.: 1970, Molec. Phys. 19, 553.
14. McLean, A.D. and Chandler, G.S.: 1980, J. Chem. Phys. 72, 5639.
15. Mayer, W.. Hariharan, P.C. and Kutzelnigg, W.: 1980, J. Chem. Phys. 73, 1880.
16. Kutzelnigg, W.: 1980, J. Chem. Phys. 73, 343.
17. Muntz, H.: 1914, Festschrift, H.A. Schwartz, 303.
18. Szasz, O.: 1926, Math. Ann. 77
19. Higgins, J.R.: 1977, *Completeness and basis properties of sets of special functions*, Cambridge University Press.
20. van Duijneveldt, F.: 1971, IBM Techn. Res. Rept. RJ948.
21. Čársky, P. and Urban, M., 1980, *Ab initio* calculations. Methods and applications in chemistry, Springer-Verlag.
22. See, for example, Ebbing, D.D.: J. Chem. Phys. 1962, 36 , 1361.
23. Starykh, V.V. and Kapyshev, V.V.: 1980, J. Chem. Phys. 72, 2713.
24. Mogilevsky, O.A., Ponomerev, L.I. and Vukajlovic, F.R.: 1980, J. Phys. B: At. Mol. Phys. 13, 2489 and references therein.
25. Silverstone, H.J., Carroll, D.P. and Silver, D.M.: 1978, J. Chem. Phys. 68, 616
26. Hall, G.G.: 1959, Rept. Progr. Phys. 22, 1.
27. Burden, F.R. and Wilson, R.M.: 1973, Adv. Phys. 21, 825.
28. Silver, D.M.: 1970, Chem. Phys. Letters. 7, 511.
29. Dunning Jr., T.H. and Hay, P.J.: 1977, *Methods in Electronic Structure Theory*, edited by H.F. Schaefer III, Plenum, New York.
30. Ahlrichs, R. and Taylor, P.: 1981, J. chim, Phys. 78, 315.
31. Wilson, S.: 1983, *Electron correlation in molecules*, Clarendon Press, Oxford.
32. See, for example, Collins, J.B., van R. Schleyer, P., Binkley, J.S. and Pople J.A., 1976, J. Chem. Phys. 64, 5142; Tatewaki, H, and Huzinaga, S. 1980, J. Chem. Phys. 72, 399; Tavouktsoglou, A. and Huzinaga, S, 1980, J. Chem. Phys. 72, 1385; see also Binkley, J.S. Pople, J.A. and Hehre, W.J. 1980, J. Am. Chem. Soc. 102, 939 for a discussion of the defiencies of the STO-3G basis set of Pople and co-workers.
33. Sadlej, A.J.: 1977, Chem. Phys. Letters, 47, 50; 1978 Acta Phys. Polon. 53, 297; see also Wilson, S, and Sadlej, A.J. 1981, Theoret. chim. Acta. 60, 19.
34. Ahlrichs, R: 1974, Theoret. Chim. Acta. 33, 157.
35. Cade, P.E. and Huo, W.M.: 1967, J. Chem. Phys. 47, 614, 649.
36. Reeves, C.M.: 1963, J. Chem. Phys. 39, 1.

37. Ruedenberg, K., Raffenetti, R,C, and Bardo, R.: 1973, in Energy, Structure and Reactivity. Proceedings of the 1972 Boulder Conference on Theoretical Chemistry, Wiley, New York.
38. Raffenetti, R.C. and Ruedenberg, K: 1973, J. Chem. Phys. 59, 5978.
39. Raffenetti, R.C.: 1973, J. Chem. Phys. 58, 4452; 59 5936.
40. Bardo, R.D. and Ruedenberg, K.: 1973, J. Chem. Phys. 59 5956, 5966; 1974 ibid 60, 918.
41. Guest, M.F. and Wilson, S.: 1981, *Supercomputers in Chemistry*, edited by P. Lykos and I. Shavitt, American Chemical Society Washington D.C.
42. Silver, D.M., Wilson, S. and Nieupoort, W.C.: 1978, Int. J. Quantum Chem. 14, 635.
43. Silver, D.M. and Nieupoort, W.C.: 1978, Chem. Phys. Letters 57, 421.
44. Silver, D.M. and Wilson, S.: 1978, J. Chem. Phys. 69, 3787.
45. Wilson, S. and Silver, D.M.: 1979, Chem. Phys. Letters. 63. 367.
46. Mezey, P.: 1979, Theoret. Chim. Acta. 53, 187
47. Wilson, S. and Silver, D.M.: 1982, J. Chem. Phys. 77 (1 Oct).
48. Bender, C.F. and Davidson, E.R.: 1966, J. Phys. Chem. 70, 2675.
49. Bender, C.F. and Davidson, E.R.: 1969, Phys. Rev. 183, 23
50. Meyer, W. and Rosmus, P.: 1975, J. Chem. Phys. 63, 2356.
51. Wilson, S. and Silver D.M.: 1977, J. Chem. Phys. 66, 5400.
52. Werner, H.-J. and Reinsch, E.-A.: 1981, Proc. 5th Seminar on Computational Problems in Quantum Chemistry, ed. P.Th. van Duijnen and W.C. Nieuwpoort, Groningen.
53. Wilson, S.: 1981, Specialist Periodical Reports: Theoretical Chemistry 4, 1.
54. Langhoff, S. and Davidson, E.R.: 1974, Int. J. Quantum Chem. 8, 61.
55. Wilson, S. and Silver, D.M.: 1977, J. Chem. Phys. 67, 1689.
56. Siu, A.K.Q. and Davidson, E.R.: 1970, Int. J. Quantum Chem. 4, 223.
57. Bartlett, R.J., Wilson, S. and Silver, D.M.: 1977, Int. J. Quantum Chem. 13, 737.
58. Wilson, S, Silver, D.M. and Bartlett, R.J.: 1977, Molec. Phys, 33, 1177.
59. Schwartz, C.M.: 1963, Methods in Computational Physics, 2, 241.
60. Schmidt, M.W. and Ruedenberg, K: 1979, J. Chem. Phys. 71, 3951.
61. Feller, D.F. and Ruedenberg, K: 1979, Theoret, Chem. Acta. 52, 231.
62. Wilson, S.: 1980, Theoret. Chim. Acta. 57, 53.
63. Hartree, D.R.: 1948, Proc. Cambridge, Philos. Soc. 45, 230.
64. Roothaan, C.C.J. and Bagus, P.S.: 1963, Methods in Computational Physics, 2, 47.
65. Wilson, S.: 1980, Theoret. Chim. Acta. 58, 31.
66. Cooper, D.L. and Wilson S.: 1982, J. Chem. Phys. (in press).
67. Cooper, D.L. and Wilson, S.: 1982, J. Chem. Phys. (in press).
68. Wilson, S.: 1981, Chem. Phys. Letters, 81, 467

69. Cooper, D.L. and Wilson S.: 1982, J. Phys. B: At. Mol. Phys. 15, 493.
70. Cooper, D.L. and Wilson, S.: 1982, J. Chem. Phys. 76, 6088.
71. Schwartz, C.M.: 1962, Phys. Rev. 126, 1015.
72. Carroll, D.P., Silverstone, H, and Metzger, R.M.: 1979, J. Chem. Phys. 71, 4142.
73. Jankowski, K. and Malinowski, P.: 1980, Phys. Rev. A22, 51.
74. Wilson, S.: 1982, J. Phys. B: At. Mol. Phys. 15, L191.
75. Eggarter, E. and Eggarter, T.P.: 1978, J. Phys. B: At. Mol. Phys. 11, 1157.
76. Eggarter, E. and Eggarter, T.P.: 1978, J. Phys. B: At. Mol. Phys. 11, 2069.
77. Eggarter, E. and Eggarter, T.P.: 1978, J. Phys. B: At. Mol. Phys. 11, 2969.
78. Eggarter, T.P. and Eggarter, E.: 1978, J. Phys. B: At. Mol. Phys. 11, 3635.
79. Jankowski, K. and Malinowski, P.: 1978, Chem. Phys. Lett. 54, 68.
80. Jankowski, K., Malinowski, P. and Polasik, M.: 1979, J. Phys. B: At. Mol. Phys. 12, 345.
81. Jankowski, K., Malinowski, P. and Polasik, M.: 1979, J. Phys. B: At. Mol. Phys. 12, 3157.
82. Jankowski, K., Malinowski, P. and Polasik, M.: 1980, J. Phys. B: At. Mol. Phys. 13, 3909.
83. Jankowski, K., Malinowski, P. and Polasik, M.: 1980, Phys. Rev. A22, 51.
84. Jankowski, K., Rutkowska, D. and Rutkowski, A.: 1982, J. Phys. B: At. Mol. Phys. 15, 1137.
85. Poirier, R.A., Daudel, R., Mezey, P.G. and Csizmadia, I.G.: 1980, Int. J. Quantum Chem. 18, 715.
86. McDowell, K.: 1979, J. Chem. Phys. 70, 3149.
87. McDowell, K.: 1979, Int. J. Quantum Chem. Symp. 13, 51
88. McDowell, K.: 1980, Int. J. Quantum Chem. 17, 895.
89. Hockney, R.W. and Jesshope, C.R., 1981. *Parallel Computers*, Adam Hilger, Bristol.
90. Mflops = Millions of floating point operators per second.

MATRIX EIGENVECTOR METHODS

Ernest R. Davidson

Chemistry Dept., University of Washington

Widespread use of the linear variation method in quantum mechanics leads to the need for efficient matrix eigenvalue methods for real symmetric matrices. Efficiency must be judged, however, in terms of the ability to solve a variety of eigenvalue problems on computers of widely different architecture, memory size, and data transfer rates. To further confuse the situation, cost rather than system throughput, is usually of paramount concern. Additionally, algorithms vary widely in ease of programming, simplicity, reliability, portability and availability as part of standard packages. Consequently there are a wide variety of eigenvalue algorithms in use, and a typical quantum laboratory will incorporate several of these into their program package.

As an illustration one may consider the following situations leading to matrix eigenvalue problems:

A. Non-orthogonal configuration interaction expansions (valence bond or r_{ij} configurations) for a few electrons.

This leads to a generalized eigenvalue problem

$$\mathbf{Hc} = E\mathbf{Sc} \tag{1}$$

where usually only the lowest few roots are wanted. Often **S** is ill-conditioned due to near linear-dependency in the configuration basis. The dimension is normally less than 1000.

G. H. F. Diercksen and S. Wilson (eds.), Methods in Computational Molecular Physics, 95–113.

B. Non-orthogonal Hartree-Fock calculations.

This leads to the generalized eigenvalue problem

$$\mathbf{Fc} = E\mathbf{Sc} \tag{2}$$

where all eigenvalues are usually wanted and the dimension is usually below 300. These equations must be solved iteratively with a succession of **F** matrices which change little between iterations. The solution from one iteration provides an orthogonalizing transformation which nearly diagonalizes **F** for the next iteration.

C. Orthogonal configuration interaction expansions for many electrons.

This leads to the ordinary eigenvalue problem

$$\mathbf{Hc} = E\mathbf{c} \tag{3}$$

with the dimension usually in the range 10^3 to 10^6. The matrix is usually randomly sparse and diagonally dominant. Normally only a few eigenvalues are wanted. In direct CI methods, **H** is not stored but is reconstructed as needed.

Eigenvalue methods are necessarily selected based on the limiting computer resource. If n is the dimension, some methods are n^3 methods (i.e. require a number of computer operations which grows like n^3) while others are kN (where k is the number of eigenvalues wanted and N is the number of non-zero elements in the matrix). The n^3 methods are easily implemented in n^2 words of random access memory with little input-output (I/O) while the kN methods can require as few as 2n words of memory but may require about 20kN words of I/O. Direct CI methods may reduce the I/O requirements of the kN methods by reconstructing the **H** matrix on each interaction from many fewer than N primitive integrals. For moderate size problems where N words of memory are available and kN is much less than n^3, a kN method implemented in core without I/O may be more efficient than an n^3 method.

SMALL MATRIX METHODS

Most small matrix methods are n^3 algorithms which bring the matrix into a simplified standard form by a sequence of similarity transformations. The Jacobi method is the oldest of these but it has fallen into disfavor with numerical analysts because it is about three times slower than the modified Givens method (1).

The Jacobi method still has two features which make it competitive for SCF calculations. In the SCF method one must solve a sequence of generalized eigenvalue problems

$$\mathbf{F}_n\mathbf{c} = \mathrm{E}\mathbf{S}\mathbf{c} \tag{4}$$

for all of the eigenvalues of fairly small matrices. After a few iterations, one has available the eigenvectors from the previous iteration which are orthogonal and can be arranged in a matrix $\mathbf{C}_{n-1}$ with

$$\mathbf{C}_{n-1}{}^{T}\mathbf{S}\mathbf{C}_{n-1} = 1. \tag{5}$$

If this matrix is used to transform $\mathbf{F}_n$,

$$\mathbf{G} = \mathbf{C}_{n-1}{}^{T}\mathbf{F}_n\mathbf{C}_{n-1} \tag{6}$$

then

$$\mathbf{c} = \mathbf{C}_{n-1}\mathbf{b} \tag{7}$$

where

$$\mathbf{G}\mathbf{b} = \mathrm{E}\mathbf{b}. \tag{8}$$

Of all the small-matrix methods, only Jacobi can take advantage of the fact that $\mathbf{G}$ is nearly diagonal near convergence. In fact, it can be shown that, for small off-diagonal elements, the Jacobi method converges quadratically.

The other advantage of the Jacobi method is that the eigenvectors are generated recursively by

$$\mathbf{B}_k = \mathbf{B}_{k-1}\mathbf{R}_k \tag{9}$$

where $\mathbf{B}$ contains the eigenvectors $\mathbf{b}$ at convergence and the $\mathbf{R}_k$ are elementary plane rotations. In the standard Jacobi method $\mathbf{B}_0$ is chosen to be a unit matrix. Without additional cost, $\mathbf{B}_0$ can be chosen to be $\mathbf{C}_{n-1}$ so that the converged matrix $\mathbf{B}$ contains the matrix $\mathbf{C}_n$. This eliminates the $\mathbf{C}_{n-1}\mathbf{B}$ multiplication which otherwise requires n^3 operations.

While these considerations may still not make Jacobi the best choice, it is not as bad for this application as the straight numerical analysis would indicate.

A. Jacobi

The Jacobi method operates by constructing a sequence of unitary transformations which reduce the size of the off-diagonal elements (1). To be explicit, if one is solving the equation

$$\mathbf{G}\mathbf{b} = E\mathbf{b} \tag{10}$$

one seeks a sequence of real orthogonal matrices $\mathbf{R}(i,j)$ which differ from unit matrices only in the elements

$$R_{ii} = R_{jj} = c \tag{11}$$

$$R_{ji} = -R_{ij} = s$$

$$c^2 + s^2 = 1.$$

Application of this transformation to $\mathbf{G}$ gives a new matrix $\mathbf{G}' = \mathbf{R}^T\mathbf{G}\mathbf{R}$ with unchanged elements except for

$$g_{ip}' = g_{pi}' = g_{pi}c - g_{pj}s \tag{12}$$

$$g_{jp}' = g_{pj}' = g_{pi}s + g_{pj}c$$

$$g_{ij}' = g_{ji}' = (g_{ii} - g_{jj})sc + g_{ij}(c^2 - s^2)$$

$$g_{ii}' = g_{ii}c^2 - 2g_{ij}sc + g_{jj}s^2$$

$$g_{jj}' = g_{ii}s^2 + 2g_{ij}sc + g_{jj}c^2.$$

If c and s are chosen so that $g_{ij}' = 0$, the measure w of the off-diagonal elements

$$w(\mathbf{G}) = \sum_{k<l} g_{kl}^2 \tag{13}$$

is reduced by

$$w(\mathbf{G}) - w(\mathbf{G}') = g_{ij}^2 \tag{14}$$

If the sequence of $\mathbf{R}(i,j)$ rotations are chosen correctly, $w(\mathbf{G})$ will converge to zero even though each successive rotation will destroy zero elements produced earlier. If the off-diagonal elements are small enough, the convergence is quadratic in the sense that, after a complete pass through $\mathbf{G}$ annihilating every off-diagonal element once, the largest remaining off-diagonal element will be proportional to the square of the maximum element before the pass.

There are many suggestions in the literature of formulae for c and s. One stable choice is

$$u = (g_{jj} - g_{ii})/(2g_{ij}) \tag{15}$$

$$t = \text{sign}(u) \,/[\,|u| + (1 + u^2)^{1/2}]$$

$$c = (1 + t^2)^{-1/2}$$

$$s = ct.$$

One advantage of this solution is that $|t| \leq 1$ and $|s| \leq c$; so g_{ii}' and g_{jj}' remain close to g_{ii} and g_{jj} respectively when g_{ij} is small.

The only opportunity for savings in the Jacobi algorithm is in the selection of the g_{ij} to be annihilated. Clearly it is wasteful to annihilate small elements. On the other hand, searching for the largest element requires n^2 operations while a rotation requires only n operations. Hence the usual technique is a compromise in which all elements above a certain threshold are annihilated on each sweep through the matrix. Several suggestions have been made for choosing the threshold (1-4). These vary from the simple choice $(w/N)^{1/2}$ where $N = n(n-1)/2$ (3) to the least squares optimized choice suggested by Birge and Hubbard (4).

B. Givens (Householder-QL-Wilkinson)

Givens' original method (5) was based on plane rotations to reduce a matrix **A** exactly to tridiagonal form by a sequence which did not destroy zeros once they were computed. This scheme used the $\mathbf{R}(i,j)$ transformation to annihilate $a_{i-1,j}$ rather than $a_{i,j}$ as in the Jacobi method. Givens then computed the eigenvalues from a Sturm sequence and the eigenvectors by substitution into the tridiagonal linear equations. Finally these vectors were back-transformed to give vectors for the original matrix. While this outline is still followed, each step of his algorithm has been replaced by a more efficient procedure.

In the Householder transformation to tridiagonal form (6) the starting matrix **A** is transformed by a sequence of unitary matrices

$$\mathbf{A}_{k+1} = \mathbf{U}_{k+1}{}^T \mathbf{A}_k \mathbf{U}_{k+1} \tag{16}$$

where $\mathbf{U}_k$ brings row k of **A** to tridiagonal form. Consider a partially transformed **A**

$$\mathbf{A}_k = \begin{pmatrix} d_1 & b_1 & & & & & \\ b_1 & d_2 & b_2 & & & & \\ & b_2 & \ddots & \ddots & & & \\ & & \ddots & \ddots & \ddots & & \\ & & & \ddots & d_k & b_k & \\ & & & & b_k & d_{k+1} & \mathbf{c}^T \\ & & & & & \mathbf{c} & \mathbf{M} \end{pmatrix} \tag{17}$$

and a partitioned $\mathbf{U}_{k+1}$,

$$\mathbf{U}_{k+1} = \begin{pmatrix} \mathbf{1}_{k+1} & \mathbf{0} \\ \mathbf{0} & \mathbf{X} \end{pmatrix}. \tag{18}$$

Then $\mathbf{A}_{k+1}$ will have the same block structure as $\mathbf{A}_k$ with the lower diagonal block becoming

$$\begin{pmatrix} d_{k+1} & \mathbf{c}^T\mathbf{X} \\ \mathbf{X}^T\mathbf{c} & \mathbf{X}^T\mathbf{M}\mathbf{X} \end{pmatrix}. \tag{19}$$

Hence if

$$\mathbf{X}^T\mathbf{c} = b_{k+1}\mathbf{e}_1 \tag{20}$$

(where $\mathbf{e}_i$ is a vector with elements $e_{ji} = \delta_{ij}$) then $\mathbf{A}_{k+1}$ will be reduced.

To accomplish this, Householder (1, 6, 7) suggested using

$$\mathbf{X} = \mathbf{1}_k - f\mathbf{w}\mathbf{w}^T \tag{21}$$

$$f = 2/\mathbf{w}^T\mathbf{w}$$

so that

$$\mathbf{X} = \mathbf{X}^T = \mathbf{X}^{-1}. \tag{22}$$

Then

$$\mathbf{X}^T\mathbf{c} = \mathbf{c} - f\mathbf{w}(\mathbf{w}^T\mathbf{c}) = b_{k+1}\mathbf{e}_1. \tag{23}$$

Clearly, except for the first element, we must have $\mathbf{w} = \mathbf{c}$, so let

$$\mathbf{w} = \mathbf{c} + h\mathbf{e}_1. \tag{24}$$

Substitution gives

$$h^2 = \|\mathbf{c}\|^2 \tag{25}$$

$$b_{k+1} = -h$$

$$f^{-1} = h(h + c_1).$$

Calculation of f^{-1} and $\mathbf{w}$ is most stable if h is chosen as

$$h = \|\mathbf{c}\| \operatorname{sign}(c_1). \tag{26}$$

Calculation of $\mathbf{X}^T\mathbf{M}\mathbf{X}$ is facilitated by defining

$$\mathbf{p} = \mathbf{M}\mathbf{w}/f^{-1} \tag{27}$$

$$\mathbf{q} = \mathbf{p} - \mathbf{w}(\mathbf{w}^T\mathbf{p})/(2f^{-1})$$

$$\mathbf{X}^T\mathbf{M}\mathbf{X} = \mathbf{M} - \mathbf{w}\mathbf{q}^T - \mathbf{q}\mathbf{w}^T.$$

This algorithm requires $2/3n^3$ multiplications to bring $\mathbf{A}$ to tridiagonal form.

Computation of the eigenvalues of a tridiagonal matrix by the QL algorithm requires only $9n^2$ operations (1, 8). The QL algorithm is based on the fact that the eigenvector corresponding to the zero eigenvalue of a singular matrix is orthogonal to all the columns of the matrix. If one writes

$$\mathbf{B} = \mathbf{A} - u\mathbf{1} = \mathbf{Q}\mathbf{L} \tag{28}$$

where u is almost an eigenvalue of $\mathbf{A}$, and the columns of $\mathbf{Q}$ are Schmidt orthonormalized columns of $\mathbf{B}$ (computed in the order

$$\mathbf{z}_n = \mathbf{b}_n, \tag{29}$$

$$\mathbf{q}_n = \mathbf{z}_n/\|\mathbf{z}_n\|,$$

$$\mathbf{z}_{n-1} = \mathbf{b}_{n-1} - \mathbf{q}_n(\mathbf{q}_n, \mathbf{b}_{n-1}),$$

$$\mathbf{q}_{n-1} = \mathbf{z}_{n-1}/\|\mathbf{z}_{n-1}\|,$$

etc.), then $\mathbf{L}$ will be lower triangular. Further the first row of $\mathbf{L}$ will be almost zero and the first column of $\mathbf{Q}$ will be almost the eigenvector of $\mathbf{A}$ with eigenvalue u. The transformation

$$\mathbf{Q}^T\mathbf{A}\mathbf{Q} = \mathbf{L}\mathbf{Q} + u\mathbf{1} = \mathbf{A}' \tag{30}$$

preserves the tridiagonal form of $\mathbf{A}$ and tends to reduce b_1' in $\mathbf{A}'$ to zero and d_1' to u. If u is only approximately an eigenvalue of $\mathbf{A}$, b_1' will be bounded by (1)

$$|b_1'| \leq \|\mathbf{B}^{-1}\|^{-1}. \tag{31}$$

In practice one usually chooses to estimate u as the eigenvalue nearest d_1 of the leading 2 x 2 submatrix of $\mathbf{A}$.

The actual QL transformation is carried out by a simpler scheme than indicated above. It can be shown (1) that the transformation $\mathbf{Q}$ which reduces a matrix $\mathbf{A}$ to tridiagonal form is uniquely specified by giving the last column of $\mathbf{Q}$. The transformation **Q** which "reduces" the tridiagonal **A** to the tridiagonal **A′** is hence uniquely specified by the choice given above for $\mathbf{q}_n$. If one writes **Q** as the product of the two unitary matrices **U** and **V** defined as

$$\mathbf{Q} = \mathbf{UV} \tag{32}$$

$$\mathbf{U} = \begin{pmatrix} \mathbf{1}_{n-2} & \mathbf{0} \\ \mathbf{0} & \mathbf{G} \end{pmatrix} \qquad \mathbf{V} = \begin{pmatrix} \mathbf{M} & \mathbf{0} \\ \mathbf{0} & 1 \end{pmatrix}$$

$$\mathbf{G} = \begin{pmatrix} d_n - u & b_{n-1} \\ -b_{n-1} & d_n - u \end{pmatrix} [\, b_{n-1}{}^2 + (d_n - u)^2)]^{-1/2}$$

then the last column of **Q** will be $\mathbf{q}_n$. If **A** is transformed by **U**, $\mathbf{U}^T\mathbf{AU}$ will differ from a tridiagonal matrix only by non-zero elements in locations (n, n-2) and (n-2, n). The matrix **V** can then be constructed as the product of a sequence of Givens plane rotations **R**(n-1,n-2), **R**(n-2,n-3),... which successively annihilate elements (n, n-2), (n-1,n-3),... That is, the rotations "chase the bulge" introduced by **U** and bring **A** back to tridiagonal form without changing $\mathbf{q}_n$. It will be noticed that u enters this calculation explicitly only in the calculation of **G**.

A new estimate of u is then made and the cycle repeated. After convergence the first row and column of the transformed **A** are discarded and the process repeated until all roots are found. Many modifications of this procedure have been discussed by Parlett (1).

The QL transformations are usually not accumulated to give the eigenvectors. Wilkinson (8) showed that the solution of

$$(\mathbf{A} - u\mathbf{1})\mathbf{x} = \mathbf{y} \tag{33}$$

will be close to the eigenvector associated with u provided **y** is not orthogonal to the desired vector. This procedure, called inverse iteration, has trouble with nearly degenerate eigenvalues. An alternative procedure can be based on repeating the QL algorithm starting with the original tridiagonal **A** matrix and the known eigenvalues. If the rotations are accumulated, this will yield, in one QL cycle per vector, a list of orthogonal vectors.

It should be noted that in both this modified method and the Jacobi method, orthogonality of the vectors is achieved but the vectors will still be inaccurate for closely spaced eigenvalues. The best which can be said is that the vectors span the nearly degenerate subspace.

In the Givens method the eigenvectors, $\mathbf{x}$, of the tridiagonal matrix must still be transformed back to the original problem by the Householder matrices

$$\prod (\mathbf{1} - f\mathbf{w}\mathbf{w}^T)\mathbf{x} \tag{34}$$

which requires another n^3 multiplications.

LARGE MATRIX METHODS

There are three methods in common use for computing a few eigenvalues of large matrices (9). No efficient method is available for finding all of the eigenvalues. All of the methods operate by selecting a set of vectors on the nth iteration, $\{\mathbf{b}_1^n \text{ --- } \mathbf{b}_K^n\}$ in which the eigenvectors are expanded as

$$\mathbf{x}_p^n = \sum_i^K c_{ip}\, \mathbf{b}_i^n. \tag{35}$$

The coefficients are chosen by the linear variation method. That is, if

$$\mathbf{A}\mathbf{x}_p = E_p \mathbf{x}_p \tag{36}$$

is the problem whose solutions are desired, then the coefficients are given by

$$\mathbf{B} = (\mathbf{b}_1^n \text{ --- } \mathbf{b}_K^n) \tag{37}$$

$$[\mathbf{B}^T\mathbf{A}\mathbf{B}]\mathbf{c}_p = d_p[\mathbf{B}^T\mathbf{B}]\mathbf{c}_p. \tag{38}$$

If the d_p are numbered in increasing order, then

$$d_p \geqslant E_p. \tag{39}$$

Further, as d_p approaches E_p, $\mathbf{x}_p^n$ will converge to $\mathbf{x}_p$.

The methods differ from each other in their choice of the set $\{\mathbf{b}\}$. The Lanczos method is most studied by numerical analysts and hence is the only textbook method. This method chooses the $\mathbf{b}'$s as an orthonormalized Krylov sequence generated by repeatedly applying $\mathbf{A}$ to a starting vector (so $K = n$). Alternatively block-Lanczos methods have been suggested which generate $\mathbf{b}'$s by applying $\mathbf{A}$ to r different vectors each iteration (sc $K = rn$).

Relaxation methods keep K fixed at some small number (usually 2). The vector $\mathbf{b}_1^n$ is the approximate eigenvector $\mathbf{x}^{n-1}$ and $\mathbf{b}_2$ is a unit vector. Higher eigenvectors are found by root shifting or by using a larger value for K. Block methods can be devised which use several unit vectors at once.

The Davidson method, like the Lanczos method uses a variable value of K and the set $\{\mathbf{b}\}$ is built up by appending a new vector each iteration. The added vector, however, is chosen by perturbation theory rather than the Krylov sequence.

A. Nesbet

Suppose $\mathbf{b}$ is the current guess to the ground state eigenvector and $\mathbf{e}$ is a unit vector. Then

$$\mathbf{x}^n = c_1\mathbf{b} + c_2\mathbf{e} \tag{40}$$

has coefficients given by

$$\begin{pmatrix} \mathbf{b}^T\mathbf{A}\mathbf{b} & \mathbf{b}^T\mathbf{A}\mathbf{e} \\ \mathbf{e}^T\mathbf{A}\mathbf{b} & \mathbf{e}^T\mathbf{A}\mathbf{e} \end{pmatrix} \begin{pmatrix} c_1 \\ c_2 \end{pmatrix} = d \begin{pmatrix} \mathbf{b}^T\mathbf{b} & \mathbf{b}^T\mathbf{e} \\ \mathbf{e}^T\mathbf{b} & \mathbf{e}^T\mathbf{e} \end{pmatrix} \begin{pmatrix} c_1 \\ c_2 \end{pmatrix}. \tag{41}$$

If $\mathbf{b}$ is normalized and $\mathbf{e} = \mathbf{e}_i$, then

$$d_0 = \mathbf{b}^T\mathbf{A}\mathbf{b} \tag{42}$$

is the eigenvalue estimate from the previous iteration and

$$R_i = \mathbf{e}_i^T(\mathbf{A} - d_0\mathbf{1})\mathbf{b} \tag{43}$$

is the residual error. Let

$$s = A_{ii} - d_0 - 2b_iR_i \tag{44}$$

$$t = 4(1 - b_i^2)R_i^2$$

then

$$d - d_0 = -2R_i^2\{s + (s^2 + t)^{1/2}\}^{-1} \tag{45}$$

$$c_2/c_1 = -[R_i - (d - d_0)b_i]/[A_{ii} - d].$$

Near convergence this solution will approach the perturbation theory result

$$d - d_0 = -R_i^2/(A_{ii} - d_0) \tag{46}$$

$$c_2/c_1 = -R_i/(A_{ii} - d_0).$$

Nesbet (10) actually used this approximate form for c_2/c_1 while Shavitt et al. (11) used the variational result. Shavitt (12) has discussed how to organize the calculation to use an $\mathbf{A}$ matrix stored in lower triangular form. The method is not easily adapted to direct CI procedures since it needs access to elements of **A** in a particular pattern in order to form the R_i sequentially.

Convergence of this method is monotonic. If the lowest diagonal element is the dominant term in the lowest eigenvalue, convergence is good when the first iteration uses the corresponding **e** as the initial guess to **x**. Convergence can be slow if the initial guess is bad or if there is near degeneracy. Convergence directly to higher eigenvalues is generally not possible.

Extension of the Nesbet method to allow computation of excited states has been made in two ways. Shavitt et al. (11) suggested shifting the lower roots, after they were found, by defining

$$\mathbf{A}_k\mathbf{b} = \mathbf{A}\mathbf{b} + \sum_i^{k-1} q_i\mathbf{x}_i(\mathbf{x}_i^T\mathbf{b}). \tag{47}$$

The q_i are chosen so that $\mathbf{A}_k$ has d_k as its lowest eigenvalue. This method is subject to accumulation of errors due to the inexactness of the $\mathbf{x}_i$. Its main advantage is that only 2n words of core memory are required.

Alternatively, when 2kn words of memory are available, k eigenvalues can be found simultaneously by methods developed by Shavitt (13) or Raffennetti (14). In the Shavitt method, an initial set of vectors $\{\mathbf{b}_1 \text{ --- } \mathbf{b}_K\}$ is found so that $\mathbf{B}^T\mathbf{A}\mathbf{B}$ is diagonal and $\mathbf{B}^T\mathbf{B}$ is a unit matrix. Each vector is then relaxed independently with the root shifted matrix

$$\mathbf{A}_j' = \mathbf{A} + \sum_i^{j-1} q_i\mathbf{b}_i\mathbf{b}_i^T \tag{48}$$

used to prevent collapse of $\mathbf{b}_j$ to a lower eigenvector. At the end of each sweep through all the $\mathbf{e}_i$, the matrix eigenvalue problem

$$(\mathbf{B}^T\mathbf{A}\mathbf{B})\mathbf{c} = d(\mathbf{B}^T\mathbf{B})\mathbf{c} \tag{49}$$

is solved to give starting **b**'s for the next iteration.

One could easily devise a method which avoided root shifting by simply expanding each $\mathbf{x}_p$ as

$$\mathbf{x}_p = \sum_j^K y_{jp}\mathbf{b}_j + z_p\mathbf{e}_i \tag{50}$$

at each step. This gives a bordered eigenvalue problem which is easily solved:

$$\begin{pmatrix} \mathbf{d}^o & \mathbf{g}^T \\ \mathbf{g} & A_{ii} \end{pmatrix} \begin{pmatrix} \mathbf{y}_p \\ z_p \end{pmatrix} = d_p \begin{pmatrix} \mathbf{1} & \mathbf{s}^T \\ \mathbf{s} & 1 \end{pmatrix} \begin{pmatrix} \mathbf{y}_p \\ z_p \end{pmatrix} \tag{51}$$

$\mathbf{d}^o = \mathbf{B}^T\mathbf{AB}$ = diagonal matrix

$\mathbf{g} = \mathbf{e}_i{}^T\mathbf{AB}$

$\mathbf{s} = \mathbf{e}_i{}^T\mathbf{B}.$

So

$$d_p - d_p{}^0 = \sum_j (g_j - d_p b_{ij})^2 (d_p - d_p{}^o)/(d_p - A_{ii})(d_p - d_j{}^o) \tag{52}$$

can easily be solved by iteration to yield an estimate for d_p and then

$$y_{jp}/z_p = (g_j - d_p b_{ij})/(d_p - d_j{}^0) \tag{53}$$

will give the coefficients. Near convergence, these equations give the perturbation estimates, to lowest order in the residuals R_{ij},

$$d_p - d_p{}^o = -R_{ip}{}^2/(A_{ii} - d_p{}^o) \tag{54}$$

$$z_p/y_{pp} = -R_{ip}/(A_{ii} - d_p{}^o)$$

$$y_{jp}/y_{pp} = 0 \qquad j \quad p$$

B. Lanczos

The Lanczos algorithm (15) was originally conceived as a scheme for reducing a matrix to tridiagonal form. Consider

$$\mathbf{B}^T\mathbf{AB} = \begin{pmatrix} d_1 & f_1 & & \\ f_1 & d_2 & f_2 & \\ & f_2 & d_3 & \ddots \\ & & \ddots & \ddots \end{pmatrix} = \mathbf{T} \tag{55}$$

where $\mathbf{B}^T\mathbf{B} = \mathbf{1}$. Then

$$\mathbf{AB} = \mathbf{BT} \tag{56}$$

so

$$\mathbf{Ab}_i = f_{i-1}\mathbf{b}_{i-1} + d_i\mathbf{b}_i + f_i\mathbf{b}_{i+1}. \tag{57}$$

This can be rewritten as

$$\mathbf{r}_{i+1} = \mathbf{A}\mathbf{b}_i - f_{i-1}\mathbf{b}_{i-1} - d_i\mathbf{b}_i \tag{58}$$
$$f_i = \| \mathbf{r}_{i+1} \|$$
$$\mathbf{b}_{i+1} = \mathbf{r}_{i+1}/f_i$$
$$d_{i+1} = \mathbf{b}_{i+1}{}^T\mathbf{A}\mathbf{b}_{i+1}$$

which provides a recursion relation for d_i, f_i, and $\mathbf{b}_i$. The recursion may be started by selecting for $\mathbf{b}_1$ any normalized vector which has a non-zero projection on all desired eigenvectors.

This scheme for generating the $\mathbf{b}$ vectors is equivalent to the Schmidt orthogonalization of the Krylov sequence $\mathbf{A}^k\mathbf{b}_1$, k = 0, 1, 2.... The eigenvectors of a truncated $\mathbf{T}$ are equivalent to expansion of the eigenvectors of $\mathbf{A}$ in the set $\{\mathbf{b}_i\}$ or in the set $\{\mathbf{A}^k\mathbf{b}_1\}$. As the power method for eigenvalues shows, the vectors $\mathbf{A}^k\mathbf{b}$ approach $\mathbf{x}_1$ or $\mathbf{x}_n$, and hence for large matrices become nearly linearly dependent. In finite precision arithmetic the higher $\mathbf{b}_i$ generated by this recursion relation will gradually lose orthogonality to earlier members of the sequence. In fact the $\mathbf{b}_i$ will eventually repeat the vector space spanned by previous members of the set and the matrix $\mathbf{T}$ will have repeated eigenvalues. At the point where f_i becomes small compared to $\mathbf{A}\mathbf{b}_i$ the first eigenvalue has in fact converged and continuation of the recursion beyond that point has difficulty in improving the approximation to other roots. Explicit orthogonalization of each $\mathbf{b}_i$ to all lower ones is possible, but this destroys the tridiagonal form of $\mathbf{T}$.

The Lanczos method is closely related to the gradient method. Consider the Rayleigh quotient

$$r(\mathbf{x}) = \mathbf{x}^T\mathbf{A}\mathbf{x}/\mathbf{x}^T\mathbf{x}. \tag{59}$$

The gradient of this for a normalized $\mathbf{x}$ is

$$1/2 \text{ grad } r = (\mathbf{A} - r\mathbf{1})\mathbf{x} = \mathbf{R}(\mathbf{x}). \tag{60}$$

In the gradient method one considers at each step the 2 dimensional problem with $\mathbf{b}_1 = \mathbf{x}^n$ and $\mathbf{b}_2 = \mathbf{A}\mathbf{x}^n$ or $\mathbf{R}(\mathbf{x}^n)$ as the expansion basis. Essentially, the Lanczos method just goes one step further by appending $\mathbf{A}\mathbf{x}$ to the whole current set $\{\mathbf{b}\}$ rather than truncating $\{\mathbf{b}\}$ back to $\mathbf{x}$. Notice that $\mathbf{A}\mathbf{x}$, $\mathbf{R}(\mathbf{x})$ and $A^k\mathbf{b}_1$ are equivalent after orthogonalization to $\{\mathbf{b}\}$.

The time limiting step of the Lanczos algorithm is forming $\mathbf{A}\mathbf{b}_i$ from $\mathbf{b}_i$. For large matrices the rest of the algorithm takes negligible time as the method normally truncates while the dimension of **T** is small. As a scheme to reduce matrices to tridiagonal form, the Lanczos method is too unstable to be useful; but it is a powerful method for finding several eigenvalues of large modulus. There is a large body of unpublished mathematical literature on this method which has been summarized by Parlett (1) and Cullum (16). Moler (9) includes an extensive bibliography of published papers including explicit orthogonalization (17) and block versions (18).

For applications to molecular structure calculations, it is best to shift the spectrum of **A** so that the highest eigenvalues are near zero. This ensures that the lowest eigenvalues are also the ones of largest modulus.

C. Davidson

In the Davidson scheme (19) the next expansion vector is chosen by perturbation theory. If the approximate eigenvector at the n'th iteration is

$$\mathbf{x}_p = \sum_i^K c_{ip}\mathbf{b}_i \tag{61}$$

$$(\mathbf{B}^T\mathbf{A}\mathbf{B})\mathbf{c}_p = d_p\mathbf{c}_p \tag{62}$$

and the residual vector is

$$\mathbf{R}_p = (\mathbf{A} - d_p\mathbf{1})\mathbf{x}_p, \tag{63}$$

then the first order correction to x_{ip} is

$$q_{ip} = R_{ip}/(d_p - A_{ii}). \tag{64}$$

This equation agrees with the perturbation estimates given above for the Nesbet scheme. Hence this method may be regarded as a modification of the Nesbet method in which first order corrections to all elements of the vector are estimated simultaneously rather than sequentially, and then the correction is added variationally.

This correction may also be viewed as the Gauss-Seidel approximation to the Newton-Raphson equation for minimizing the Rayleigh quotient (1). It differs from the correction used in the gradient method by approximately taking into account the second derivatives of the Rayleigh quotient. Consequently it will give a correction vector which is better targeted to improving a particular eigenvalue when **A** is diagonally dominant with a wide range of diagonal elements. As noted previously, the Lanczos method is equivalent to using the residual as the correction, and after orthogonalization of $\mathbf{R}_p$ to the previously found {**b**} the result is independent of p. Orthogonalization of one of the $\mathbf{q}_p$ to the {**b**} to get the next expansion vector produces a correction which depends strongly on which vector is to be corrected. This allows preferential convergence directly to excited states.

The correction vector is explicitly orthogonalized to the previous set {**b**} and normalized before being included as the next **b**. If orthogonalization via

$$\mathbf{q}' = \prod_i (1 - \mathbf{b}_i\mathbf{b}_i^T)\mathbf{q} \tag{65}$$

reduces the norm of **q** appreciably, this step is repeated to ensure accurate orthogonality. At convergence the loss of significant figures in **R** become severe, but this does not cause any numerical problems as long as the {**b**} remain orthonormal. Unlike the Lanczos method, this method will not give repeated eigenvalues and will terminate with the exact eigenvectors if the dimension of the set {**b**} ever equals the dimension of the matrix. If the denominators are omitted in the definition of **q** the method reduces to the explicitly orthogonalized Lanczos method.

Of course, the method is designed to have the dimension of {**b**} always much less than the dimension of the matrix **A**. In practice the number of **b** vectors is only allowed to be about 10-15 larger than the number of eigenvectors wanted. When this maximum occurs, the set {**b**} is replaced by the truncated set {**x**} of current approximations to the eigenvectors of interest. Simultaneously the file containing {**Ab**} is replaced by {**Ax**} computed as

$$(\mathbf{Ax}_p) = \sum_i^K c_{ip}(\mathbf{Ab}_i) \tag{66}$$

and the matrix $\mathbf{B}^T\mathbf{AB}$ is replaced by

$$(\mathbf{X}^T\mathbf{AX})_{ij} = d_i\delta_{ij}. \tag{67}$$

If {**b**} is replaced by {**x**} every iteration, this becomes the often rediscovered method of variationally adding the first order correction.

The next iteration of the algorithm proceeds by forming $\mathbf{Ab}_K$ which is always assumed to be the time-limiting step. Then the previous $\mathbf{B}^T\mathbf{AB}$ is bordered by $\mathbf{B}^T\mathbf{Ab}_K$ and diagonalized to give the next estimate of **x**. Both {**b**} and {**Ab**} are retained on external memory so that **R** can be formed rapidly from

$$\mathbf{R}_p = \sum_i^K c_{ip}[(\mathbf{Ab}_i) - d_p(\mathbf{b}_i)]. \tag{68}$$

Several tests have been suggested for convergence but the

most reliable is the size of $|c_{Kp}|$. Since $\mathbf{b}_K$ is normalized, the largest element of $\mathbf{b}_K$ is less than unity and $|c_{Kp}|$ gives an estimate of the largest change in $\mathbf{x}_p$. As in any variational scheme, the error in the eigenvalue is quadratic in the error in the vector. The norm of the residual provides a pessimistic lower bound to the exact eigenvalue,

$$d_p - \|\mathbf{R}_p\| \leq E_p \leq d_p. \tag{69}$$

Unfortunately $\|\mathbf{R}\|$ is proportional to $\|\mathbf{A} - d\mathbf{1}\|$ so it measures absolute rather than relative error, and hence does not provide a suitable test for convergence of the eigenvector.

In practice, convergence is rapid and is not bothered by near degeneracy to the same extent as the Nesbet method. Even faster convergence can be obtained by the block method (see later) for nearly degenerate eigenvalues. Convergence will also be more rapid, to the targeted vector, than the Lanczos method provided $\mathbf{q}$ is a better approximation than $\mathbf{R}$ to the Newton-Raphson correction $(d_p\mathbf{1} - \mathbf{A})^{-1}\mathbf{R}_p$. Convergence of vectors other than the targeted one will be less rapid than Lanczos since the Lanczos scheme provides a simultaneous gradient correction to all vectors. This distinction has caused some confusion in the literature (20,21).

In using the Davidson method to compute several eigenvectors, the subscript p labeling the target vector must be varied during the calculation. Usually p is selected according to one of the following strategies:

(1) Start p at 1 and increment each time a vector is converged.

(2) Vary p cyclically through the list of eigenvectors desired.

(3) Choose p to track on an eigenvector with a desired pattern of coefficients (22).

(4) Choose p to correspond to the largest $|c_{Kj}|$ or $\|\mathbf{R}_j\|$ from the previous iteration (among the desired eigenvectors).

In addition to the choice of p, there is some discretion in the definition of $\mathbf{q}$. In practice, division by zero is avoided by replacing very small values of $d_p - A_{ii}$ by some fixed number. Also the A_{ii} are often replaced by E_i^{di} from a zeroth order hamiltonian such as the Fock operator (23) or a spin-coupling averaged energy. This is particularly convenient for direct CI calculations which compute $\mathbf{Ab}$ directly from integrals without forming $\mathbf{A}$.

The original method has been extended in two very useful ways. Butscher and Kammer (22) suggested the root-homing method which converges to an eigenvector of the desired form without accurate knowledge of even the number of lower eigenvalues. In this case there is no variational principle if lower eigenvalues are missing, but a small residual vector still indicates an accurate vector.

Liu (9) has suggested a block-Davidson method which adds correction vectors to the set {**b**} for several vectors simultaneously. This requires additional memory to be most effective since two vectors must be held in core for each **Ab** which is formed in one pass through **A**. The block method, however, may greatly reduce the number of passes through **A** by working on corrections to several target vectors at once and may improve convergence of nearly degenerate eigenvalues.

Weber et al (24) have published an alternative derivation and a FORTRAN program based on the Davidson method with careful orthogonalization.

Attempts have been made to extend the method for use with the non-symmetric and/or non-orthogonal (generalized) eigenvalue problem. The non-symmetric case (25, 26) can be approached by introducing a basis $\{\underline{\mathbf{b}}\}$ for the left-hand vectors. Then the expansion coefficients are found by solving the small non-symmetric problem

$$(\underline{\mathbf{B}}^T\mathbf{A}\mathbf{B})\mathbf{c} = d\mathbf{c} \tag{70}$$

$$\underline{\mathbf{c}}^T(\underline{\mathbf{B}}^T\mathbf{A}\mathbf{B}) = d\underline{\mathbf{c}}^T$$

where $\underline{\mathbf{B}}^T\mathbf{B} = \mathbf{1}$. As in the symmetric case, $\{\underline{\mathbf{b}}\}$ and $\{\mathbf{b}\}$ can be bordered by constructing perturbation corrections $\underline{\mathbf{q}}$ and $\mathbf{q}$ and orthogonalizing these by

$$\mathbf{b}_{n+1} = \prod_i (1 - \mathbf{b}_i\underline{\mathbf{b}}_i^T)\mathbf{q} \tag{71}$$

$$\underline{\mathbf{b}}_{n+1} = \prod_i (1 - \underline{\mathbf{b}}_i\mathbf{b}_i^T)\underline{\mathbf{q}}$$

and normalizing so that

$$\mathbf{b}_i^T\mathbf{b}_i = \underline{\mathbf{b}}_i^T\mathbf{b}_i = 1. \tag{72}$$

In general the eigenvalues are complex and root-tracking is used to drive convergence to the desired solution.

Attempts to apply the Davidson method to the generalized eigenvalue problems encountered in non-orthogonal CI calculations have not been very successful (27, 28). At present block Nesbet or conjugate gradient methods are more effective. Other methods based on rewriting

$$\mathbf{Hc} = E\mathbf{Sc} \tag{73}$$

as

$$(\mathbf{H} - d\mathbf{S})\mathbf{c} = (E - d)\mathbf{Sc} \tag{74}$$

where E - d is small seem to hold some promise, expecially where (**H** - d**S**) is diagonally dominant (29) or **S** on the right hand side can be approximated by a unit matrix (30) for the purpose of computing a new estimate of d. These latter methods have not been tried for the nearly linear dependent case typical of non-orthogonal CI matrices.

REFERENCES

1. Parlett, B. N.: 1980, "The Symmetric Eigenvalue Problem", Prentice-Hall, Englewood Cliffs, N.J.
2. Wilkinson J. H. and Reinsch, C.: 1971, "Handbook for Automatic Computation, Volume II--Linear Algebra", Springer-Verlag, New York.
3. Corneil, D.: 1965, "Eigenvalues and Orthogonal Eigenvectors of Real Symmetric Matrices", Masters Thesis, University of Toronto.
4. Birge, R. R. and Hubbard, L. M.: 1978, J. Comput. Phys. 29, pp. 199-207.
5. Givens, W.: 1954, "Numerical Computation of the Characteristic Values of a Real Symmetric Matrix", ORNL-1574, Oak Ridge National Laboratory.
6. Wilkinson, J. H.: 1960, Comp. J. 3, pp. 23-27.
7. Householder, A. S.: 1958, J. Soc. Ind. Appl. Math. 6, pp. 189-195.
8. Wilkinson, J. H.: 1965, "The Algebraic Eigenvalue Problem", Oxford University Press, New York.
9. Moler, C. and Shavitt, I.: 1978, "Numerical Algorithms in Chemistry: Algebraic Methods", LBL-8158 Lawrence Berkeley Laboratory.
10. Nesbet, R. K.: 1965, J. Chem. Phys. 43, pp. 311-312.
11. Shavitt, I., Bender, C. F., Pipano, A., and Hosteny, R. P.: 1973, J. Comput. Phys. 11, pp. 90-108.
12. Shavitt, I.: 1970, J. Comput. Phys. 6, pp. 124-130.
13. Shavitt, I.: 1977, unpublished report.
14. Raffenetti, R. C.: 1979, J. Comput. Phys. 32, pp. 403-419.
15. Lanczos, C.: 1950, J. Res. Natl. Bur. Stand. 45, pp. 255-282.
16. Cullum, J. and Willoughby, R. A.: 1981, J. Comput. Phys. 44, pp. 329-358.
17. Paige, C. C.: 1970, BIT 10, p. 183; 1972, J. Inst. Math. Applic. 10, pp. 373-381.

18. Cullum, J. and Donath, W. E.: 1974, "A Block Generalization of the Symmetric S-Step Lanczos Algorithm", IBM Watson Research Center RC-4845; Saad, Y.: 1980, Soc. Ind. Appl. Math. J. Num. Anal. 17, p. 687.
19. Davidson, E. R.: 1975, J. Comput. Phys. 17, pp. 87-94.
20. Kalamboukis, T. Z.: 1980, J. Phys. A: Math. Gen. 13, pp. 57-62.
21. Davidson. E. R.: 1980, J. Phys. A: Math. Gen. 13, pp. 179-180.
22. Butscher, W. and Kammer, W. E.: 1976, J. Comput. Phys. 20, pp. 313-325.
23. Seeger, R. Krishman, R., and Pople, J. A.: 1978, J. Chem. Phys. 68, pp. 2519-2521.
24. Weber, J., Lacroix, R., and Wanner, G.: 1980, Computers and Chemistry 4, pp. 55-60.
25. Rettrup, S.: 1982, J. Comput. Phys. 45, pp. 100107.
26. Hirao, K., and Nakatsuji, H.: 1982, J. Comput. Phys. 45, pp. 246-254.
27. Cheung, L. M., and Bishop, D. M.: 1977, Computer Phys. Comm. 13, pp. 247-250.
28. Gallup, G.: 1982, J. Comput. Chem. 3, pp. 127-129.
29. Ruhe, A.: 1975, J. Comput. Phys. 19, pp. 110-120.
30. Tsunematsu, T., and Takeda, T.: 1978, J. Comput. Phys. 28, pp. 287-293.

GROUP THEORY APPLIED TO CI METHODS

B. T. Sutcliffe

Department of Chemistry, University of York, York,
YO1 5DD, England

1. CI-THE PROBLEM AND SOME NOTATION

The basic approximation in CI theory is the linear combination of configurations

$$\Psi_K = \sum_{i=1}^{M} C_{Kj}\Phi_j \qquad (1.1)$$

This approximation leads to the usual M-dimensional secular problem

$$\underset{\sim}{H}\,\underset{\sim}{C} = \underset{\sim}{M}\,\underset{\sim}{C}\,\underset{\sim}{E} \qquad (1.2)$$

where, in conventional notation,

$$H_{ij} = \langle\Phi_i|\hat{H}|\Phi_j\rangle\ , \qquad M_{ij} = \langle\Phi_i|\Phi_j\rangle$$

$$(\underset{\sim}{C})_{Kj} = C_{Kj}\ , \qquad (\underset{\sim}{E})_{ij} = \delta_{ij}E_i \qquad (1.3)$$

and $\hat{H}$ is the usual spinless Hamiltonian.

We imagine that each configuration Φ_j consists of an anti-symmetrized product of N spin-orbitals where N is the number of electrons in the problem. We shall assume that there are n orbitals

$$\phi_1(\underset{\sim}{r}),\ \phi_2(\underset{\sim}{r}),\ \ldots\ ,\ \phi_n(\underset{\sim}{r}) \qquad (1.4)$$

using $\underset{\sim}{r}$ to denote the space coordinates of an electron. There are

G. H. F. Diercksen and S. Wilson (eds.), Methods in Computational Molecular Physics, 115–159.

therefore 2n possible spin-orbitals

$$\phi_1(\underset{\sim}{r})\alpha(s),\ \phi_1(\underset{\sim}{r})\beta(s),\ \ldots\ \phi_n(\underset{\sim}{r})\beta(s)$$

or

$$\psi_1(\underset{\sim}{x}),\ \psi_2(x)\ ,\ \ldots\ \psi_{2n}(\underset{\sim}{x}) \qquad (1.5)$$

where s is used to denote the spin-coordinate and $\underset{\sim}{x}$ is used to denote the space and spin coordinates collectively.

The central problem in any CI method is to evaluate the matrix elements H_{ij} and M_{ij}. From general considerations we know that any matrix element of $\hat{H}$ can be written formally as

$$H_{ij} = \sum_{r,s=1}^{2n} D^{1ij}_{sr} \langle\psi_r|\hat{h}|\psi_s\rangle + \tfrac{1}{2} \sum_{rstu}^{2n} D^{2ij}_{tu,rs} \langle\psi_r\psi_s|g|\psi_t\psi_u\rangle \qquad (1.6)$$

where D^{1ij}_{rs} is an element of the one-particle transition density matrix and $D^{2ij}_{tu,rs}$ is an element of the two-particle transition density matrix. The notation for the integrals is the conventional one and it should be noticed that it refers to <u>spin-orbitals</u>.

Similarly a matrix element of the overlap matrix can be written as

$$M_{ij} = N^{-1} \sum_{rs=1}^{2n} D^{1ij}_{sr} \langle\psi_r|\psi_s\rangle \qquad (1.7)$$

In fact the spin integration is quite formal and the matrix element (1.6) can be re-written as

$$H_{ij} = \sum_{rs=1}^{n} P^{ij}_{sr} \langle\phi_r|\hat{h}|\phi_s\rangle + \tfrac{1}{2} \sum_{rstu=1}^{n} Q^{ij}_{tu,rs} \langle\phi_r\phi_s|g|\phi_t\phi_u\rangle \qquad (1.8)$$

where P^{ij}_{sr} is an element of the one-particle spinless transition density matrix and $Q^{ij}_{tu,rs}$ is an element of the two-particle spinless transition density matrix. In this formula the notation for the integrals is again conventional but now the implied integration is over space only.

It is possible to write these matrix elements in a different way using second quantized notation. As is well known we can replace the operator $\hat{H}$ with an effective operator operating on the spin-orbital space directly, thus

$$\hat{H} \Rightarrow \sum_{rs} \langle\psi_r|\hat{h}|\psi_s\rangle a_r^+ a_s + \tfrac{1}{2} \sum_{rstu} \langle\psi_r\psi_s|g|\psi_t\psi_u\rangle a_r^+ a_s^+ a_u a_t \qquad (1.9)$$

where the creation (annihilation) operators a_r (a_r^+) are assumed to act on the spin-orbitals and to obey the anti-commutation relations.

$$[a_r^+, a_s]_+ \equiv a_r^+ a_s + a_s^+ a_r = \langle\psi_r|\psi_s\rangle$$

$$[a_r, a_s] = [a_r^+, a_s^+]_+ = 0 \qquad (1.10)$$

It is usual in the CI problem to assume that the spin-orbital basis is orthonormal and we shall in fact assume this from now on so that

$$\langle\psi_r|\psi_s\rangle = \delta_{rs} \qquad (1.11)$$

Thus to be perfectly safe it should be assumed that none of the results which follow can be taken over into a theory based on non-orthogonal orbitals.

We can re-write (1.9) by introducing the operators

$$E'_{rs} = a_r^+ a_s \qquad (1.12)$$

as

$$\sum_{rs} \langle\psi_r|\hat{h}|\psi_s\rangle E_{rs} + \tfrac{1}{2} \sum_{rstu} \langle\psi_r\psi_s|g|\psi_t\psi_u\rangle [E'_{rt}E'_{su} - \delta_{st}E'_{ru}] \qquad (1.13)$$

where the operators E_{rs} satisfy the <u>commutation</u> relations

$$[E'_{rs}, E'_{tu}] \equiv E'_{rs}E'_{tu} - E'_{tu}E'_{rs}$$

$$= \delta_{st}E'_{ru} - \delta_{ru}E'_{ts} \qquad (1.14)$$

At this stage we are still in a spin-orbital basis, but we can remove spin here just as we did in passing from (1.6) to (1.8). In this effective operator this is achieved by summing over the spin index for each orbital index. Let us write, formally,

$$E'_{rs} \equiv E'_{r\sigma,s\sigma} \qquad (1.15)$$

where σ denotes the two possible spin indices then we define

$$E_{rs} = \sum_{\sigma} E'_{r\sigma,s\sigma} \tag{1.16}$$

Using these operators the effective operator (1.13) becomes

$$\sum_{rs} \langle\phi_r|\hat{h}|\phi_s\rangle E_{rs} + \tfrac{1}{2} \sum_{rstu} \langle\phi_r\phi_s|g|\phi_t\phi_u\rangle [E_{rt}E_{su} - \delta_{st}E_{ru}] \tag{1.17}$$

where the integrals are now over orbitals and the operators obey the commutation relations which are just like (1.14) namely

$$[E_{rs}, E_{tu}] = \delta_{st}E_{ru} - \delta_{ru}E_{ts} \tag{1.18}$$

Thus in order to do CI we need to know before starting any calculation formulas for the P^{ij}_{sr} and $Q^{ij}_{tu,rs}$ for all configurations ij that might arise and for all orbitals rstu involved. Equivalently we must know the matrix elements of E_{rs} and $[E_{rt}E_{su} - \delta_{st}E_{ru}]$ on the orbital space appropriate to the configuration sets chosen.

It must be understood quite clearly that if one worked a calculation using both (1.17) and (1.8) then either formulation would lead to exactly the same answer. If it does not then we have got something wrong.

Now in practice CI calculations lie between two extremes of formulation. At one extreme one has the conventional or configuration driven CI in which it is assumed that one has a given set of Φ_i and the object is to find which P^{ij}_{sr}, $Q^{ij}_{tu,rs}$ are non-zero and what their values are. (Or, equivalently, matrix elements of E_{rs}, etc.) At the other extreme one has direct or integral driven CI in which one has implicitly a standard list of the Φ_i and one wants to know with which P^{ij}_{sr} the integral $\langle\phi_r|\hat{h}|\phi_s\rangle$ can be associated to produce a contribution to all the correctly identified H_{ij}. (Equivalently, the appropriate contribution of E_{rs} to all the H_{ij} to which it contributes.) Similarly for the two-electron integrals.

As a rough and ready guide it is reasonable to say that the formalism (1.8) is well suited to conventional CI while the formalism (1.17) is well suited to integral drived CI. But which one uses in fact is a matter of taste and in which it is found easiest to see the required results. A mixture of both styles is perfectly possible if you wish to prove your virtuosity.

Now group theory can help in structuring the CI problem. In particular it is possible using certain groups to obtain explicit expressions for the density matrix elements. (Equivalently, the E_{rs} matrix elements.) Thus the use of group theory may help one to see an efficient coding strategy. But on the other hand it may not. There is no intrinsic virtue in the use of group theory in the CI problem. It should be remembered that the earliest practical formulations of conventional CI were made entirely without benifit of group theory, as were the earliest practical formulations of direct CI. Subsequent to these practical achievements, clever people realized the relevance of group theory and were able to make suggestions on this basis about how the codes could be cleaned up and made more efficient.

Thus the situation at present is that every well-brought up quantum chemist has to understand some of the relevant group theory in order to understand the logic behind many modern programs. But it is certainly not necessary to take any kind of loyalty-oath to abstract group theory before initiation into the arcane mysteries of quantum chemistry.

Now in practice group theory gets in on the act in two quite distinct ways.

(A) Use of the invariances of $\hat{H}$ under transformations in the coordinate space (KINEMATIC INVARIANCE)

Among such invariances are the usual point-group symmetry operators. More important for CI generally among such invariances are the permutation of electron variables which enables one to use the group of permutations of N ojects, the symmetric group of degree N, denoted usually as S_N.

(B) Use of the invariance of the roots of the secular problem under linear transformations of the orbital space (DYNAMICAL INVARIANCE)

This is a more unusual, but in fact very "deep" kind of symmetry and it makes possible the use of the unitary group in n-dimensions, denoted usually as U(n). More strictly one should consider the special unitary group SU(n). The unitary group in n-dimensions consists of all unitary n by n matrices. The special unitary group consists of all such matrices with determinant +1. The unitary group is a continuous group dependent upon n^2 parameters. This simply means that any unitary matrix can be written down in terms of n^2 real numbers, for examples, the real parts of the elements of the matrix elements of U(n). The special unitary group has n^2-1 parameters, since there is 1 extra requirement to be imposed by virtue of requiring $|\underset{\sim}{U}(n)| = +1$ for all $\underset{\sim}{U}(n)$ in SU(n).

2. POINT GROUP SYMMETRY

Imagine a fixed basis in which electrons are described by coordinates. For definiteness let us imagine this basis to be a right-handed cartesian basis with unit vectors $\hat{e}_x$, $\hat{e}_y$, and $\hat{e}_z$

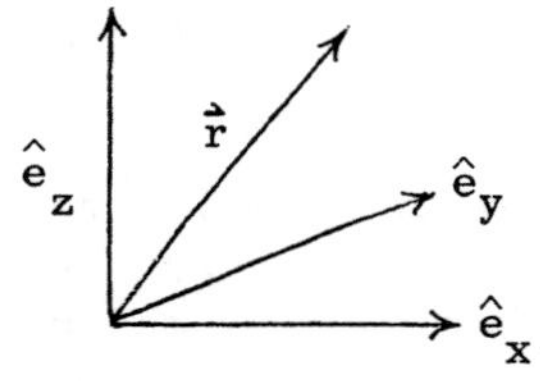

$$\vec{r} = \hat{e}_x r_x + \hat{e}_y r_y + \hat{e}_z r_z$$

$$= (\hat{e}_x \hat{e}_y \hat{e}_z) \begin{pmatrix} r_x \\ r_y \\ r_z \end{pmatrix} \equiv \underset{\sim}{\hat{e}}\,\underset{\sim}{r} \qquad (2.1)$$

For the electrons one would have $\underset{\sim}{r}_1, \underset{\sim}{r}_2 \ldots \underset{\sim}{r}_N$.

Now imagine an orthogonal coordinate transformation in this basis, thus

$$\vec{r} \rightarrow \vec{r}\,' = \underset{\sim}{\hat{e}}\,\underset{\sim}{r}' , \qquad \underset{\sim}{r}' = \underset{\sim}{R}\,\underset{\sim}{r} , \qquad \underset{\sim}{R}\,\underset{\sim}{R}^T = \underset{\sim}{E}_3 \qquad (2.2)$$

For example,

$$\underset{\sim}{R} = \begin{pmatrix} \cos\phi & -\sin\phi & 0 \\ \sin\phi & \cos\phi & 0 \\ 0 & 0 & 1 \end{pmatrix}$$

as a rotation about $\hat{e}_z$ by $+\ \phi$.

Now imagine the operator $\hat{H}$ made up in a set of transformed coordinates

$$\underset{\sim}{r}' = \underset{\sim}{R}\,\underset{\sim}{r} \qquad (2.4)$$

where $\underset{\sim}{R}$ is some as yet unspecified 3 by 3 orthogonal matrix. Let us call the transformed operator $\hat{H}'$ and it is made up just as is $\hat{H}$ but with r_x, etc., wherever r_x, etc., occurs in H. It is then a matter of simple algebra to show that

$$\nabla^2(i') \equiv \nabla^2(i) \qquad (\text{i.e. } \nabla^2(i')g(\underset{\sim}{r}_i') \equiv \nabla^2(i)g(\underset{\sim}{r}_i))$$

and

$$r'_{ij} \equiv r_{ij}$$

$$(\text{where } r_{ij} = ((r_{xy}-r_{xi})^2 + (r_{yi}-r_{yj})^2 + (r_{zi}-r_{zj})^2)^{\frac{1}{2}} \qquad (2.5)$$

These relations hold for absolutely any 3 by 3 orthogonal transformation whatsoever, so that obviously the kinetic energy and electron repulsion parts of the Hamiltonian are invariant under any operation in the group of all such matrices. This group is again a continuous group often denoted O(3). It is the full orthogonal group in three-dimensions and contains all possible rotations and reflections.

However this may be, it is easy to show too that in general, the electron-nucleus distance operator is not invariant. That is

$$r'_{ai} \not\equiv r_{ai} \tag{2.6}$$

However, there are certain situations in which the nuclei are so fixed that for a particular choice of orthogonal transformations

$$r'_{ai} \equiv r_{bj} \quad \text{and} \quad Z_a = Z_b \tag{2.7}$$

that is the operation effectively "swaps" nuclei of equal charge. In this case

$$\sum_{a=1}^{N_n} Z_a / r'_{ai} \equiv \sum_{a=1}^{N_n} Z_a / r_{ai} \tag{2.8}$$

If there is a set of matrices that have the property (2.8) then they form a representation of an invariance group of the molecule which is usually called the point group of the molecule. Clearly the point group will be the point group of a given clamped nucleus configuration.

A little more algebra then enables one to show that a molecular point group is always isomorphic with a sub-group of the symmetric (permutation) group of the nuclei with identical charges. In practice we usually find molecular point groups by looking for this isomorphic group, but it should be remembered that it is not the nuclear positions that we are actually monkeying around with in this approach. We are changing the description of the electronic variables.

Now any non-singular transformation of coordinates induces a transformation in a function space built on this coordinate space. The rule is

$$f(\underset{\sim}{r}) \xrightarrow{\underset{\sim}{R}\underset{\sim}{r}} \bar{f}(\underset{\sim}{r}) = f(\underset{\sim}{R}^{-1}\underset{\sim}{r}) = f(\underset{\sim}{R}^{T}\underset{\sim}{r}) \tag{2.9}$$

That is to say, make up the function or o erator on the function space by replacing r_x, etc., by $(\underset{\sim}{R}^T\underset{\sim}{r})_x$ wherever it occurs.

Now we can think of the operator $\underset{\sim}{R}$ inducing an operation O_R in the function space so that we can re-write the left-hand part of (2.9) as

$$O_R f = \bar{f} \tag{2.10}$$

Then if

$$\hat{H} f = g$$

$$O_R \hat{H} f = O_R g$$

therefore

$$O_R \hat{H} O_R^{-1} O_R f = O_R g$$

or

$$O_R \hat{H} O_R^{-1} \bar{f} = \bar{g} \tag{2.11}$$

We now interpret $O_R \hat{H} O_R^{-1}$ as the operator $\hat{H}$ but made up appropriately for the transformed space (i.e., with $(\underset{\sim}{R}^T \underset{\sim}{r})_x$, etc., in place of r_x, etc.).

And as we have seen there are certain operations under which $\hat{H}$ is invariant so that in such cases we can say

$$O_R \hat{H} O_R^{-1} = \hat{H} \tag{2.12}$$

or

$$O_R \hat{H} = \hat{H} O_R$$

or

$$[O_R, \hat{H}] = 0$$

That is, the induced symmetry operator commutes with $\hat{H}$.

Now suppose that we knew the exact eigen-solutions of $\hat{H}$ and suppose that we look at a p-fold degenerate one then

$$\hat{H} f_i = E f_i \ , \qquad i = 1,2,\ldots p \tag{2.13}$$

and

$$O_R \hat{H} f_i = \hat{H} O_R f_i = \hat{H} \bar{f}_i = E \bar{f}_i \tag{2.14}$$

and from this it follows at once that $\bar{f}_i$ can be at most a linear combination of the f_i so that

$$O_R f_i = \sum_{j=1}^{p} f_j (\underset{\sim}{O}_R)_{ji} \tag{2.15}$$

where the matrix $\underset{\sim}{O}_R$ is a non-singular matrix. Now it is very easy to show that if

$$O_R O_S = O_p \quad \text{then} \quad \underset{\sim}{O}_R \underset{\sim}{O}_S = \underset{\sim}{O}_p \tag{2.16}$$

so that the $\underset{\sim}{O}_R$ form a representation of the O_R. In fact if the solutions are exact the representation is usually irreducible. Now for any finite group a fundamental theorem assures us that we may choose irreducible representations to be unitary matrices and in future we shall assume this done. We say that the eigenfunctions f_i form or provide a basis for the unitary irreducible representations $\underset{\sim}{O}_R$. Now another fundamental theorem assures us that functions which form a basis for unitary irreducible representations (unitary irreps for short) are such that

$$\langle f_i^{(\mu)} | f_j^{(\nu)} \rangle = \delta_{\mu\nu}\delta_{ij} \tag{2.17}$$

where μ and ν label the irreps and i and j the rows of the irrep matrix. It also assures us that

$$\langle f_i^{(\mu)} | \hat{H} | f_j^{(\nu)} \rangle = \delta_{\mu\nu}\delta_{ij} \langle f^{(\mu)} || \hat{H} || f^{(\mu)} \rangle \tag{2.18}$$

where the matrix element on the right-hand side of (2.18) is called the reduced matrix element and is independent of i and j.

Thus if one constructs trial functions for a CI calculation that form a basis for unitary irreps of the molecular point group, one has got, at least to some extent, a handle on the matrix elements, but it is not usually enough to specify completely the required matrix element. This is essentially because, assuming that the orbitals themselves form a basis for irreps of the point group, the symmetry of the configuration functions is determined completely by the orbitals. Usually in point groups the dimension of the irreps is small, often only 1-D. Thus there are many distinct configurations with the same point group symmetry designation but constructed from very different orbitals. Thus the density matrix elements are not usually determined in any "fine" way by the point group specification of the function and orbital symmetry.

Now this is not to say that point group symmetry is useless. It is useful if it is conveniently possible to use orbitals which form a basis for the point group, or for a sub-group of the point group, with only one-dimensional irreps. In such a case it is easy to show that

$$\langle\phi_r^{(\mu)}|\hat{h}|\phi_s^{(\nu)}\rangle = \delta_{\nu\mu}\langle\phi_r^{(\mu)}|\hat{h}|\phi_s^{(\mu)}\rangle \tag{2.19}$$

and that

$$\langle\phi_r^{(\nu)}\phi_s^{(\mu)}|g|\phi_t^{(\sigma)}\phi_u^{(\lambda)}\rangle = 0 \quad \text{unless the } \nu,\ \mu, \text{ etc. are equal in pairs} \tag{2.20}$$

Thus in that case one knows that certain density matrix elements vanish for certain sets of orbital numbers, and this is useful. One could of course make a "finer" classification by using whatever two-, three- and four-dimensional irreps occurred in the point group, but such use is rare in practice. One can understand this when one realizes that the point group symmetry is a contingent symmetry. Not all molecules have significant point group symmetries and if one is moving the nuclei then the actual point group symmetry will change from case to case. Thus in a general code one will not build in point group symmetry too deeply because it might not be relevant in any given calculation. Incorporating point group symmetry must effectively carry an overhead in the code. One has to consider whether the return on average is worth the overhead. It is fair to say that in special codes for atoms and linear molecules it is sensible to build in the equivalent of point group summetry pretty deeply, because there really is a pay off there, but for a completely general code it is a matter of judgement.

Roughly speaking it seems that some advantage can be gained by incorporating symmetry recognition at any rate into integral evaluating parts of the code. It seems too, that one can win in the integral transformation step. Most people incorporate limited symmetry into the density matrix construction step in conventional CI. As far as the direct method is concerned, if the integral $\langle\phi_r|\hat{h}|\phi_s\rangle$ is not present then E_{rs} never gets looked at so the advantage is obvious here. It is quite tricky both in the conventional and direct approaches to deal with other than one-dimensional irreps and it is only quite recently that this problem has been studied systematically.

It might be thought that one could take advantage of the O(3) symmetry of the operators ∇^2 and $1/r_{ij}$ and in the original version of Polyatom, to which I was a contributor, we did try to do so. But in practice nowadays the use of that symmetry is confined to a sort of zero-value recognition procedure at integral evaluation time, if it is used at all.

However, there is one area in which O(3) symmetry is very useful. Every transformation $\underset{\sim}{R}$ in coordinate space induces a transformation not only in space-function space, but also in spin-function space. If we denote that transformation by O'_R then, since there is not spin in $\hat{H}$,

$$[O'_R, \hat{H}] = 0 \ , \qquad \text{all } \underset{\sim}{R} \text{ in } O(3) \tag{2.21}$$

This implies that

$$[\hat{S}^2, \hat{H}] = [\hat{S}_\alpha, \hat{H}] = 0 \ , \qquad \alpha = x, y \text{ or } z \tag{2.22}$$

and in fact

$$[\hat{S}^2, \hat{S}_\alpha] = 0 \tag{2.23}$$

Thus we can choose our configurations to be spin-eigenfunctions and our spin-eigenfunctions must be a basis for irreps of the continuous group O(3). Now in fact O(3) is a rather rotten group, because it has two sheets, that is it has two distinct manifolds, one of which consists of all the operators in SO(3), that is all orthogonal matrices with determinants +1, and the other sheet consists of those with determinants -1. As those who live alone will know, changing sheets ia really a dreadful bore, for one has then to do some washing, so it is convenient to stick to one sheet and say that our spin-eigenfunctions provide a basis for irreps SO(3). However, this too is a pretty nasty group, since it is doubly connected and it turns out that the spin-eigenfunctions want to get between the sheets again and absolutely refuse to get out and choose a basis for a nice clean irrep of SO(3). In fact they provide a basis for the universal covering group of SO(3) which is the special unitary group of dimension 2, called SU(2). This group is the group of all two-dimensional unitary matrices with determinant +1. Now although this all sounds pretty indecent, in fact when we come to the unitary group it will turn out to be a blessing in disguise. The spin functions (α,β) can at most span a two-dimensional space and the most that coordinate transformation in three-dimensions can induce in one particle spin-function space is a transformation of one particle spin-functions so that their scalar products remain constant. That is, so that

$$\langle \sigma'_1 | \sigma'_2 \rangle = \langle \sigma_1 | \sigma_2 \rangle \ , \qquad \sigma_i = \alpha \text{ or } \beta \tag{2.24}$$

A little thought shows that this is possible if

$$(\alpha\beta) = (\alpha\beta)\underset{\sim}{U}_2 \tag{2.25}$$

where $\underset{\sim}{U}_2$ is any two by two unitary matrix so at least it is plausible that spin gets mixed up with the unitary group.

But it turns out that many paticle spin-functions have a more immediate group-theoretical relevance in the context of the symmetric group, and it is to this that we turn next.

3. THE SYMMETRIC GROUP OF N ELECTRONS, S_N

For this section we are still thinking of kinematic symmetry, for $\hat{H}$ really is invariant under any permutation of electron space and spin variables and the set of all such permutations do form a group. May I also remind you that we are after the D^{ij}_{sr}, etc., or their equivalents. The point of using the symmetric group here is that every electronic wave function is a basis function for a special one-dimensional irrep of S_N. This wave function can be expressed as a sum of products of space and spin parts. The spin-parts in turn must form a basis for a very special set of irreps of S_N and these spin-based irreps in their turn determine the possible irreps for the space part. If one can find the spin irreps then it turns out that you have effectively found the density matrix elements.

Before beginning a more systematic discussion of matrix elements it is necessary to introduce some notation and definitions. First the notation for an abstract permutation of N objects. Such a permutation is written as

$$P = \begin{pmatrix} 1 & 2 & 3 & \dots & N \\ i_1 & i_2 & i_3 & \dots & i_N \end{pmatrix} \qquad (3.1)$$

This is interpreted as the object 1, going into i_1, 2 going into i_2 and so on. We shall imagine that there is some suitable definition from an operator on the coordinate space of the system and we shall assume that there is an induced operator O_p in a function space built on this coordinate space such that

$$O_p f(x_1, x_2 \dots x_N) = f(x_{i_1}, x_{i_2}, \dots x_{i_N}) \qquad (3.2)$$

This is a pretty conventional definition of the effect of O_p and agrees with the one offered in Pauncz's book on Spin Eigenfunctions. With this definition on the definition of permutations (3.1) then it is easy to show that the O_p form a representation of S_N as required.

S_N is a group of order N! and each permutation operator in the group can be characterized by its cycle structure. To give an example, the permutation

$$\begin{pmatrix} 1 & 2 & 3 \\ 2 & 3 & 1 \end{pmatrix} \text{ may be written as } (123) \qquad (3.3)$$

on the understanding that the cycle notation on the right means $1 \to 2$, $2 \to 3$, $3 \to 1$. Somewhat more generally,

$$\begin{pmatrix} 1 & 2 & 3 & 4 & 5 & 6 \\ 2 & 4 & 5 & 1 & 3 & 6 \end{pmatrix} = (124)(35)(6) \qquad (3.4)$$

In fact <u>any</u> permutation can be decomposed <u>uniquely</u> into cycles of disjoint elements, the cycles commute and by convention the one of greatest length is written first. This is exhibited in (3.4) above, where there is first a cycle of length three, then one of length two and then a cycle of length one. A cycle of length two is called a <u>transposition</u> and every permutation can actually be decomposed into a product of transpositions. This decomposition is <u>not</u> unique and neither are the transpositions generally of disjoint elements. However, the <u>number</u> of transpositions into which a permutation can be decomposed, is unique. This number is either even or odd and if even, the permutation is said to have <u>even parity</u> and if odd, <u>odd parity</u>. The point is generally denoted by $(-1)^p$ where p is the number of transpositions in the decomposition.

Now it is reasonably easy to show that all permutations with the same cycle structure are in the same class so that by determining the cycle structure of the permutations in a given S_N we can determine its class structure. But a fundamental theorem of finite groups assures us that there are as many irreps of a group as there are classes, so that in determining the class structure we also determine the irrep structure. Let us take the example of S_4. Here the possible cycle structures are

1) $(i_1 i_2 i_3 i_4)$

2) $(i_1 i_2 i_3)(i_4)$

3) $(i_1 i_2)(i_3 i_4)$ $\qquad i_j$ in 1,2,3,4

4) $(i_1 i_2)(i_3)(i_4)$

5) $(i_1)(i_2)(i_3)(i_4)$ (3.5)

So there is in 2), for example, one cycle of length 3 and one of length 1 and so on. In general if there are m cycles in any decomposition we denote the cycle lengths by $\lambda_1, \lambda_2 \ldots \lambda_m$. Obviously

$$\lambda_1 + \lambda_2 + \ldots \lambda_m = N , \qquad \text{for } S_N \tag{3.6}$$

and by convention we always choose

$$\lambda_1 \geqslant \lambda_2 \geqslant \ldots \lambda_m > 0 \tag{3.7}$$

We usually symbolize the cycle structure by writing the λ_i as a vector

$$[\lambda] \equiv (\lambda_1 \lambda_2 \ldots \lambda_m) \tag{3.8}$$

and denote r repeats of a given cycle length by λ^r. Thus for (3.5) above

$$1)\ [4] \qquad 2)\ [3,1] \qquad 3)\ [2,2] \equiv [2^2]$$

$$4)\ [21^2] \qquad 5)\ [1^4] \tag{3.9}$$

Now there is a much more fruitful notation than that used in (3.5) above to represent the cycle structure. This notation is due to Young and is called the Young diagram (pattern or shape). The Young diagrams corresponding to (3.5) above are

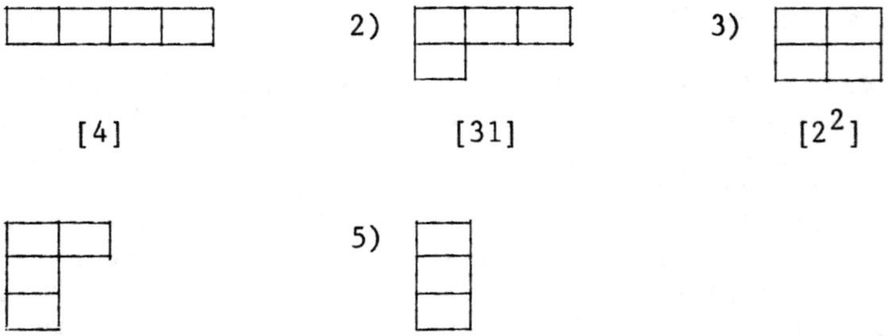

1) [4] 2) [31] 3) $[2^2]$

4) $[21^2]$ 5) $[1^4]$ (3.10)

In (3.10) it can be seen that the boxes in a given row represent a cycle and each row is a new cycle.

The reason why this notation is fruitful is because Young was able to show that by putting the numbers 1,2...N in a given order into the boxes of a given pattern gives the dimension of irrep that is associated with that pattern.

The rule is to put the numbers 1,2...N into the boxes so that when you read across any row or down any column the numbers for a strictly increasing sequence. Thus to take S_4 again, the possibilities are

$[1^4]$ ⇒ 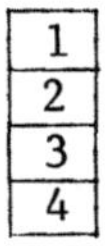 $[4]$ ⇒ 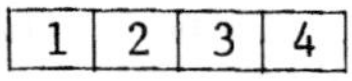

(3.11)

so that there is only one assignment possible here. We say that each diagram generates only one <u>tableau</u>.

However

$[31]$ ⇒ 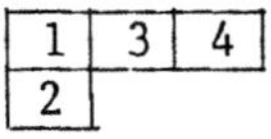 or 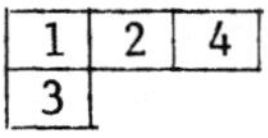 or 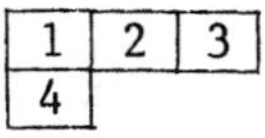 (3.12)

so this diagram generates three tableaux and is associated with an irrep of dimension three.

Now the notation $[\lambda]$ is used to represent both class structure and irrep structure. When it is used to represent class structure then obviously the permutation $[1^N]$ in S_N is just the class containing the identity permutation alone, while $[N]$ is the class containing the permutation that interchanges all elements of S_N. When the notation is used to represent irreps the convention is that

$[N]$ ⇒ TOTALLY SYMMETRIC IRREP OF S_N
(ALL OPS. REP. BY 1)

$[1^N]$ ⇒ ANTI-SYMMETRIC IRREP OF S_N
(ALL EVEN PARITY OPS BY 1
ALL ODD PARITY OPS BY −1)

Generally $[\lambda]$ stands for a given class or a given irrep, but notice that the diagram associated with $[\lambda]$ occur in pairs. Thus

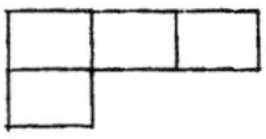 and 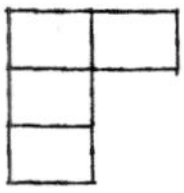

$[31]$ $[21^2]$ (3.13)

or

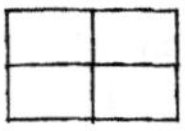 and itself

$[2^2]$

(3.14)

The pairs of diagrams that can be turned into one another by "transposition" of rows and columns in the patterns are called conjugate (adjoint, dual, associated) diagrams. A diagram like (3.14) is called self-conjugate. If a pattern is represented by $[\lambda]$ then the convention is that the conjugate pattern is written $[\tilde{\lambda}]$.

Now if we imagine that we have found an irrep of the group S_N in which all the transpositions in S_N are represented by $(ij) \to \underset{\sim}{P}(ij)$ and that this irrep is $[\lambda]$ then there will be another irrep in which the transpositions are represented by $(ij) \to -\underset{\sim}{P}(ij)$ and this irrep is in fact $[\tilde{\lambda}]$ the irrep conjugate to $[\lambda]$. In general the irreps will differ only in the sign they assign to irreps of permutations of odd parity.

Now the Pauli principle assures us that any wave function must be antisymmetric under the permutation of any pair of electron (space and spin) variables. In the present context this tells us that an electronic wave function for N-electrons must be a basis function for the irrep $[1^N]$ of S_N, where the operators of S_N are understood to be appropriate to electron space and spin variable space. It is convenient to denote an operator induced from such a permutation of variables into ordinary function space by $O^{(r)}$ and into spin-function space by $O^{(s)}$.

Now what we want to do is start off with some product function of the form

$$\Phi_i(r_1,r_2\ldots r_N)\Theta_{\mu i}(s_1 s_2\ldots s_N) \tag{3.15}$$

where Φ_i is an orbital product and $\Theta_{\mu i}$ is a spin-function product, and adapt this product to be a basis function for $[1^N]$ and use this adapted product to find matrix elements H_{ij} and M_{ij}. In fact what we shall be able to show is that if one chooses the spin-functions with care then the matrix element H_{ij} can be written as

$$H_{ij} = \sum_{P \varepsilon S_N} (-1)^P \langle\Phi_i|\hat{H}|O_P^{(r)}\Phi_j\rangle\langle\Theta_{\mu i}|O_P^{(s)}|\Theta_{\mu j}\rangle \tag{3.16}$$

where the matrix $\underset{\sim}{D}(P)$ of spin-function expectation values, such that

$$(\underset{\sim}{D}(P))_{\mu_i\mu_j} = \langle\Theta_{\mu_i}|O_P^{(s)}|\Theta_{\mu_j}\rangle \tag{3.17}$$

is actually an irrep of S_N. What this means is that if one knows the irreps of S_N then one actually has the required density matrix elements, since the space expectation value in (3.16) yields the required orbital integrals and the spin term will yield their proper coefficients.

In fact, as we shall see, the irreps for which the spin functions are a basis must be conjugate to the irreps for which the space functions are a basis so the whole matrix element business is very tightly sewn up by the symmetric group.

That this is possible depends upon the following results.

RESULT (1)

If you can find a set of spin functions that are a basis for $[\lambda]$ then in order to get a total function that is a basis for $[1^N]$ it can be combined only with a space function that is a basis for $[\tilde{\lambda}]$. Furthermore the bases may be so ordered as to force a 1-1 correspondence between the functions in the space and spin bases so that we may write

$$\psi^{[1^N]} \Rightarrow \sum_i \Phi_i^{[\tilde{\lambda}]} \Theta_{\mu i}^{[\lambda]} \qquad (3.18)$$

where the sum goes over all functions in the basis for the irrep $[\lambda]$ ($[\tilde{\lambda}]$).

This result arises from the standard direct product theorem, which it will be remembered states that the direct product of two irreps is in general reducible into sums of irreps. For S_N a special result says that in this reduction the irrep $[1^N]$ can arise in one way and one way only. Thus (3.18) is a sort of Clebsch-Gordan or vector coupling result for S_N.

The result (1) actually holds for particles of any spin whatsoever but the next result is absolutely peculiar to particles of spin-1/2.

RESULT (2)

For particles of spin-1/2 <u>any</u> set of spin-eigenfunctions $\Theta_{S,M_S,i}$ form a basis for irreps of S_N of the form $[\lambda_1, \lambda_2]$ (that is, two row diagrams). No other irreps can occur. Furthermore λ_1 and λ_2 are such that

$$\lambda_1 + \lambda_2 = N$$

$$\lambda_1 - \lambda_2 = 2S$$

so

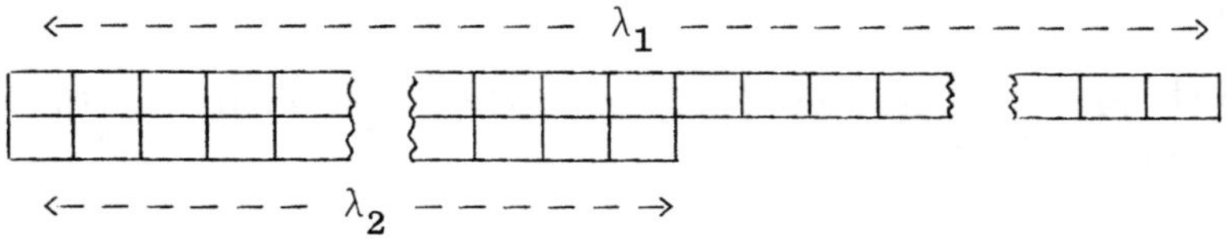

Thus any allowed set of orbital products must form a basis for $[\tilde{\lambda}]$.

This result arises basically because there are only two kinds of spin-functions for spin-1/2 particles. Thus lots of the permutations that could arise do not arise in practice because they are the same if they involve, let us say, only α spin functions. However, that the spin-eigenfunctions are a basis for irreps of S_N is a rather deeper result and takes a bit of proving. The irreps of S_N are actually indifferent to the M_S values of the spin-eigenfunctions and it is the custom to stick to the $M_S = S$ value. That there are exactly as many spin-eigenfunctions with the same S value as the dimension of the irrep $[\lambda_1\lambda_2]$ of S_N is an enormous blessing and makes our life much easier than it might otherwise be. When this is not the case the whole business becomes extremely painful as those of you who have tried working out statistical weights for nuclear rotation-vibration wave functions will know.

Thus for electrons the whole business of the symmetric group boils down to finding sets of suitable spin-eigenfunctions. Now any set of spin-eigenfunctions will do, for all spin-eigenfunctions are equal, but some are more equal than others. In particular it is found that the Yamanouchi-Kotani spin eigenfunctions, that is, those which are constructed by the usual branching diagram procedure, are a basis for a standard (Young-Yamanouchi) unitary irrep of S_N. The standard set of Gelfand-Tsetlin spin eigenfunctions, that is those which are constructed as a standard basis for SU(2), are also a basis for the standard (Young-Yamanouchi) unitary irreps of S_N. In fact they differ from the Yamanouchi-Kotani functions only by a phase factor.

However, the spin-eigenfunctions constructed by the Rumer diagrams method, which are equivalent to those which are generated in the Boys bonded-function CI are not a basis for a unitary irrep of S_N. Of course they are a basis for a non-unitary irrep, but what this means is that the Rumer spin-eigenfunctions do not form an orthonormal set so that one often has M_{ij} matrix elements in the secular problem even when one has orthonormal orbitals, whereas in the standard unitary bases such matrix elements will not arise.

Finally we must return to the space part of the problem to complete our job and for this we need this result.

RESULT (3)

If $\Phi_i^{[\tilde{\lambda}]}$ and $\Phi_j^{[\tilde{\lambda}]}$ are products of orthonormal orbitals in some standard order then

$$\langle \Phi_i^{[\tilde{\lambda}]} | \hat{H} | O_P^{(r)} \Phi_j^{[\tilde{\lambda}]} \rangle$$

vanishes unless the function on the right-hand side above differs from the function on the left-hand side in two or fewer orbitals.

What this result means is that if we work in a basis of orthonormal orbitals then we need consider the irreps $\underset{\sim}{D}(P)$ (see (3.17)), for P at most a single transposition.

If this were not the case then we should have to know the $\underset{\sim}{D}(P)$ for all N! operators in the problem. For N greater than about 10 this is a job that verges upon the impossible. (At least it has not been done yet.) Thus, as a rough and ready rule, one can say that it is orthonormal orbitals or one is dead for any significant problem. This is not completely true as those of you familiar with pair-function methods will know, but such methods are possible as Professor Ahlrichs will doubtless demonstrate, because the number of permutations that one actually has to consider is still a reasonably small number.

To finish this section let us just consider the case where $\Phi_i^{[\tilde{\lambda}]}$ consists of a product of distinct orthonormal orbitals then using (3.16)

$$\langle \Phi_i^{[\tilde{\lambda}]} | \hat{H} | \Phi_i^{[\tilde{\lambda}]} \rangle \Rightarrow \sum_{r=1}^{n} \langle \phi_r | \hat{h} | \phi_r \rangle + \sum_{r,s=1}^{n}{}' \langle \phi_r \phi_s | g | \phi_t \phi_u \rangle \tag{3.19}$$

($r \neq s$ in double sum). Then these integrals are obviously associated with $\underset{\sim}{D}_{\mu_i \mu_i}(1)$ that is the diagonal element of the spin irrep and this will therefore make the same contribution to each integral. Clearly we are free to call this contribution 1 and this is the relevant contribution to the density matrix elements.

If we now consider the single transposition we get again using (3.16)

$$- 1 \sum_{r,s}{}' \langle \phi_r \phi_s | g | \phi_s \phi_r \rangle \tag{3.20}$$

and the contribution from $\underset{\sim}{D}(P_{ij})$ is whatever it is, depending on the basis, but again it is the same for all integrals of this kind and so determines a contribution to the two-particle transition density matrix and so on.

Thus one can quite easily build up the required density matrix contributions if one starts with a knowledge of the required $\underset{\sim}{D}$ matrices and the relevant orbital rules. In fact it is usually easier to combine the orbital rules with a rule for constructing the elements of $\underset{\sim}{D}$ and Dr Diercksen has exhibited such a rule system in the context of constructing symbolic elements in the bonded-function CI approach. As you heard in that lecture, most of the required coefficients are 0 or 1, but a few are slightly odd involving powers of 2 and so on. In a sort of homage to the late S. F. Boys the coefficients in this context are called projective reduction coefficents rather than density matrix elements, because that was what Boys called them.

I think that it is now pretty easy to see now how one can also incorporate point group symmetry for one-dimensional irreps. One simply knocks out the vanishing integrals from the general equivalents of the right-hand side of (3.19) or (3.20). (Of course in (3.19) and (3.20) themselves none of the integrals will vanish, but in one orbital difference cases and so on, some of them may vanish.)

You can see also why it is rather a tricky job to cope with many-dimensional irreps of the point group at this stage.

[Note added in proof]

Dr Rettrup kindly informed me that he, in collaboration with Dr Sarma and Professor Dahl, had come up with a method for managing many dimensional irreps of the point group. The relevant reference is Int. J. Quant. Chem. 22, 127 (1982).

4. THE UNITARY GROUP IN A BASIS PROVIDED BY 2n SPIN-ORBITALS AND IN A BASIS PROVIDED BY n ORBITALS

In this section we move into dynamical groups. The operator $\hat{H}$ is in no way invariant under any sort of super unitary coordinate transformation in an arbitary number of dimensions. What is invariant however are the roots of the secular problem (1.2) under a unitary transformation of the complete orbital space. One must however be a bit careful here. If one considers a one-electron problem and a set of 2n spin orbitals ψ_r $r = 1,2,\ldots 2n$, then the one electron secular problem in this basis has the same roots as the secular problem solved in the basis

$$\bar{\psi}_r = \sum_{p=1}^{2n} \psi_p U_{pr}$$

$$\underset{\sim}{\psi} = \underset{\sim}{\psi}\,\underset{\sim}{U}\,, \qquad \underset{\sim}{U}\,\underset{\sim}{U}^{+} = \underset{\sim}{E}_{2n} \tag{4.1}$$

where $\underset{\sim}{U}$ is a 2n by 2n unitary matrix.

Now suppose that we wish to solve a two electron problem then we can obviously use a basis constructed from all possible products of pairs of orbitals, that is all possible $\psi_r(\underset{\sim}{x}_1)\psi_s(\underset{\sim}{x}_2)$. Put symbolically we can solve the two electron problem in the direct product basis

$$\underset{\sim}{\psi}(\underset{\sim}{x}_1) \otimes \underset{\sim}{\psi}(\underset{\sim}{x}_2) \tag{4.2}$$

If we perform a unitary transformation like (4.1) on the spin orbital basis then we induce in the two-particle function space a transformation

$$\begin{aligned} \bar{\underset{\sim}{\psi}}(\underset{\sim}{x}_1) \otimes \bar{\underset{\sim}{\psi}}(\underset{\sim}{x}_2) &= (\underset{\sim}{\psi}(\underset{\sim}{x}_1)\underset{\sim}{U}) \otimes (\underset{\sim}{\psi}(\underset{\sim}{x}_2)\underset{\sim}{U}) \\ &= (\psi(\underset{\sim}{x}_1) \otimes \psi(\underset{\sim}{x}_2))(\underset{\sim}{U} \otimes \underset{\sim}{U}) \end{aligned} \tag{4.3}$$

Now we can say in this case that if we solve the secular problem using the $\bar{\underset{\sim}{\psi}} \otimes \bar{\underset{\sim}{\psi}}$ basis we shall get the same roots as we would have done if we had solved it in the basis $\underset{\sim}{\psi} \otimes \underset{\sim}{\psi}$. And clearly we can go on for 3, 4, 5, etc., particles. However, we know that in fact we only want to work with a sub-set of all possible orbital products, namely those that are an antisymmetric basis for S_N so that we have to be able to divide the space of orbital products into invariant sub-spaces such that we can perform unitary transformations only within that space which is a basis also for $[1^N]$ of S_N. If we can do this then we can make use of the unitary invariances of the secular problem with the sub-space. There is another snag too. In any real problem, from a given spin orbital basis we can generate many more orbital products even adapted to $[1^N]$ of S_N, than we can possibly solve a secular problem for. Thus although we can be sure of the basic theoretical invariance of the secular problem we have to show that we can get useful results using this sort of unitary symmetry in any practical case where we cannot use all possible configurations. It is also appropriate to remind ourselves here that whatever happens we absolutely must get the same answers using unitary symmetry as using any other sort of symmetry. It is just that looking at the problem from the point of view of unitary symmetry is suggestive and useful for a particular way of doing CI.

Now before we bury ourselves in the details of the unitary group I propose to make a huge digression into the theory of continuous matrix groups, because I think that the apparatus of such group theory is not yet a common-place of quantum chemistry and, to be truthful, I enjoy it. Now it may not turn everyone on, and all that I can offer is the advice formerly offered by the mothers of aristocratic young English women to their daughters on their wedding nights. They used to say "I do not suppose that you enjoy it, my dear, but lie back and think of the Empire". So here goes.

The set of all possible non-singular matrices of dimension n form a group under matrix multiplication. This is pretty obvious, I think, with the unit matrix $\underset{\sim}{E}_n$ being the unit of the group and every element in the group being a matrix with a unique inverse. This group is called the general linear group symbolized by Gl(n). As a very special case of this group let us consider the set of all possible 2 by 2 matrices of the form

$$\underset{\sim}{R}(\omega) = \begin{pmatrix} \cos\omega & -\sin\omega \\ \sin\omega & \cos\omega \end{pmatrix} \qquad 0 \leqslant \omega \leqslant 2\pi \tag{4.4}$$

Here the matrices are 2 by 2 but depend upon just one parameter, in this case the angle ω. The set of all possible matrices of this kind obviously form a group thus

$$\underset{\sim}{R}(\omega')\underset{\sim}{R}(\omega) = \underset{\sim}{R}(\omega''), \qquad \omega'' = \omega + \omega' \pmod{2\pi}$$

$$\underset{\sim}{R}(0)\underset{\sim}{R}(\omega) = \underset{\sim}{R}(\omega)\underset{\sim}{R}(0) = \underset{\sim}{R}(\omega)$$

$$\underset{\sim}{R}^{-1}(\omega)\underset{\sim}{R}(\omega) = \underset{\sim}{R}(0), \qquad \underset{\sim}{R}^{-1}(\omega) = \underset{\sim}{R}(-\omega) \tag{4.5}$$

and so on.

However we cannot label the group operations by a discrete parameter set as we can in point groups (e.g., C_3, $\bar{C}_3$, etc.) or as in S_N (e.g., (321), (123)(45)(6)). We can label the operations only by the continuous label ω. The group that we have chosen as an example is a particular case of the special orthogonal group in n dimensions, labelled SO(n) which consists of all orthogonal n by n matrices with positive determinant. In general the matrices of this group depend upon (n(n-1)/2 real parameters and in our special case SO(2), the matrices depend only on 1 parameter and the group SO(2) is abelian, though no other group in the set is.

Now let us treat the matrices as though they transformed a set of variables. (This is just a fictional set of variables, they should not be thought of a necessarily related to electrons or anything.) So

$$\begin{pmatrix} x_1' \\ x_2' \end{pmatrix} = \underset{\sim}{R}(\omega) \begin{pmatrix} x_1 \\ x_2 \end{pmatrix} \tag{4.6}$$

or

$$\underset{\sim}{x}' = \underset{\sim}{R}(\omega)\underset{\sim}{x}$$

or, quite often written as

$$x_i' = f_i(\underset{\sim}{x},\omega) \quad \text{or} \quad \underset{\sim}{x}' = \underset{\sim}{f}(\underset{\sim}{x},\omega)$$

where by definition f_i is analytic in ω.

Now one can get to $\underset{\sim}{x}'$ in two ways. The first way is,

$$\underset{\sim}{x}' = \underset{\sim}{f}(\underset{\sim}{x},\omega) , \quad \text{for some } \omega \tag{4.7}$$

and the trivial second way is,

$$\underset{\sim}{x}' = \underset{\sim}{f}(\underset{\sim}{x}',0) , \quad \text{"near the identity"} \tag{4.8}$$

But because by construction f_i is analytic in ω, we can expand (4.7) in a Taylor series about $\omega = 0$ so that

$$x_i' = f_i(\underset{\sim}{x},0) + (\partial f_i(\underset{\sim}{x},\omega)/\partial\omega)_{\omega=0})\omega+\ldots \tag{4.9}$$

but using the form (4.8) for $f(\underset{\sim}{x},0)$ we get

$$dx_i = x_i' - x_i = (\partial f_i(x,\omega)/\partial\omega)_{\omega=0}\delta\omega \tag{4.10}$$

if $\delta\omega$ is sufficiently small in relation to the functional form f_i.

In our example the form (4.7) is

$$x_1' = x_1\cos\omega - x_2\sin\omega \equiv f_1(\underset{\sim}{x},\omega)$$

and

$$\begin{aligned}(\partial f_1/\partial\omega)_{\omega=0} &= (-x_1\sin\omega - x_2\cos\omega)_{\omega=0} \\ &= -x_2\end{aligned} \tag{4.11}$$

and so on. Thus we can write for SO(2)

$$\begin{aligned} dx_1 &= -x_2 d\omega \\ dx_2 &= x_1 d\omega \end{aligned} \tag{4.12}$$

where we can change from $\delta\omega$ to $d\omega$ because the relation between successive ω values is an analytic relation, namely

$$\omega'' = \omega + \omega' \tag{4.13}$$

(compare with (4.5) above). Such a passage is not always possible in more complicated cases.

To consider now this sort of move in a general problem where one has n x n matrices each depending on p parameters, so that one has

$$x_i' = f_i(\underset{\sim}{x}, \omega_1, \omega_2 \ldots \omega_p) \; , \qquad i = 1,2 \ldots n \tag{4.14}$$

and one gets

$$dx_i = \sum_{r=1}^{p} U_{ir}(\underset{\sim}{x}) \, \delta\omega_r \tag{4.15}$$

where

$$U_{ip}(\underset{\sim}{x}) = (\partial f_i / \partial \omega_p)_{\underset{\sim}{\omega}=\underset{\sim}{0}} \tag{4.16}$$

Now if there is an analytic relationship between the various parameter values (and there must be <u>some</u> kind of relationship between their values, it just need not be analytic) then one can also write, using a Taylor series

$$d\omega_r = \sum_{s=1}^{p} V_{rs}(\underset{\sim}{\omega}) \, \delta\omega_s \tag{4.17}$$

where

$$V_{rp} = (\partial \phi_r / \partial \omega_p')_{\underset{\sim}{\omega}'=\underset{\sim}{0}} \tag{4.18}$$

where $\phi_r(\underset{\sim}{\omega}, \underset{\sim}{\omega}')$ is the assumed analytic relation between related values of the parameters $\underset{\sim}{\omega}$. In our example this relation is (4.13) and obviously here $\delta\omega = d\omega$. In general if a relation like (4.17) exists, it is invertable and gives $\delta\omega_r$ in terms of the $d\omega_r$. This is important in considerations of integrating over the group manifold for orthogonality theorems.

Now let us consider an analytic function F built on the variable space $x_1, x_2 \ldots x_n$ then the total differential

$$dF = \sum_{i=1}^{n} (\partial F / \partial x_i) dx_i \tag{4.18}$$

But, using (4.15) for dx_i, we get at once

$$dF = \sum_{i=1}^{n} \sum_{r=1}^{p} U_{ir} \partial F/\partial x_i \delta\omega_r$$
$$= \sum_{r=1}^{p} I_r F \delta\omega_r \qquad (4.19)$$

where

$$I_r = \sum_{i=1}^{n} U_{ir} \partial/\partial x_i \qquad (4.20)$$

The operators I_r are called the infinitessimal operators (IOs) or infinitessimal generators of the problem.

To consider their form in our example, is of course to consider a problem with only one such operator. Clearly from (4.12)

$$U_{11} = -x_2 , \qquad U_{21} = x_1 \qquad (4.21)$$

so that

$$I_1 \equiv I_\omega = x_1 \partial/\partial x_2 - x_2 \partial/\partial x_1 \qquad (4.22)$$

So that I_ω is pretty clearly connected with angular momentum, a fact that will not come as a great surprise to you.

In the general case there is an amazing result due to Lie that can be proved for these operators. (It is not very easy to prove but is well worth while struggling with.) This result is that the commutator of any pair of them can be written so that

$$[I_t, I_u] = \sum_{s=1}^{p} C^s_{tu} I_s \qquad (4.23)$$

where the C^s_{tu} are genuine constants (not functions of anything) called the structure constants and they satisfy the relation

$$C^s_{tu} = -C^s_{ut} \qquad (4.24)$$

Let me just give you an example which arises from SO(3). In this case the IOs are

$$I_1 = x_3 \partial/\partial x_2 - x_2 \partial/\partial x_3$$
$$I_2 = x_1 \partial/\partial x_3 - x_3 \partial/\partial x_1$$
$$I_3 = x_2 \partial/\partial x_1 - x_1 \partial/\partial x_2 \qquad (4.25)$$

so that

$$[I_1, I_2] = I_3 , \quad [I_2, I_3] = I_1 , \quad [I_3, I_1] = I_2 \qquad (4.26)$$

or

$$[I_i, I_j] = \varepsilon_{ijk} I_k , \quad \varepsilon_{ijk} = +1$$

ijk, 123 and cyclic

$$\varepsilon_{ijk} = -1$$

ijk = 213 and cyclic

$$\varepsilon_{ijk} = 0 \text{ otherwise} \qquad (4.27)$$

Again these infinitessimal operators are essentially the angular momentum operators as might be expected since SO(3) is the pure rotation group in 3-dimensions.

Now we must see if all this can help us to get irreducible representations of continuous groups, which is what we are really after. Now from (4.18) we see that what the I_p actually do to a function is to induce in it an infinitessimal change under the infinitessimal transformations $\delta\omega_p$ in parameter space. Thus we can see for our SO(2) example that the infinitessimally transformed function

$$\bar{F} = F + dF$$

can be written as

$$\bar{F} = (1 - i/\hbar \, d\omega J_z)F \qquad (4.28)$$

where we have written

$$J_z = \hbar/i \; I_\omega \qquad (4.29)$$

and chosen $d\omega$ so that we induce in the function space the correct transformation. Now can we integrate this infinitessimal relation to find the result for the finite transformation? Well here we pretty clearly can, by regarding the integration as sequence of applications of the infinitessimal transformation (4.28) and passing to the limit. Thus

$$\bar{F} = (1 - (i\omega/\hbar k)J_z)(1 - (i\omega/\hbar k)J_k)\ldots(1 - (i\omega/\hbar k)J_z)F \quad (4.30)$$

where k is sufficiently large for $d\omega = \omega/k$ to be sufficiently small. Assuming that we can pass to the limit k in some suitable way we get, using $\bar{F}$ to denote the result of some finite transformation

$$\bar{F} = \lim_{k\to\infty}(1 - (i\omega/\hbar k)J_z)^k F = \exp(-i\omega J_z/\hbar)F \quad (4.31)$$

using the power series definition of the exponential.

Thus on some suitable space for the operation of J_z we can represent the finite operation (clearly some sort of rotation) originally represented by the matrix $\underset{\sim}{R}(\omega)$ by the operator

$$R(\omega) = \exp(-i\omega J_z/\hbar) \quad (4.32)$$

Now this may seem a perverse thing to have done, but it is not really. If we can find in the space on which J_z operates, eigenfunctions of J_z, say

$$J_z u_m = \hbar m u_m$$

then if these functions are "square integrable" in some sense we have actually found by this process a representation of SO(2). To see this, consider

$$R(\omega)u_m = \exp(-i\omega J_z/\hbar)u_m = \exp(-i\omega J_z u_m/\hbar) \quad (4.33)$$

where this last step would not be correct if u_m were not an eigenfunction of J_z and from this last result it follows that

$$\begin{aligned}\langle u_n|R(\omega)|u_m\rangle &= \exp(-i\omega\langle u_n|J_z|u_m\rangle/\hbar)\\ &= \exp(-i\omega m\delta_{nm})\\ &= (\underset{\sim}{U})_{nm}\end{aligned} \quad (4.34)$$

so that the matrix with elements U_{nm} which is the right-hand side of (4.34) is a unitary matrix representation of SO(2) in the basis u_m such that

$$R(\omega)\underset{\sim}{u} = \underset{\sim}{u}\,\underset{\sim}{U} \quad (4.35)$$

Now this is a bit of a swindle in this case because a little thought shows that $\underset{\sim}{U}$ is a matrix of infinite dimension, but this is not because there is anything intrinsically wrong with the argument, it merely reflects the fact that SO(2) is a rather nasty group which, as a matter of fact, does not have <u>any</u> finite dimensional irreducible representations. If I had worked with SO(3) I could have used I_3 in exactly the same way as I_ω and there, there would have been only a finite range of m values. But to show that would have been to get a bit ahead of myself.

What I want to persuade you of is that it is not implausible to suppose that one can get at representations of continuous groups via the eigenfunctions of their IOs. Furthermore if one can find a set of commuting IOs then one can find simultaneous eigenfunctions for this set and thus get a very "fine" representation of the group by the exponential trick used in (4.33). In fact for most of the groups of interest in theoretical physics and all that I know of in quantum chemistry if one finds such a set of commuting operators then a specification of their eigenfunction is sufficient to specify all the unitary irreducible representations of finite dimensions of continuous group.

If you believe me, then you will see that it is worthwhile to consider the abstract algebra of the IOs to determine the largest set that we can of commuting operators. The algebra of interest is the Lie algebra of the operators which is concerned with the commutation properties of operators that consist of real linear combinations of IOs. Now for the purpose of studying this algebra we forget about the function space on which the IOs can be thought of as acting. For the moment that is not important. The space that we are interested in here, is the p-dimensional space spanned by the p independent IOs considered as abstract "vectors" in the algebra.

Consider the operators

$$A = \sum_{r=1}^{p} a_r I_r , \qquad S = \sum_{r=1}^{p} s_r I_r \tag{4.36}$$

where the a_r, s_r, etc., are real numbers. Then, because we can obviously write

$$[A,S] = T \tag{4.37}$$

where the coefficients of the IOs in T are some proper combination of the structure constants and the a_r and the s_r. There are of course at most p operators like A, S, T, etc., that are in the algebra. It is perhaps helpful to think of (4.37) as analogous to the result in ordinary vector algebra

$$\vec{a} \times \vec{b} = \vec{c}$$

and indeed ordinary vectors in 3-dimensions do form a Lie algebra under vector product as the operator.

Now consider the operator denoted F(A) (or sometimes ad(A)) such that

$$F(A)S \equiv [A,S] \qquad (4.38)$$

then, by definition

$$F(A)I_q = \sum_{r,s=1}^{p} a_r C^s_{rq} I_s \qquad (4.39)$$

and by definition also

$$F(B)F(A)S = [B,[A,S]]$$

so that

$$F(B)F(A)I_q = \sum_{r,s,1,t=1}^{p} b_1 a_r C^s_{rq} C^t_{1s} I_t$$
$$= \sum_{t=1}^{p} P^{BA}_{tq} I_t \qquad (4.40)$$

where

$$P^{BA}_{tq} = \sum_{r,s,1=1}^{p} b_1 a_r C^s_{rq} C^t_{1s}$$
$$\equiv (\underset{\sim}{P}^{BA})_{tq} \qquad (4.41)$$

Now $\underset{\sim}{P}^{BA}$ is a matrix of real numbers relevant to the algebra and as we know from general experience one can milk a given algebra for a lot more results than otherwise is possible if one can turn it into a normed vector space. That means getting a rule for constructing a scalar product in the algebra. And it is the case that in Lie algebras $\underset{\sim}{P}^{BA}$ sometimes provides us with this possibility. If one considers the trace of $\underset{\sim}{P}^{BA}$ one gets

$$\mathrm{tr}\underset{\sim}{P}^{BA} \equiv \sum_t P^{BA}_{tt} = \sum_{1,r=1}^{p} b_1 a_r \sum_{s,t=1}^{p} C^s_{rt} C^t_{1s}$$
$$= \sum_{1,r=1}^{p} b_1 a_r g_{1r} \qquad (4.42)$$

where the matrix $\underset{\sim}{g}$ with elements $g_{1r} = g_{r1}$ is constructed entirely from the structure constants of the IOs and thus is characteristic of the basic relations between the IOs rather than some particular combinations. It is a symmetric matrix as can easily be shown by remembering that $C^s_{rt} = -C^s_{tr}$ and so on and obviously (4.42) looks a bit like

$$\underset{\sim}{b}^T \underset{\sim}{g} \underset{\sim}{a} \tag{4.43}$$

where $\underset{\sim}{a}$, $\underset{\sim}{b}$, etc., are column matrices of coefficients. Thus one is very tempted to say that one has found a scalar product to turn the algebra into a normed vector space with the scalar product defined as

$$\langle A|B\rangle = \mathrm{tr}\underset{\sim}{P}^{BA} \equiv \underset{\sim}{b}^T \underset{\sim}{g} \underset{\sim}{a} \tag{4.44}$$

Now the matrix g is called the Killing form for the algebra, but alas it can have eigenvalues which are zero. In this case it is no good as a metric because no unambiguous definition of the norm of vector can result, since there could well be a perfectly nice vector but of norm zero, merely because it belonged to a zero-eigenvalue of $\underset{\sim}{g}$. However there are a large number of algebras for which $\underset{\sim}{g}$ has no zero-eigenvalues and such algebras are called semi-simple Lie algebras. One can recognize them from the algebra of the IOs in that semi-simple algebras have no invariant abelian sub-algebras (or no abelian ideals). Since SO(2) has a Lie algebra with only one operator in it it cannot be semi-simple (g has one element and it is zero) so one would expect SO(2) to be a rather nasty group. Indeed as far as Lie algebras are concerned SO(2) has the same Lie algebra as the Lie algebra of the uniform translation group in one-dimension, which is a well-known pig of a group. However if the algebra is semi-simple then $\underset{\sim}{g}$ has no zero eigenvalues and we can regard it as a metric and as such it is called the Cartan-Killing metric. Of the groups of interest to us SO(n), $n > 2$ yields a semi-simple algebra as does U(n), $n \geqslant 2$.

There is also another very desirable property for metrics, namely that all the eigenvalues of a metric should have the same sign. Unfortunately this is not the case for the general semi-simple Lie algebra, but only for a sub-set of such algebras which are called compact algebras. Cartan showed that for compact algebras, $\underset{\sim}{g}$ is negative definite. A compact algebra certainly arises from a group which has a finite number of parameters only, and in which the parameters have only a finite range, and this recognition is sufficent for most purposes. By definition then a compact algebra must be a semi-simple algebra. The groups referred to above yield, in fact, compact algebras, but the group Gl(n) does not. This is reasonable when one relaizes that the parameters of Gl(n) can be taken as the $2n^2$ real and complex parts of the matrix elements $(\underset{\sim}{G})_{ij}$ where $\underset{\sim}{G}$ is any non-singular matrix. Clearly these matrix elements are not bounded even though there are only a finite number of them.

Now if we are dealing with a semi-simple algebra then the Cartan-Killing metric $\underset{\sim}{g}$ is certainly invertable and it can be shown that the operator

$$C = \sum_{r,q} g^{-1}_{rq} I_r I_q \tag{4.45}$$

which is called the Casimir operator is an operator that commutes with <u>all</u> the IOs in the algebra. Thus in (4.45) we have found the first of our commuting operators. To give an example of a Casimir operator consider SO(3). A little monkey business with (4.27) will convince you that for the Lie algebra of SO(3), then

$$g_{rs} = -2\delta_{rs} \tag{4.46}$$

and so the Casimir operator is just

$$-\tfrac{1}{2} \sum_{r=1}^{3} I_r^2 \tag{4.47}$$

And this operator is clearly essentially identical to our old friend J^2 from angular momentum theory. The Casimir operator is clearly the sort of principal designator of an irrep achieved from the algebra. One should perhaps say that there are other Casimir-like operators that are definable in certain cases, but we shall not need to use them and so I shall neglect them. In fact we will not even have to use the Casimir operator in the unitary group, because there we can latch on to the irreps completely by using a device for getting commuting operators invented by Cartan.

Cartan saw that if one could define a suitable eigenvalue problem in the operator algebra then all the operators in the algebra that belonged to the same eigenvalue would commute. For this aspect of the problem one does not have to think in terms of the metric space. The eigenvalues here have nothing to do with the eigenvalues of the Cartan-Killing metric. What Cartan saw was that if one could actually find eigensolutions to the problem

$$F(A)S \equiv [A,S] = \rho S \tag{4.48}$$

then if one could find a ρ that had several different operators S associated with it then this set of S would commute with each other and with the operator A.

Rewriting (4.48) in terms of the IO as

$$\sum_{rl} a_r s_l [I_r, I_l] = \sum_{r,l,t} a_r s_l C^t_{rl} I_t = \rho \sum_t s_t I_t \tag{4.49}$$

it follows that one has a p by p secular problem to solve, with secular equations

$$\sum_l ((\sum_r a_r C^t_{rl}) - \rho\delta_{tl}) s_l = 0 \qquad l = 1,2,\ldots p$$

or

$$|(\sum_r a_r C^t_{rl}) - \rho\delta_{tl}| = 0 \tag{4.50}$$

This equation determines the ρ and hence the s_l for any given set of a_r.

Now of course there is no guarantee that this equation has any solutions at all and it usually will not have p solutions for which the coefficients s_r are all real. So even in the best case we shall have to expand our Lie algebra a bit to make use of them. However Cartan was able to show a remarkable result if one allows complex s_r. The result (which is not easy to show) is that, for any semi-simple Lie algebra there is a particular choice of the a_p for which the secular problem (4.50) has a maximum number of distinct roots. With this choice of A, then there is degeneracy only for the root $\rho = 0$. With this choice there are then $l < p$ commuting operators and this is the most there can be. The algebra is said then to be of rank l and the set of commuting operators are said to constitute a Cartan sub-algebra of the full algebra.

By convention the distinct operators are denoted by E with suitable subscripts and the operators of the Cartan sub-algebra are denoted $H_1, H_2, \ldots H_l$.

Let us look at this result in SO(3). Here from (4.27)

$$\begin{aligned}\sum_{i=1}^{3} a_i C^k_{ij} &= \sum_i a_i \varepsilon_{ijk} \\ &= 0 \quad j = k \\ &= a_3 \quad k = 1, \quad j = 2 \\ &= -a_2 \quad k = 1, \quad j = 3 \\ &\text{etc.}\end{aligned} \tag{4.51}$$

so that the secular problem (4.50) is just

$$\begin{vmatrix} \rho & a_3 & -a_2 \\ -a_3 & \rho & a_1 \\ a_2 & -a_1 & \rho \end{vmatrix} = 0 \qquad (4.52)$$

The solutions here are given as roots of

$$\rho(\rho^2 + a_1^2 + a_2^2 + a_3^2) = 0$$

and are obviously

$$\rho = 0$$

$$\rho = \pm i(a_1^2 + a_2^2 + a_3^2)^{\frac{1}{2}} \qquad (4.53)$$

Clearly here the Cartan sub-algebra is of dimension 1 = 1 and no amount of monkey business with the a_i will alter that fact. In this case one can choose the a_i for convenience and we always choose $a_1 = a_2 = 0$ and $a_3 = 1$ so that the other roots are

$$\rho = \pm\alpha , \qquad \alpha = ia_3 \qquad (4.54)$$

Thus the relevant operators are therefore

$$S_0 = s_3 I_3$$

and

$$S_{\pm\alpha} = s_1(I_1 \pm iI_2) \qquad (4.55)$$

by convention S_0 is written H_1 and $S_{\pm\alpha}$ and $E_{\pm\alpha}$.

In this case pretty obviously

$$H_1 \rightarrow J_z$$

$$E_{\pm\alpha} \rightarrow J_{\pm} \qquad (4.56)$$

Now the move from H_1 to J_z in (4.56), which is made by multiplying by $\hbar/i$, has produced the operator J_z which we know to be hermitian on a space of decent enough functions. The move from $E_{\pm\alpha}$ to $J_{\pm}$ has similarly produced a pair of normal operators. Now we know that J_z, like all proper hermitian operators, has a complete eigenvalue spectrum on a suitable function space so that we actually can construct its eigenfunctions and hence reps of SO(3) by means of the exponential trick (4.32). Further if we can find eigenfunctions of the hermitian equivalent of the Casimir operator, namely J^2, which are simultaneously eigenfunctions of J_z then the reps that we produce using the exponential trick are, we know, irreps. Furthermore we need only find eigenfunctions for particular eigenvalues of J^2 and J_z because we can use the $J_\pm$ operators to get the rest.

Now the really remarkable thing is that one can do something like this for any semi-simple Lie algebra and hence for any continuous group that generates such an algebra. It is not, of course, usually as simple as it is for SO(3) but there are no essential differences. The operators in the Cartan sub-algebra can always be used to get Hermitian operators and hence eigenfunctions in a suitable space. Similarly the operators not in the Cartan sub-algebra always act, in some way, as raising and lowering operators.

From this discussion it should be clear that although groups that generate semi-simple Lie algebras (semi-simple groups) have finite dimensional unitary irreps, in general there are an infinite number of such irreps and in general the dimension of these irreps, though remaining countable, increases without limit.

It must also be clear that in some sense the eigenvalues of the operators in the Cartan-sub-algebra actually provide a complete labelling of the irreps. It is usual to put these eigenvalues together to make a vector of length p called the weight of an irrep. Thus if one has

$$H_i |m_{ij}\rangle = m_{ij} |m_{ij}\rangle \qquad i = 1,2\ldots l \tag{4.57}$$

then the weight is

$$(m_{1j}, m_{2j} \ldots m_{lj}) \tag{4.58}$$

for some agreed choice of j. It can be shown that one can make a choice of j such that there is one weight, the so-called highest weight in terms of which the irrep can be uniquely specified. This specification does not make explicit use of the Casimir operator, but of course its eigenvalue is implicitly contained in the weight. The idea of highest weight states is made use of in the Gelfand-Tsetlin theory of the irreps of U(n). In the case of the group SO(3) the idea is almost trivial. Here the highest weight state specified by the eigenvalue of J_z for which m = j where j is the eigenvalue of J^2.

Now the nice thing about functions chosen to provide a basis for a highest weight irrep of a continuous group is that of course the operators H_i are diagonal in such a basis and the $E_{\pm\alpha}$ have relatively simple matrix elements. These properties have been exploited with great skill and ingenuity by Paldus in his account of the CI method using the unitary group. He noticed that the operators E_{rr} of (1.16) satisfied the commutation relations, specializing (1.18),

$$[E_{rr}, E_{ss}] = 0 \tag{4.59}$$

and that they were essentially the operators of the Cartan subalgebra of the Lie algebra generated by the unitary group U(n). They are usually called the weight operators. Furthermore the operators $E_{r(r\pm1)}$ play the role of elementary raising and lowering operators and that in fact all E_{rs} can be expressed in terms of these.

Before going on perhaps I ought to remind you what the operators E_{rs} and E_{rs} actually mean in the context that we are going to use them, since our consideration of them has been rather brief and very formal. For us these operators operate upon a space of orbital products and the exclusion principle is taken care of by the anticommutation relations of the creation and annihilation operators. Let us denote a product of spin orbitals by $|m\rangle$ then

$$E'_{rs}|m\rangle = a_r^+ a_s|m\rangle = 0 \qquad \text{unless } \psi_s \text{ is in } |m\rangle$$

$$= |m(r\to s)\rangle \qquad \text{if } \psi_s \text{ is in } |m\rangle$$

That is it replaces the orbital ψ_s by the orbital ψ_r in the product. It thus raises (or lowers) one of the spin orbital numbers in the product. Looking now at E_{rs} this obviously works on orbital products and if there were a pair of spin orbitals with the same orbital number (but different spins) in the original orbital product one would have (using (1.16))

$$E_{rs}|m_0\rangle = 2|m_0(r\to s)\rangle$$

where $|m_0\rangle$ denotes an orbital product and we have assumed that the paired orbital was s. If s had been unpaired then we should just have got 1.

I hope that this gives enough of an idea but the best way to get it is to fiddle about yourselves for a bit.

To return now to the main theme. I do not want to come directly at the Gelfand-Tsetlin basis for the unitary group, instead I want to use a different basis, to so-called Weyl tableaux basis because with its aid one can see rather more clearly how the symmetric group gets in on the act. To do this let us return to (4.3) and notice that if we applied a permutation operator for two particles to the left and right-hand side of (4.3) then we could write (remember $\tilde{U}$ is of dimension 2n).

$$O_p(\underset{\sim}{\bar\psi}(\underset{\sim}{x}_1)\otimes\underset{\sim}{\bar\psi}(\underset{\sim}{x}_2)) = (O_p(\underset{\sim}{\psi}(\underset{\sim}{x}_1)\otimes\underset{\sim}{\psi}(\underset{\sim}{x}_2))\underset{\sim}{U}\otimes\underset{\sim}{U} \qquad (4.60)$$

However the effect of O_p in the spin-orbital basis is effectively to interchange the order of the indexing on the orbital pairs, keeping the electron labels the same. Denoting this operator that effects the labelling change by $\bar{O}_p$ and suppressing the variable symbols in $\underset{\sim}{\psi}$ (4.60) is effectively

$$\bar{O}_p(\underset{\sim}{\bar\psi}\otimes\underset{\sim}{\bar\psi}) = \bar{O}_p((\underset{\sim}{\psi}\otimes\underset{\sim}{\psi}))\underset{\sim}{U}\otimes\underset{\sim}{U} \qquad (4.61)$$

so that we can classify the basis by the way that its constituents transform under the operations of S_N where N is the number of electrons.

Thus for two electrons there will only be two types of basis function

$$\psi_{\boxed{i}\boxed{j}} \Rightarrow \psi_i(\underset{\sim}{x}_1)\psi_j(\underset{\sim}{x}_2) + \psi_j(\underset{\sim}{x}_1)\psi_i(\underset{\sim}{x}_2)$$

and

$$\psi_{\begin{smallmatrix}\boxed{i}\\ \boxed{j}\end{smallmatrix}} \Rightarrow \psi_i(\underset{\sim}{x}_1)\psi_j(\underset{\sim}{x}_2) - \psi_j(\underset{\sim}{x}_1)\psi_i(\underset{\sim}{x}_2) \qquad (4.62)$$

because S_2 has only two irreps.

Now this partitioning of the basis is obviously into two invariant sub-spaces because no linear transformation (and hence a a fortiori a unitary transformation) of the basis set will produce from the set of functions labelled $\psi_{\boxed{i}\boxed{j}}$, any function that lies in the space of functions labelled $\psi_{\begin{smallmatrix}\boxed{i}\\ \boxed{j}\end{smallmatrix}}$ all i,j.

Thus if one adapts the matrix $\underset{\sim}{U}\otimes\underset{\sim}{U}$ to product functions made up like (4.62) (rather than to simple products like $\psi_i(\underset{\sim}{x}_1)\psi_j(\underset{\sim}{x}_2)$ as in the ordinary direct product basis for (4.3)) then the matrix $\underset{\sim}{U}\otimes\underset{\sim}{U}$ goes into a matrix which can still be chosen unitary but which consists simply of diagonal blocks. One block is the representation of U(2n) in the basis of symmetric combinations $\psi_{\boxed{i}\boxed{j}}$ and the other is a representation in the basis of antisymmetric combination $\psi_{\begin{smallmatrix}\boxed{i}\\ \boxed{j}\end{smallmatrix}}$.

Obviously we can generalise (4.62) to deal with three-fold, four-fold and so on up to N-fold products, classifying the products that block the N-fold direct product matrix $\underset{\sim}{U}\otimes\underset{\sim}{U}\otimes\cdots\otimes\underset{\sim}{U}$ according to the irreps of S_N.

Now in fact we have been a little bit too special in confining our attention to unitary transformations of basis. In fact it is pretty clear that this sort of blocking can be managed under only general linear transformation, and the blocking that results in a linear transformation actually reduces the resulting direct product matrices to irreducible form. That is to say a set of irreps of Gl(2n) can be found by considering any element $\underset{\sim}{G}$ of Gl(2n) and constructing for it a basis of 2,3,4,... and so on (ad infinitum) function products such that the relevant direct product matrices are reduced to block-diagonal form. The functions that do this according to the relevant irreps of S_p where p is the order of the product, are called (linear) tensors and are actually usually the chosen means of dealing with irreps of Gl(2n) which is not a compact group.

This may seem an aside but it is not totally without relevance, because it is the case that operators E'_{rs} actually obey the commutation relations relevant to Gl(2n) rather than to U(2n). To see this we simply notice that if we think up a fictitious variable space $\underset{\sim}{q}$ such that

$$\underset{\sim}{q}' = \underset{\sim}{G}\,\underset{\sim}{q}$$

and treat the real parts of the elements of $\underset{\sim}{G}$, g_{ij} say and the complex parts, $\bar{g}_{ij}$ say as the parameters of the group then

$$(\partial q_j/\partial g_{ji})_{\underset{\sim}{G}=\underset{\sim}{0}} = q_i \;, \qquad \text{all } j \tag{4.63}$$

(and similarly for $\bar{g}_{ji}$). Thus from (4.63) the IOs appropriate to the g_{ji} are

$$I_{ij} = q_i\partial/\partial q_j \tag{4.64}$$

and obviously

$$[I_{ij},I_{kl}] = \delta_{jk}I_{il} - \delta_{il}I_{kj} \tag{4.65}$$

which is just the relation (1.14) (or (1.18)). Thus the IO of the general linear group Lie algebra generated by the real (or complex) parameters are the ones with the right commutation relations for our purposes. Unfortunately

$$I_{ij} \neq I^{+}_{ji} \tag{4.66}$$

and we require for our E'_{rs} (or E_{rs}) that

$$E'_{rs} = E'^{+}_{sr} \qquad (\text{or } E_{rs} = E^{+}_{sr}) \tag{4.67}$$

However we can, with a little care make sure (4.67) is satisfied and actually use the linear group operators as though they were the unitary group operators.

This is possible because of a very remarkable result: this is, that the irreps of Gl(2n) remain distinct irreps of U(2n) and do not become reducible on passing to the sub-group. It is also the case that the irreps of U(2n) remain irreps of SU(2n) but in this case they do not necessarily remain distinct. (This is the end of the sequence if one went from SU(2n) to SO(2n), the irreps SU(2n) would in general be reducible in SO(2n).)

To return now to the unitary group we must now split up the space and spin parts of the spin orbital basis ψ. To do this we write

$$\underset{\sim}{\psi} = \underset{\sim}{\phi} \otimes (\alpha, \beta) \equiv \underset{\sim}{\phi} \otimes \underset{\sim}{\eta} \tag{4.68}$$

and make use of (2.25) to see that in fact the unitary transformation U(2n) is really the transformation

$$\underset{\sim}{U}(\underset{\sim}{n}) \otimes \underset{\sim}{U}(2) \tag{4.69}$$

where U(n) is relevant to the orbital space and U(2) to the spin space. From this it follows that the group of interest is <u>not</u> actually U(2n) but the sub-group $U(n) \otimes U(2)$ which consists of all matrices like (4.69). We thus need irreps of this sub-group, but we do not need all of them. We only need a special sub-set of irreps that correspond to a basis of properly antisymmetrized orbital products, regarded as the tensors like (4.62) and its more general successors.

To get the right irreps we notice that since the spin functions transform according to U(2) then the spin-eigenfunctions that form a tensor basis for U(2) must transform like irreps of U(2). (Strictly SU(2), but this point is not important here.) But we also know that these spin-eigenfunctions are a basis for S_N. We have also established that if the spin-eigenfunctions are a basis for $[\lambda_1 \lambda_2]$ of S_N then the orbital product functions must be a basis for $[\tilde{\lambda}_1 \tilde{\lambda}_2]$ of S_N. Thus we need only concern ourselves with a basis of orbital tensor products that are a basis for $[\tilde{\lambda}_1 \tilde{\lambda}_2]$ when we consider a tensor product basis for U(n). Strictly speaking this tensor product basis is a basis for an irrep of Gl(n) but as we have seen it is possible to use it as a basis for an irrep of U(n).

At this stage I think an example is mandatory. Consider the problem in which one has 3 electrons (N = 3) and n orbitals, where n is not yet decided upon. Then the possible spin-eigenfunction irreps of S_3 are

 and 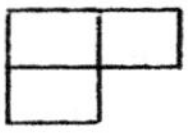

[3] [21] (4.70)

and remembering that $\lambda_1 + \lambda_2 = N$, $\lambda_1 - \lambda_2 = 2S$ then [3] is $2S = 3$ and [21] is $2S = 1$.

Thus the only possible classification of the space part is

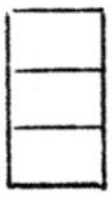 or 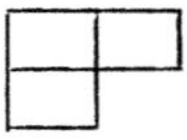

For $S = 3/2$ For $S = 1/2$ (4.71)

Thus for $S = 3/2$ states we can only have a tensor basis (compare (4.52)) denoted ψ 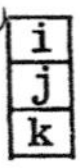

or for $S = 1/2$ a tensor basis like ψ 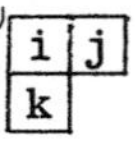

where in each case i,j,k denote orbital numbers.

Now let us introduce a bit of notation relevant to the unitary group. It is the custom to denote the number of two-column rows in a Young diagram for the space part by a and the number of one-column rows by b.

Thus for the $S = 3/2$ pattern $a = 0$, $b = 3$ and for the $S = 1/2$ pattern $a = 1$, $b = 1$. Pretty obviously

$$2a + b = N , \quad b = 2S \tag{4.72}$$

or

$$\lambda_1 = a + b , \quad \lambda_2 = a$$

By now you must be wondering how the orbital numbers get in on the act, but the clue lies in the way in which the orbital symbols occur in the tensor designation. The actual result due to Weyl is as follows. If you put the orbital numbers in the cells of the space part Young diagram so that they are non-decreasing along a row and strictly increasing down a column then there are as many Weyl tableaux produced by the process as the dimension of the required irrep of U(n) associated properly with U(2) to give $S = b/2$. A tensoral set of orbital products is a basis for this irrep.

Unfortunately this tensoral product set is not a basis that has the nice properties with respect to the operators of the Cartan sub-algebra of the Lie algebra of the unitary group. It is not, that is, the Gelfand-Tsetlin basis. It is in fact the bonded function (Rumer valence bond) basis. This is rather a pity, but in practice as we shall see it does not matter.

Let us now consider the Weyl tableaux for the case S = 1/2 for 3 electrons and now specialize n to be 4. The relevant tableaux are

$$\begin{array}{ccccc}
\begin{array}{|c|c|}\hline 1&1\\\hline 2\\\cline{1-1}\end{array} &
\begin{array}{|c|c|}\hline 1&2\\\hline 2\\\cline{1-1}\end{array} &
\begin{array}{|c|c|}\hline 1&1\\\hline 3\\\cline{1-1}\end{array} &
\begin{array}{|c|c|}\hline 1&2\\\hline 3\\\cline{1-1}\end{array} &
\begin{array}{|c|c|}\hline 2&2\\\hline 3\\\cline{1-1}\end{array} \\[2ex]
\begin{array}{|c|c|}\hline 1&3\\\hline 2\\\cline{1-1}\end{array} &
\begin{array}{|c|c|}\hline 1&3\\\hline 3\\\cline{1-1}\end{array} &
\begin{array}{|c|c|}\hline 2&3\\\hline 3\\\cline{1-1}\end{array} &
\begin{array}{|c|c|}\hline 1&1\\\hline 4\\\cline{1-1}\end{array} &
\begin{array}{|c|c|}\hline 1&2\\\hline 4\\\cline{1-1}\end{array} \\[2ex]
\begin{array}{|c|c|}\hline 2&2\\\hline 4\\\cline{1-1}\end{array} &
\begin{array}{|c|c|}\hline 1&2\\\hline 4\\\cline{1-1}\end{array} &
\begin{array}{|c|c|}\hline 2&3\\\hline 4\\\cline{1-1}\end{array} &
\begin{array}{|c|c|}\hline 3&3\\\hline 4\\\cline{1-1}\end{array} &
\begin{array}{|c|c|}\hline 1&4\\\hline 2\\\cline{1-1}\end{array} \\[2ex]
\begin{array}{|c|c|}\hline 1&4\\\hline 3\\\cline{1-1}\end{array} &
\begin{array}{|c|c|}\hline 2&4\\\hline 3\\\cline{1-1}\end{array} &
\begin{array}{|c|c|}\hline 1&4\\\hline 4\\\cline{1-1}\end{array} &
\begin{array}{|c|c|}\hline 2&4\\\hline 4\\\cline{1-1}\end{array} &
\begin{array}{|c|c|}\hline 3&4\\\hline 4\\\cline{1-1}\end{array}
\end{array} \qquad (4.73)$$

so that the irrep of U(4) adapted properly to antisymmetric spin-eigenfunctions for 3 electrons is of dimension 20.

The general dimension formula can be shown to be

$$t = \frac{2S+1}{n+1}\begin{pmatrix} n+1 \\ N/2-S \end{pmatrix}\begin{pmatrix} n+1 \\ n-S-N/2 \end{pmatrix} \qquad (4.74)$$

which is clearly a very nasty large number for reasonable n and N.

To complete our specification of the representation we need a way of specifying the orbital number involved and it turns out to be convenient to do this using the number

$$c = n - N/2 - S$$

so that generalizing the [λ] notation from the symmetric group we can specify a useful irrep of U(n) by

$$[2^a 1^b 0^c]$$

thus for our S = 1/2 case, it would be

$$[2^1 1^1 0^2]$$

Obviously the correspondence

a = "doubly occupied spin paired orbitals"
b = "singly occupied orbitals"
c = "unoccupied orbitals"

makes some sort of sense.

We are now on the home straight and to reach the winning post we need only a couple more results. The first is that the irreps of U(n-1) can be got from the irreps of U(n) by striking out from any tableau the box containing n. If n does not occur in a set of boxes then these tableaux already specify a basis for an irrep of U(n). One may go on down to U(1) so that any irrep of U(n) may be specified completely by this process. In the context of the numbers a, b and c this means that we can start with a given a, b, c and call it a_n, b_n and c_n, then we can construct a_{n-1}, b_{n-1}, c_{n-1} and so on to generate a tableau of three columns that completely specifies the basis function or Weyl tableau in the irrep. Thus in the case of our example $a_n = 1$, $b_n = 2$ and $c_n = 2$ the Paldus or abc tableau corresponding to the first Weyl tableau in (4.17) is

n	a	b	c
4	1	1	2
3	1	1	1
2	1	1	0
1	1	0	0
0	0	0	0

(4.75)

where the last row n = 0 is a fictitious row added because it is computationally convenient. The significance of c = 1 in row 3 is that the orbital 4 will never be used, similarly c = 0 in row 2 means that orbital 3 will never be used. That b = 0 in row 1 signifies the passing of orbital 2 from the irrep and that is the end.

Now we come to the big swindle. Clearly one does not actually need the Young tableaux to construct the Paldus tableaux. The direct rules for the construction of Paldus tableaux are obviously pretty easy as can be seen from (4.75). So suppose that we regard the Paldus tableau as primary then we might as well regard it as specifying not a basis function associated with a given Weyl tableau but a basis function in the Gelfand-Tsetlin basis. That it is possible to do this is clear because there must be exactly the same number of basis function for any form of an irrep of a group. Whether it is a useful thing or not to do is of course not clear. But Paldus was able to show that it was useful. By making this correspondence Paldus showed for example that the matrix element of E_{rr} between two functions specified by Paldus tableaux (abc) and (a'b'c') was given by

$$\langle(a'b'c')|E_{rr}|(abc)\rangle = \delta_{(a'b'c')(abc)} n_r(abc) \tag{4.76}$$

meaning that it vanishes unless the functions are identical, an expected reflection of the fact that the basis is adapted to the Cartan sub-algebra. When the functions are identical then

$$\begin{aligned} n_r(abc) &= 2(a_r - a_{(r-1)}) + b_r - b_{(r-1)} \\ &= 2\Delta a_r + \Delta b_r \end{aligned} \tag{4.77}$$

Clearly if the orbital r does not occur in the tableau as occupied at the level of row r in the tableau, then $n_r = 0$. Otherwise clearly all that one is giving is some combination of 0; 1 and 2, to give possible values 0, 1 or 2 for n_r. A dead simple result. There are other rather nastier results for the E_{rs}, but relatively simple results for $E_{r(r\pm1)}$.

And that is it. If one has a CI calculation using n orbitals for N electrons to produce a state of a given S value, all one has to do is to imagine that each configuration in the calculation is symbolized by a Paldus tableau, then let the hammer drop and accumulate the relevant density matrix elements using (4.76) and its congeners in the context of (1.17).

If we did this in a conventional CI then one would end up with exactly the same density matrix elements as one would get by using the same orbital configuration and Yamanouchi-Kotani spin coupling. Thus it is doubtful if a unitary group approach would have been worth inventing if one could only do conventional CI with it.

Where it scores, however, is in direct CI where it is a "natural" because the E_{rr}, E_{rs} and so on effectively specify integrals and "point at" configurations in the sense of Paldus tableaux by virtue of rules like (4.76) and so on. In fact there is an ordering of Paldus tableaux, the so called lexical ordering that makes this pointing very easy. Furthermore it is possible to specify this ordering in terms of "walks" through a particular kind of graph called the Shavitt digraph (each walk corresponds to a Paldus tableau) and thus makes it possible to program the "pointing" rules with only considerable ingenuity instead of inhumanly fiendish ingenuity.

There is however a snag and it is the one that I alluded to at the beginning of this section; namely, one can only be sure of the unitary group invariance if one deals with all possible antisymmetric configurations that result from a given set of orbitals. But this is generally quite out of the question as a quick look at (4.74) will assure you. Now the fact that one cannot use such a set does not mean that all that has gone before is useless. The matrix element formulae and so on do not depend in any way on this invariance. What it does mean however is that the roots of the secular problem will, as usual, depend on a particular orbital choice and more importantly, if one uses the E_{rs}, etc., as pointing devices to point to configurations then one has to make jolly sure that the configuration pair that is pointed to, really is in the space of configurations that one wants to consider. Fortunately the Shavitt digraph makes this possible in the most important case (all singles and doubles in a given reference state) and the ingenuity of Professor Siegbahn has made it possible in a lot more cases, but that is something about which he will doubtless talk to you.

To end now on a rather pathetic if not bathetic note. It is the case that following Professor Roos' invention of direct CI in a special case, for which he did not use group theory explicitly, Professor Shavitt fell upon the general problem. He managed to get all the formulas relevant to the unitary group and invent the digraph using purely algebraic techniques. It must have been a labour of Hercules! He was pipped at the post in publication by Professor Paldus to whom Professor Shavitt's endeavours were unknown. I suppose that this shows both that you can get away without any group theory at all if you have a mind to, even in the most difficult problems, but on the other hand, if you do know some group theory then you may be able to get to the solution of a problem faster than might otherwise be the case.

NOTE ADDED IN PROOF

After I had finished giving the lectures of which the foregoing is the manuscript, Dr Saunders pointed out that I had overlooked a whole method of doing the CI problem utilizing continuous groups. In Section (3) I concentrated on the irreps of S_N in a basis of spin-eigenfunctions (a basis for SO(3)). But the discussion can be "inverted". One can think of the irreps of SO(3) in a basis for S_N. If one thinks of the problem this way round then it becomes one in angular momentum coupling and re-coupling to get the required basis for the spin irreps of SO(3). This approach also makes a very useful connection with pair-theory mathods.

Dr Saunders told me that this approach was the one that he had used to check the results of Paldus and Boyle (Physica Scripta 31, 295 (1980)) and also to prepare the projective reduction coefficients for the new ATMOL direct CI.

I can only apologize for my oversight and perhaps explain it as a Freudian slip. Many years ago I wrote a paper on CI in which I got the projective reduction coefficients by exactly this approach. To my shame, however, in that paper I succeeded in re-inventing a rather special case of the Jucys spin diagram technique to solve the angular momentum problem, because I was too young and ignorant to realize that it had been done before.

BIBLIOGRAPHY

M. Hamermesh, Group Theory, Addison-Wesley.

R. Pauncz, Spin Eigenfunctions, Plenum Press.

C. D. H. Chisholm, Group Theoretical Techniques in Quantum Chemistry, Academic Press.

J. Paldus, Many-electron Correlation Problem. A Group Theoretical Approach in "Theoret. Chem., Advances & Perspectives", Vol. 2, p. 131.

J. D. Louck, The Unitary Group, Amer. J. Phys., 38, 3, 1970.

I. Shavitt, Int. J. Quant. Chem., Symp., 12, 5, 1978.

J. Hinze (Ed.), The Unitary Group in the Evaluation of Electronic Energy Matrix Elements, Springer-Verlag.

B. G. Wybourne, Classical Groups for Physicists, Wiley.

B. Schutz, Geometric Methods of Mathematical Physics, C.U.P.

THE MULTICONFIGURATIONAL (MC) SCF METHOD

Björn O. Roos
Division of Physical Chemistry 2
Chemical Centre
P.O.B. 740
S-220 07 LUND
Sweden

This series of lectures will treat the methods employed to solve the orbital and CI optimization problem for multiconfigurational wave functions. The main emphasis will be put on the recent development of quadratically or near-quadratically convergent procedures such as the Newton-Raphson and the super-CI methods. The equations for a general second order Newton-Raphson procedure will be derived in a form suitable for large scale applications. Different approximations to the general procedure will be discussed, such as the augmented Hessian and the super-CI methods.

G. H. F. Diercksen and S. Wilson (eds.), Methods in Computational Molecular Physics, 161–187.

1. INTRODUCTION

In this series of lectures (4x45 min) we shall discuss the multiconfiguration (MC) self-consistent field method (SCF) and its computational implementation. Due to the limited time available, it will not be possible to give a complete survey of the method, and the discussion will be restricted to the basic features of the method with the emphasis on the recent development of efficient optimization procedures and the modern implementations, which have made the MCSCF method a practical tool for theoretical studies of a wide class of problems in chemistry.

The MCSCF method dates back to the early days of quantum chemistry /1/, and has been used for calculations on atoms and small molecules since the beginning of the 60's (for a compilation of atoms and molecules studied with the MCSCF method from this time up to the end of the 70's, see the report by Detrich and Wahl in /3/). The method was, however, difficult and time-consuming to use due to bad convergence behaviour of the (mainly Fock matrix based) first order optimization procedures in use, and sever limitations to the size of the MC expansion which could be used. This situation has drastically changed during the last three or four years due to the development of efficient second order MCSCF techniques /4,5/ and the use of density matrix formulated procedures which in conjunction with the direct CI method /6/ has made it possible to deal with very long MC expansions /7/. As a result of this progress the MCSCF method is today used for a wide variety of problems in chemistry and molecular physics.

Accurate quantum chemical calculations on molecular systems are today normally performed in two steps: first a set of molecular orbitals and a zeroth order approximation to the wave function are obtained. This wave function is then used as the reference in a subsequent treatment of correlation effects, either by means of a configuration interaction calculation or by many-body perturbation theory. The zeroth order approximation has traditionally been obtained from single configuration SCF calculations. The recent progress in multiconfiguration SCF procedures has now made it possible to extend the zeroth order treatment to a much higher level of accuracy. The MCSCF method is today routinely used to obtain a qualitatively correct description of the electronic structure for a wide variety of molecular systems, including potential energy surfaces for chemical reactions in ground and excited states, ionization processes, accurate studies of molecular properties, etc. This represents an important step forward and extends our understanding of the electronic structure in molecules to more complex systems where the independent particle approximation breaks down. A common feature of such system is a near degeneracy of several electronic configurations, making a single configuration approximation invalid. Such situations obtain in most chemical reactions but also in many molecules at their equilibrium geometries. A drastic example is the weak multiple bonds appearing in transition metal dimers, where in some cases (eg. Cr_2) the wave function comprises several thousand configurations with most of the coefficients of a similar magnitude /8/. A number of other examples can be given, also for simpler molecules containing only first row atoms. Typically near degeneracies occur in most cases where the chemical bonds are obtained as resonances between different valence structures (eg. O_3, NO_2, C_6H_6, etc.), and in weakly bonded systems (eg. F_2, N_2O_4, etc.).

This series of lectures will, however, not deal with the chemical aspects of multiconfiguration wave function, but has to concentrate on the technical aspects of the theory.

2. THE WAVE FUNCTION AND THE ENERGY EXPRESSION

The multiconfigurational wave function $|0\rangle$ - the MCSCF reference state - is obtained as a superposition of configuration state functions (CSF's),

$$|0\rangle = \sum_{\mu=1}^{M} C_\mu^{(0)} |\mu\rangle, \tag{1a}$$

where the sum runs over all CSF's$|\mu\rangle$. These will be assumed to be properly spin- and space-symmetrized electronic configurations built from a given set of molecular orbitals $\{\phi_p;p=1,n\}$. The wave function (1) represents one single electronic state of the system, but not necessarily the ground state. An orthogonal complement to $|0\rangle$ can formally be written as

$$|K\rangle = \sum_{\mu} C_\mu^{(K)} |\mu\rangle \qquad K = 1,\ldots, M-1 \tag{1b}$$

It will not be assumed, that the states $|0\rangle$ and $|K\rangle$ are eigenstates of the Hamiltonian of the system projected onto the MC sub-space, unless this is explicitly stated.

The first problem encountered when setting up an MCSCF calculation is the selection of CSF's for the expansion (1). A number of different criteria has been used through the years, in order to try to determine the most important (in some sense) CSF's which can be generated from the orbital subspace /3/. One of the serious bottlenecks of MCSCF calculations has been the severe limitations on the length, M, of the expansion (1). A glance at the compilation of MCSCF calculations made by Detrich and Wahl /3/, which covers most of the calculations performed in the period 1962-1980, shows that no MCSCF wave function published before 1980 comprised more than 61 CSF's. Normally the value of M lies in the range 2-10. Obviously such sever limitations on the wave function would limit the applicability of the MCSCF method to a rather small range of problems. This limitation has, however, now been lifted to a large extent, due to a combination of multiconfiguration SCF methods with direct CI procedures /6/.

The CSF selection problem is in most cases considered as a two step procedure. The first step consists in a partitioning of the orbital space into occupied (primary) and not occupied (secondary) subspaces. The primary subspace is further divided into inactive and active subspaces, where the inactive orbitals are assumed to be doubly occupied in all CSF's, while the active orbitals have variable occupations. Such a partitioning of the orbital space is not straightforward in non-trivial applications, but is anyway much easier than the subsequent selection of CSF's from the list generated by the active subspace. A special and very attractive variant of the MCSCF wave function is obtained by including all CSF's which can be generated by distributing the 'active' electrons among the active orbitals in all possible ways consistent with the overall spin and space symmetry of the wave function. This procedure avoids the difficult and somewhat ambigous selection of individual CSF's. The wave function obtained in this way has been named the complete active space (CAS) wave function /7/, and the computational procedure connected with it, the CASSCF method. CAS expansions can be very long (calculations with M in the range 10^4-$3x10^4$ have actually been performed) and this sets an upper limit to the size of the active subspace at around ten orbitals. Starting from a CAS configuration, space restrictions can be imposed which reduce the number of CSF's in a controlled way. One possibility, which for many cases has proven to lead to only negligible changes in the results, is to restrict the occupancies of the active orbitals in different symmetry species. Consider as an example the $X^1\Sigma g^+$ ground state of the N_2 molecule with an active subspace consisting of all valence orbitals: $2\sigma_g$, $2\sigma_u$, $3\sigma_g$, $3\sigma_u$, $1\pi_u$, and $1\pi_g$. The number of active electrons is ten. The dominant configuration of N_2 at equilibrium is $(2\sigma_g)^2(2\sigma_u)^2(3\sigma_g)^2(1\pi_u)^4$. A very good approximation to the full CAS result is obtained by including into (1) only CSF's with six electrons in σ-orbitals (g and u) and four electrons in π-orbitals. This leads to a substantial reduction of the MC space. More soft conditions can also be imposed where the occupation of a given sub-symmetry is restricted to be between a lower and an upper limit (eg. between 4 and 8 electrons in σ and 2 and 6 in π for N_2). This type of wave function can be called symmetry restricted complete active space (SRCAS). It has the advantage of leaving the computational procedure for the CASSCF method unchanged (see below). In the following discussion we shall label the inactive orbitals as i,j,k,....; the active orbitals as t,u,v,.... and secondary orbitals as a,b,c.... Orbitals not specified to any of the subspaces will be labeled p,q,r,...

The energy expression will be obtained from the Hamiltonian in second quantized form. The Hamiltonian will be assumed to be spin-independent. It can then be written in terms of the spin-averaged excitation operators (the generators of the unitary group /9/),

$$\hat{E}_{pq} = \sum_{\sigma} \hat{a}^{\dagger}_{p\sigma} \hat{a}_{q\sigma} \tag{2}$$

where $\hat{a}_{q\sigma}(\hat{a}^{\dagger}_{q\sigma})$ are the normal annihilation (creation) operators for an electron in the molecular orbital ϕ_q with spin quantum number σ. The sum extends over the two possible values of σ. The generators (2) fulfill the following commutation relation

$$\hat{E}_{pq}\hat{E}_{rs} - \delta_{qr}\hat{E}_{ps} = \hat{E}_{rs}\hat{E}_{pq} - \delta_{sp}\hat{E}_{rq} \tag{3}$$

The Hamiltonian (projected onto the finite subspace of the Hilbert space, in which we are working) is given in terms of the generators $\hat{E}_{pq}$ as,

$$\hat{H} = \sum_{p,q} h_{pq}\hat{E}_{pq} + \frac{1}{2}\sum_{p,q,r,s}(pq|rs)\{\hat{E}_{pq}\hat{E}_{rs} - \delta_{qr}\hat{E}_{ps}\} \tag{4}$$

where h_{pq} and $(pg|rs)$ are the normal one- and two-electron integrals. The total energy is now directly obtained from the expression

$$E_0 = \langle 0|\hat{H}|0\rangle \;/\; \langle 0|0\rangle \tag{5}$$

as

$$E_0 = \sum_{p,q} h_{pq} D^{(00)}_{pq} + \sum_{p,q,r,s}(pq|rs)P^{(00)}_{pqrs} \tag{6}$$

where $D^{(00)}_{pq}$ and $P^{(00)}_{pqrs}$ are the first and second order reduced density matrices which according to (4) can be obtained as

$$D^{(00)}_{pq} = \langle 0|\hat{E}_{pq}|0\rangle \tag{7a}$$

$$P^{(00)}_{pqrs} = \frac{1}{2}\langle 0|\hat{E}_{pq}\hat{E}_{rs} - \delta_{qr}\hat{E}_{ps}|0\rangle \tag{7b}$$

These density matrix elements are functions of the MC expansion coefficients $C_\mu^{(0)}$ in (1). Expanding $|0\rangle$ according to (1) we obtain

$$D_{pq}^{(00)} = \sum_{\mu,\nu} C_\mu^{(0)} C_\nu^{(0)} A_{pq}^{\mu\nu} \tag{8a}$$

$$P_{pqrs}^{(00)} = \sum_{\mu,\nu} C_\mu^{(0)} C_\nu^{(0)} A_{pqrs}^{\mu\nu} \tag{8b}$$

where $A_{pq}^{\mu\nu}$ and $A_{pqrs}^{\mu\nu}$ are the well-known one- and two-electron spin coupling constants which are needed to construct the matrix elements over the Hamiltonian in the CSF basis,

$$A_{pq}^{\mu\nu} = \langle\mu|\hat{E}_{pq}|\nu\rangle \tag{9a}$$

$$A_{pqrs}^{\mu\nu} = \frac{1}{2}\langle\mu|\hat{E}_{pq}\hat{E}_{rs} - \delta_{qr}\hat{E}_{ps}|\nu\rangle, \tag{9b}$$

with the matrix elements given as

$$H_{\mu\nu} = \langle\mu|\hat{H}|\nu\rangle = \sum_{p,q} A_{pq}^{\mu\nu} h_{pq} + \sum_{p,q,r,s} A_{pqrs}^{\mu\nu}(pq|rs) \tag{10}$$

The values of the spin coupling constants depend on the way the CSF's have been constructed. In the unitary group formulation of the CI problem by Paldus and Shavitt /11/ a Gelfand-Tsetlin CSF basis is used and the coupling constants are generated by an elegant graphical procedure. It is important that it is easy to generate the density matrises (8) from the list of coupling constants also for very long MC expansions. This is achieved by computing and storing the $A_{pqrs}^{\mu\nu}$ sequentially for each integral label p,q,r and s. Only a simple read through the list of coupling constants is then necessary for the construction of $P_{pqrs}^{(00)}$. The construction of $D_{pq}^{(00)}$ is of course trivial.

The indices p,q,r, and s in (7) obviously only run over the primary orbital subspace. Actually it is only necessary to consider explicitly density matrix elements with all indices in the active subspace, since elements corresponding to inactive orbitals are given as

$$D_{ip}^{(00)} = 2\delta_{ip} \tag{11a}$$

$$P_{ijpq}^{(00)} = \delta_{ij} D_{pq}^{(00)} - \delta_{iq}\delta_{jp} \tag{11b}$$

$$P^{(00)}_{pijq} = 2\delta_{ip}\delta_{jq} - \frac{1}{2}\delta_{ij}D^{(00)}_{pq} \tag{11c}$$

where p and q run over the entire primary subspace. Density matrix elements, containing inactive orbital indices and not included in (11), are zero.

The number of active orbitals is normally small (<20) and it is therefore possible to store the density matrices in the central memory of the computer, which greatly simplifies the data handling.

The energy expression (6) is a function of the MC expansion coefficients through the density matrix elements, according to (8), and a function of the molecular orbitals ϕ_p through the one- and two electron integrals. In an MCSCF optimization procedure both these quantities are varied until the energy reaches a stationary value ($\delta E=0$). We shall now discuss how these variations are performed in modern implementations of the MCSCF method.

3. THE GENERAL SECOND ORDER MCSCF PROCEDURE

A number of different methods to optimize the parameters of an MCSCF wave function has been suggested in the literature. Some of them are based on Fock matrices /12/ and can be viewed upon as generalizations of the methods traditionally used in Hartree-Fock calculations. Other methods are based on the annihilation of single excitations out of the MC reference state using the generalized Brillouin's theorem /13/. They have been called "Super-CI" methods, since they lead to the solution of a CI secular problem including the reference state and the singly excited Brillouin states /14,7/. The third and maybe most powerful of these methods is the second order Newton-Raphson approach, where the energy is expanded as a Taylor series in the non-redundant parameters /4,5,7,14/. The series is truncated after the second order term and updated values of the parameters are obtained by solving a system of linear equations. Cubic methods are obtained if one more term is included in the Taylor expansion /5/.

It will not be possible to cover all these methods to any depth in the present lectures. Instead we shall concentrate on the general second order scheme and derive some of the other methods from the general equations

3.1. The Non-Linear Newton Raphson Method

Before analyzing the energy expression (6) with respect to the variational parameters in detail, let us briefly review the multidimensional Newton-Raphson procedure. Assume that the energy is a function of a set of parameters λ_i, which we arrange as a column vector $\underline{\lambda}$: $E=E(\underline{\lambda})$. Make a Taylor expansion of the energy around a point $\underline{\lambda_0}$, which we arbitrarily can put equal to zero

$$E(\underline{\lambda}) = E(0)+ \sum_i \left(\frac{\partial E}{\partial \lambda_i}\right)_0 \lambda_i + \frac{1}{2} \sum_{i,j} \lambda_i \left(\frac{\partial^2 E}{\partial \lambda_i \partial \lambda_j}\right)_0 \lambda_j \ldots \tag{12}$$

or in matrix notation

$$E(\underline{\lambda}) = E(0) + \underline{g}^{\dagger}\underline{\lambda} + \frac{1}{2}\underline{\lambda}^{\dagger}\underline{\underline{h}}\,\lambda + \ldots \tag{13}$$

where the gradient vector $\underline{g}$ and the Hessian matrix $\underline{\underline{h}}$ have the elements

$$g_i = \left(\frac{\partial E}{\partial \lambda_i}\right)_0, \tag{14a}$$

and

$$h_{ij} = \left(\frac{\partial^2 E}{\partial \lambda_i \partial \lambda_j}\right)_0 \tag{14b}$$

The condition to be fullfilled for an optimized solution is $\delta E \doteq 0$ or equivalently

$$\left(\frac{\partial E}{\partial \lambda_i}\right)_0 = 0 \qquad i=1,2\ldots \tag{15}$$

An iterative procedure can now be set up by computing the set of parameters λ for which the Newton-Raphson iterative function (13) has an extremum value

$$\frac{\partial E}{\partial \lambda_i} = 0, \qquad i=1,2\ldots \tag{16}$$

which leads to the following set of linear equations.

$$\underline{g} + \underline{\underline{h}}\,\underline{\lambda} = 0, \tag{17a}$$

or

$$\underline{\lambda} = - \underline{\underline{h}}^{-1}\,\underline{g}. \tag{17b}$$

A sequence of Newton-Raphson iterations is obtained by solving equation (17); redefine the zero point ($\underline{\lambda_0}$) as the new set of parameters; recalculate $\underline{g}$ and $\underline{\underline{h}}$ and return to equation (17). Such a procedure converges quadratically, that is, the error vector in iteration n is given as a quadratic function of the error vector in iteration (n-1) /5/. This does not necessarily means that the NR procedure will converge fast or even at all. However, close to the stationary point, we can expect a quadratic behaviour. We shall return later to a more precise discussion of what 'close' means in this respect.

3.2 Exponential Unitary Operators

The vector $\underline{\lambda}$ contains two sets of parameters; one set describing variations of the molecular orbitals, and the other set the variations of the MC coefficients in (1). The energy variations are constrained by the condition that the MO's remain orthogonal to each other. This can be achieved by introducing Lagrangian multipliers, but a more frequently used approach is to introduce exponential unitary transformations to rotate the orbitals /16, 17,5/. In order to have an equivalent treatment of the variations of the MC coefficients, we shall describe these also by an exponential unitary transformation.

We can define a general unitary operator by

$$\hat{U} = e^{\hat{X}} = 1 + \hat{X} + \frac{1}{2!}\hat{X}^2 + \ldots . \tag{18}$$

If $\hat{U}$ is unitary ($\hat{U}^\dagger\hat{U} = \hat{1}$) it is always possible to find an anti-hermitian operator $\hat{X}$ such that equation (18) is fulfilled. Also if $\hat{X}$ is anti-hermitian, $\hat{U}$ is unitary. The proof of the first statement follows from the fact that the eigenvalues of $\hat{U}$ have unit modulus, that is, they can be written in the form $e^{i\alpha}$. The proof of the second statement is trivial. The operator $\hat{X}$ is thus anti-hermitian

$$\hat{X}^\dagger = - \hat{X} \tag{19}$$

For the unitary rotation of the MC reference state $|0\rangle$ we use $\hat{X}=\hat{S}$ with

$$\hat{S} = \sum_{K\neq 0} S_{KO} \left(|K\rangle\langle 0| - |0\rangle\langle K|\right). \tag{20}$$

The variational parameters for the state rotations are given as the first row of the matrix $\underline{\underline{S}}$ appearing in (20). Similarly for the orbital rotations we define a matrix $\hat{X}=\hat{T}$ with

$$\hat{T} = \sum_{p,q} T_{pq}\hat{E}_{pq} = \sum_{p>q} T_{pq}(\hat{E}_{pq} - \hat{E}_{qp}) \tag{21}$$

where $\underline{\underline{T}}$ is an antisymmetric matrix in the orbital indices p and q.

It is maybe not obvious that $e^{\hat{T}}$ operating on $|0\rangle$ corresponds to a unitary rotation of the orbitals in $|0\rangle$. In order to see that this is the case let $e^{\hat{T}}$ operate on a product of spin-orbitals and expand the operator in a prower series:

$$e^{\hat{T}} = 1 + \hat{T} + \frac{1}{2!}\hat{T}^2 + \ldots$$

Now we obtain

$$e^{\hat{T}}\{\phi_p\phi_q\ldots\} = \{\phi_p\phi_q\ldots\} + T\{\phi_p\phi_q\ldots\} + \frac{1}{2}\hat{T}^2\{\phi_p\phi_q\ldots\} + \ldots$$

$$= \{\phi_p\phi_q\ \ldots\} +$$

$$+ \{(\hat{T}\phi_p)\phi_q\ldots\} + \{\phi_p(\hat{T}\phi_q)\ \ldots\} + \ldots$$

$$+ \{(\tfrac{1}{2}\hat{T}^2\phi_p)\phi_q\ldots\} + \{(\hat{T}\phi_p)(\hat{T}\phi_q)\ldots\} + \{\phi_p(\tfrac{1}{2}\hat{T}\phi_q)\ldots + \ldots$$

$$+ \ldots$$

$$= \{(e^{\hat{T}}\phi_p)(e^{\hat{T}}\phi_q)\ \ldots\}$$

which corresponds to a unitary rotation of each orbital in the product. Since our reference state is a linear combination of spin-orbital products, the same is true there.

Our rotated reference state is given by the combined operation of $\hat{S}$ and $\hat{T}$,

$$|0'\rangle = e^{\hat{T}}e^{\hat{S}}|0\rangle \tag{22}$$

To simplyfy the notation for the following derivation of the energy expression we introduce the operator $\hat{\lambda} = \hat{T} + \hat{S}$, and write the unitary rotation of the reference state as

$$|0'\rangle = e^{\hat{\lambda}}|0\rangle, \tag{23}$$

where we must remember to always let $\hat{S}$ operate first when expanding this operator.

3.3. The Energy Expression and the Newton-Raphson Equations.

We now have the tools needed to expand the energy expression in terms of the variational parameters given by the matrices $\underline{\underline{S}}$ and $\underline{\underline{T}}$. Formally we obtain from equation (23)

$$E(\underline{\lambda}) = \langle 0'|\hat{H}|0'\rangle = \langle 0|e^{-\hat{\lambda}}\,\hat{H}e^{\hat{\lambda}}|0\rangle. \tag{24}$$

Expanding the operator $e^{\hat{\lambda}}$ in powers of $\hat{\lambda}$ we obtain for the transformed Hamiltonian

$$e^{-\hat{\lambda}}\hat{H}\, e^{\hat{\lambda}} = \hat{H} + [\hat{H},\hat{\lambda}] + \frac{1}{2!}\,[\hat{\lambda},[\hat{\lambda},\hat{H}]] + \ldots. \qquad (25)$$

where the expansion has been truncated at second order in $\hat{\lambda}$. Inserting (25) into (24) yields for the energy expression

$$E(\underline{\lambda}) = E(0) + \langle 0|[\hat{H},\hat{\lambda}]|0\rangle + \frac{1}{2!}\,\langle 0|[\hat{\lambda},[\hat{\lambda},\hat{H}]]\,|0\rangle + \ldots \qquad (26)$$

In order to compare (26) with the Taylor expansion (13) we expand $\hat{\lambda}$ according to (20) and (21), remembering that $\hat{\lambda}=\hat{T}+\hat{S}$,

$$\hat{\lambda} = \sum_{\alpha} \lambda_{\alpha}\, \hat{R}_{\alpha} \qquad (27)$$

where α runs over both the orbital rotations (21) and the state rotations (20). Equation (26) can then be rewritten as

$$E(\underline{\lambda}) = E(0) + \sum_{\alpha} \lambda_{\alpha}\, \langle 0|[\hat{H},\hat{R}_{\alpha}]|0\rangle + \frac{1}{2!}\sum_{\alpha,\beta} \lambda_{\alpha}\langle 0|[\hat{R}_{\alpha},\hat{R}_{\beta},\hat{H}]|0\rangle\, \lambda_{\beta} + \ldots \qquad (28)$$

where the second order term has been symmetrized over α and β and

$$[\hat{R}_{\alpha},\hat{R}_{\beta},\hat{H}] = \frac{1}{2}\,[\hat{R}_{\alpha},\, [\hat{R}_{\beta},\hat{H}]] + \frac{1}{2}[\hat{R}_{\beta},[\hat{R}_{\alpha}\hat{H}]]. \qquad (29)$$

Note again that $\hat{S}$ always operates before $\hat{T}$ in the mixed term. Comparing equations (28) and (13) we obtain the following expressions for the matrix elements of the gradient vector and the Hessian as,

$$g_{\alpha} = \langle 0|[\hat{H},\hat{R}_{\alpha}]|0\rangle \qquad (30a)$$

$$h_{\alpha\beta} = \langle 0|[\hat{R}_{\alpha},\hat{R}_{\beta},\hat{H}]|0\rangle \qquad (30b)$$

These experessions can be simplified by expanding the commutators, remembering that $\hat{R}$ is a real anti-Hermitian operator ($\hat{R}_{\alpha}^{\dagger} = -\hat{R}_{\alpha}$), and assuming that also the wave function is real,

$$g_{\alpha} = 2\langle 0|\hat{H}\, \hat{R}_{\alpha}|0\rangle \qquad (31a)$$

$$h_{\alpha\beta} = \langle 0|\hat{R}_{\alpha}\hat{R}_{\beta}\hat{H} - 2\hat{R}_{\alpha}\hat{H}\, \hat{R}_{\beta} + \hat{H}\, \hat{R}_{\alpha}\hat{R}_{\beta}|0\rangle \qquad (31b)$$

Finally we need to split these matrix elements into an orbital and a configuration part according to the partioning of the operators $\hat{R}_\alpha$. For the configuration part we have from equation (20),

$$\hat{R}_K^{(C)} = |K\rangle\langle 0| - |0\rangle\langle K| \tag{32a}$$

and similarly from the last part of equation (21),

$$\hat{R}_{pq}^{(O)} = \hat{E}_{pq} - \hat{E}_{qp} = \hat{E}_{pq}^{-} . \tag{32b}$$

We now obtain for the two parts of the gradient vector

$$g_K^{(C)} = 2\langle 0|\hat{H}|K\rangle, \tag{33a}$$

$$g_{pq}^{(O)} = 2\langle 0|\hat{H}\,\hat{E}_{pq}^{-}\,|0\rangle . \tag{33b}$$

For a fully optimized reference state the energy gradient is zero. The conditions to be fullfilled for an MCSCF reference state are thus given by the expressions (33) being zero. The first of the conditions states that the reference state does not interact with the orthogonal complement, that is, it is a solution to the secular problem

$$(\underline{\underline{H}} - E\,\underline{\underline{1}})\,\underline{C} = 0 \tag{34}$$

in the configuration space. The second condition is the generalized Brillouin's theorem /13/, first derived by Levy and Berthier.

The Hessian is divided into four parts: the orbital-orbital (OO), the configuration-configuration (CC) and the coupling (CO) part,

$$h_{KL}^{(CC)} = 2\{\langle K|\hat{H}|L\rangle - \delta_{KL}\,\langle 0|\hat{H}|0\rangle\}, \tag{35a}$$

$$h_{pq,rs}^{(OO)} = \langle 0|\hat{E}_{pq}^{-}\hat{E}_{rs}^{-}\,\hat{H}|0\rangle + \langle 0|\hat{H}\,\hat{E}_{pq}^{-}\hat{E}_{rs}^{-}|0\rangle$$

$$- 2\langle 0|\hat{E}_{pq}^{-}\,\hat{H}\,\hat{E}_{rs}^{-}\,|0\rangle \tag{35b}$$

and for the mixed term

$$h_{K,pq}^{(CO)} = h_{pq,K}^{(OC)} = 2\langle K|[\hat{H},\hat{E}_{pq}^{-}]|0\rangle \tag{35c}$$

These formulas were first derived by Levy in his thesis /4/.

They give the matrix elements of the gradient vector and the Hessian in terms of matrix elements of the excitation operators $\hat{E}_{pq}$. They can be evaluated in terms of one- and two-electron integrals and first- and second order reduced density matrices by inserting the Hamiltonian (4) into equations (34)-(35). Note that the transition density matrix elements $D_{pq}^{(KO)}$ and $P_{pqrs}^{(KO)}$ are needed for the evaluation of the CI coupling matrix (35c).

Using the matrices (34) and (35) we can now write the Newton-Raphson equations (17) in the following form,

$$\begin{bmatrix} \underline{\underline{a}} & \underline{\underline{b}} \\ \underline{\underline{b}} & \underline{\underline{c}} \end{bmatrix} \begin{bmatrix} \underline{S} \\ \underline{T} \end{bmatrix} = - \begin{bmatrix} \underline{v} \\ \underline{w} \end{bmatrix} , \tag{36}$$

where we have introduced the simplifying notation $\underline{\underline{a}} = \frac{1}{2} \underline{\underline{h}}^{(CC)}$, $\underline{\underline{b}} = \frac{1}{2} h^{(CO)}$, $\underline{\underline{c}} = \frac{1}{2} h^{(OO)}$, $\underline{v} = \frac{1}{2} \underline{g}^{(C)}$, and $\underline{w} = \frac{1}{2} \underline{g}^{(O)}$

The direct solution for the rotation parameters $\underline{S}$ and $\underline{T}$ from (36) is not very practical if the MC expansion is large, due to the complications in computing matrix elements over the orthogonal complement space $|K\rangle$, defined by equation (2). An M^3 transformation is needed to obtain $\underline{h}^{(CC)}$ from the Hamiltonian matrix elements $\langle\mu|\hat{H}|\nu\rangle$. Olsen and Yeager have suggested a choice of the complementary space which reduces the computational effort to an M^2 procedure, but even this procedure is unsuitable in connection with direct CI techniques /18/.

A more attractive computational procedure is obtained by rewriting equation (36) in terms of variations of the original CI coefficients C_μ in equation (1) /15/. Equation (36) can be written in the form

$$\begin{bmatrix} \underline{\underline{C}}^{\dagger} \underline{\underline{A}}\, \underline{\underline{C}} & \underline{\underline{C}}^{\dagger} \underline{\underline{B}} \\ \underline{\underline{B}}^{\dagger} \underline{\underline{C}} & \underline{\underline{c}} \end{bmatrix} \begin{bmatrix} \underline{S} \\ \underline{T} \end{bmatrix} = - \begin{bmatrix} \underline{\underline{C}}^{\dagger} \underline{G} \\ \underline{w} \end{bmatrix} , \tag{37}$$

where $\underline{\underline{C}}$ is a matrix containing the M-1 column vectors $\underline{C}_K$, $\underline{\underline{A}}$ is the $(\hat{H} - E_0 \hat{1})$ matrix in the CSF basis: $A_{\mu\nu} = \langle\mu|\hat{H}-E_0|\nu\rangle$; $B_{\mu,pq} = \langle\mu|[\hat{H},\hat{E}^-_{pq}]|0\rangle$; $G_\mu = \langle 0|\hat{H}|\mu\rangle$. The state rotation parameters $\underline{S}$ are obtained from variations of the individual CI coefficients as $\underline{S} = \underline{\underline{C}}^{\dagger}\, \delta\underline{C}$. However, the number of parameters in $\delta\underline{C}$ is M, while they are only M-1 in $\underline{S}$. This redundancy of $\delta\underline{C}$ can be removed by adding one row to equation (37) where we define the redundant rotation $S_0 = \underline{C}_0^{\dagger}\, \delta\underline{C}$ and demand that $S_0 = 0$; by introducing an arbitrary, but non-zero, constant z:

$$\begin{bmatrix} z & 0 & 0 \\ 0 & \underline{\underline{C}}^{\dagger}\underline{\underline{A}}\,\underline{\underline{C}} & \underline{\underline{C}}^{\dagger}\underline{\underline{B}} \\ 0 & \underline{\underline{B}}^{\dagger}\underline{\underline{C}} & \underline{\underline{c}} \end{bmatrix} \begin{bmatrix} \underline{C}_0^{\dagger}\delta\underline{C} \\ \underline{\underline{C}}^{\dagger}\delta\underline{C} \\ \underline{T} \end{bmatrix} = - \begin{bmatrix} 0 \\ \underline{\underline{C}}^{\dagger}\underline{G} \\ \underline{W} \end{bmatrix} \tag{37a}$$

This equation is now multiplied with the unitary rotation

$$\begin{bmatrix} \underline{\underline{U}} & 0 \\ 0 & 1 \end{bmatrix}$$

where $\underline{\underline{U}}$ is the MxM matrix $(\underline{C}_0, \underline{\underline{C}})$ which contains all the CI vectors. Introducing the projection matrices $\underline{\underline{P}} = \underline{C}_0\underline{C}_0^{\dagger}$ and $\underline{\underline{Q}} = \underline{\underline{1}} - \underline{C}_0\underline{C}_0^{\dagger} = \underline{\underline{C}}\,\underline{\underline{C}}^{\dagger}$ we obtain the final equation system

$$\begin{bmatrix} z\underline{\underline{P}} + \underline{\underline{Q}}\,\underline{\underline{A}}\,\underline{\underline{Q}} & \underline{\underline{Q}}\,\underline{\underline{B}} \\ \underline{\underline{B}}^{\dagger}\underline{\underline{Q}} & \underline{\underline{c}} \end{bmatrix} \begin{bmatrix} \delta\underline{C} \\ \underline{T} \end{bmatrix} = - \begin{bmatrix} \underline{\underline{Q}}\,\underline{G} \\ \underline{W} \end{bmatrix} \tag{38}$$

From this equation we can determine the changes in the CI coefficients, $\delta\underline{C}$, and the orbital rotations $\underline{T}$. The updated coefficients are obtained from $\delta\underline{C}$ as

$$\underline{C}_0' = \underline{C}_0 + \delta\underline{C} - \frac{1}{2}(\delta\underline{C}^{\dagger}\delta\underline{C})\,\underline{C}_0 \tag{39}$$

where we have used the fact that $\delta\underline{C} = \underline{\underline{C}}\,\underline{S}$ and employed the exponential transformation through the second order.

3.4. Computational Aspects on the NR procedure.

When setting up methods to solve the linear equation system (38) or (36) one must be aware of the dimension of the problem. A moderately big example is M=2000 and n=50 (the number of orbitals). Assuming no symmetry the dimension of $\underline{T}$ is then 1275, so the total dimension of the problem becomes 3275. It is not practical to set up an entire Hessian of this dimension. Instead we should look for iterative algorithms were the correction vector $\underline{\lambda}^{(i)}$ in iteration i is obtained from $\underline{\lambda}^{(i-1)}$ by multiplication with the Hessian $\underline{\underline{h}}$. In such a method we can apply direct CI methods and construct $\underline{\underline{h}}\underline{\lambda}$ directly from our list of one- and two-electron integrals together with first and second order density matrix elements. An iterative procedure of this kind, which seems to work well (i.e. converge fast) for the solution of NR type equations, has been designed by Pople et al. /19/, and will be briefly described below in a slightly modified form.

Write the linear equation system in the form

$$(\underline{\underline{A}}_0 - \underline{\underline{A}})\ \underline{x} - \underline{b} = 0 \tag{40}$$

where $\underline{\underline{A}}_0$ is a diagonal non-singular matrix (eg. the diagonal part of the Hessian). Multiplying by $\underline{\underline{A}}_0^{-1}$ we can write (40) in the form

$$\underline{X} = \underline{\underline{A}}_0^{-1}\ \underline{\underline{A}}\ \underline{X} + \underline{\underline{A}}_0^{-1}\ \underline{b} \tag{41}$$

It is easy to see that the solution to this equation can be given as a power series expansion

$$\underline{X} = \sum_{n=0}^{\infty} (\underline{\underline{A}}')^n \underline{b}$$

where $\underline{\underline{A}}' = \underline{\underline{A}}\ \underline{\underline{A}}_0^{-1}$.

Direct use of this equation is not advisable, however, since it is only slowly convergent. A much more efficient procedure is obtained by constructing a set of orthogonal vectors $\underline{X}_n$ as

$$\underline{X}_{n+1} = \underline{\underline{A}}'\underline{X}_n - \sum_{\ell=0}^{n} \frac{\underline{X}_\ell^\dagger \underline{\underline{A}}'\ \underline{X}_n}{\underline{X}_\ell^\dagger\ \underline{X}_\ell}\ \underline{X}_\ell \tag{42}$$

An improved vector $\underline{X}$ can be obtained by taking

$$\underline{X} = \sum_{i=0}^{k} \alpha_i\ \underline{X}_i \tag{43}$$

The parameters α_i are obtained by inserting (43) into (41) and project onto each of the $\underline{X}_n$. The process is claimed to converge in four to six iterations for the problems considered by Pople et al. /19/. Such good convergence can probably not always be obtained in solving the NR equations, especially not in near-redundant situations where the Hessian matrix has small eigenvalues. The fundamental operation to be performed, in the procedure outlined above, is the multiplication of $\underline{X}$ with $\underline{\underline{A}}'$ ($\underline{Y}=\underline{\underline{A}}'\underline{X}$) with components

$$Y_i^{(n+1)} = \sum_k A_{ik}\ X_k^{(n)}/(A_0)_{kk} \tag{44}$$

If $\underline{\underline{A}}$ is the non-diagonal part of the Hessian in (38), we obtain for the CI part of $\underline{Y}$

$$Y_{\mu}^{(n+1)} = \sum_{\nu \neq \mu} H_{\mu\nu} \, \delta C_{\nu}^{(n)} / (H_{\nu\nu} - E_0), \tag{45}$$

which is a formula equivalent to that used in solving the secular problem in the direct CI procedure (eg. when using the Davidson algorithm). Lengsfield /15/ has derived detailed expressions for the calculation of such update vectors both for the CI and the coupling part of the Hessian matrix.

The convergence of the NR procedure is only second order when the starting vector is already in the local area, that is, rather close to the minimum. For a ground state calculation a necessary condition is that the Hessian is positive definite, that is, it has only positive eigenvalues. For the n-th excited state the number of negative eigenvalues must not exceed (n-1). Quite frequently this is not the case. A straightforward application of the NR procedure will then diverge, or converge to a non-desired solution. In order to deal with these problems Yeager et al. /5/ have devised a step size and sign control method which they claim always assures convergence to the desired stationary point. We shall not describe their procedure in detail here. Essentially the sign control is obtained by reversing the sign of those rotations which corresponds to undesired negative eigenvalues of the Hessian, while the step size control consists in reducing the amplitudes of the rotations to some predetermined empirical limits. Their method has the disadvantage of requiring an a priori search for all negative eigenvalues of the Hessian matrix, a task which might not be easily undertaken in large calculations. However, it should be emphasized that knowledge about the eigenvalues of the Hessian gives valuable information about the stability of the calculation. Therefore a search for the n lowest roots of the Hessian at the end of a calculation on the n-th excited state is advisable.

One method which avoids the problem with undesired negative eigenvalues of the Hessian and also introduces an automatic damping of the rotations is the augmented Hessian (AM) method /20/. Consider again the NR equations (17),

$$\underline{\underline{h}}\, \underline{\lambda} = -\underline{g} \tag{46}$$

In the AM method one solves, instead of (46) the secular problem,

$$\begin{bmatrix} 0 & \underline{g}^{\dagger} \\ \underline{g} & \underline{\underline{h}} \end{bmatrix} \begin{bmatrix} 1 \\ \underline{\lambda} \end{bmatrix} = E \begin{bmatrix} 1 \\ \underline{\lambda} \end{bmatrix} \tag{47}$$

In order to relate (47) and (46) it is convenient to transform both equations to a basis where $\underline{\underline{h}}$ is diagonal. Thus defining the diagonal matrix $\underline{\underline{\varepsilon}} = \underline{\underline{U}}^{\dagger}\underline{\underline{h}}\ \underline{\underline{U}}$; $\underline{g}' = \underline{\underline{U}}^{\dagger}\underline{g}$; and $\underline{\lambda}' = \underline{\underline{U}}^{\dagger}\underline{\lambda}$ we obtain from equation (46)

$$\lambda_i' = - g_i'/\varepsilon_i, \tag{48a}$$

while the solution of (47) yields

$$\lambda_i' = - g_i'/(\varepsilon_i - E_i) \tag{48b}$$

According to MacDonald's theorem /21/, the eigenvalues E_i satisfy a betweenness condition: $\varepsilon_{i-1} \leqslant E_i \leqslant \varepsilon_i$ for all i. For the lowest state we have especially $E_1 \leqslant \varepsilon_1$, where ε_1 is the smallest eigenvalue of the Hessian. Equation (4 b) therefore automatically includes a sign control, since $\varepsilon_i - E_i$ is always positive even if ε_i is negative. This feature of the AM method is especially attractive in determining MCSCF wave functions for higher roots. At convergence the two sets of rotation parameters (48a) and (48b) become identical since E_i goes to zero with $\underline{g}$.

Before ending this section a few words about redundant parameters. It may happen that the set of variation parameters in the MC wave function is linearly dependent. When such a situation obtains, the Hessian will be singular and the NR equations lead to indefinite values for the rotations. A trivial example of a redundant set of parameters is the orbital rotation T_{ij}, where both indices correspond to inactive (or secondary) orbitals. The energy gradient $\partial E/ T_{ij}$ is in this case automatically zero since the wave function is invariant to such rotations. Another example is given by the CASSCF wave function. Here the energy (but not the wave function) is also invariant to first order against internal rotations of orbitals in the active subspace (T_{tu}) due to the identity,

$$\langle 0|\ \hat{H}\ \hat{E}^{-}_{tu}\ |0\rangle \equiv 0$$

It is easy to prove that also in this case will the total Hessian in (36) become singular. Redundant parameters must not be included in the NR procedure. They are trivial to exclude for the examples given above, but in more general cases a redundant variable might occur as a linear combination of $\underline{S}$ and $\underline{T}$. In such cases the Hessian will only be singular at convergence and it might be difficult to exclude them. The CASSCF method has the advantage that all parameters except those given above are non-redundant.

4. APPROXIMATIONS TO THE GENERAL SECOND ORDER MCSCF PROCEDURE

The fully coupled NR method as it is given by equation (36) or (38) shows a quadratic convergence behaviour in the local region, and is in this respect very attractive. Also satisfactory global convergence can normally be obtained by using mode damping, step size and sign control, or the augmented Hessian method. Convergence can also be improved, when far from the local region, by including cubic terms in the NR equations /4,5/. The full NR method is, however, computationally complex and is, in spite of the large scale formulation given by Lengsfield /15/ limited in size to a few thousand terms in the Hessian. Especially the calculation of the coupling matrix $\underline{\underline{B}}$ in (38) becomes time consuming for a large orbital space together with a long CI expansion. It is therefore natural to look for simplifications of the NR equations which avoid some of the computational bottlenecks of the full second order procedure.

The direct solution of equation (36) (or (38)) is in the MCSCF language called a "one-step" procedure, since both $\underline{S}$ and $\underline{T}$ are obtained simultaneously. We can formally rewrite (36) as a "two-step" procedure by solving for $\underline{S}$ first. From the first row of (36) we obtain

$$\underline{S} = - \underline{\underline{a}}^{-1}\,\underline{v} - \underline{\underline{a}}^{-1}\underline{\underline{b}}\,\underline{T} \tag{49a}$$

which inserted into the second row yields

$$(\underline{\underline{c}} - \underline{\underline{b}}^{\dagger}\underline{\underline{a}}^{-1}\underline{\underline{b}})\underline{T} = - \underline{w} - \underline{\underline{b}}^{\dagger}\underline{\underline{a}}^{-1}\underline{v} \tag{49b}$$

The matrix $\underline{\underline{c}} - \underline{b}^{\dagger}\underline{\underline{a}}^{-1}\underline{b}$ is the partitioned orbital Hessian.

Equation (49) is not very practical to use as it stands, since it involves the inverse of the CI part of the Hessian. Suppose, however, that we work in a configuration basis ($|0\rangle, |K\rangle$) where $\underline{\underline{a}}$ is diagonal, that is, we start each iteration by solving the CI problem to all orders. The matrix $\underline{\underline{a}}$ then has the elements $(E_K-E_0)\delta_{KL}$ and the corresponding gradient vector $\underline{v}$ with elements $\langle 0|\hat{H}|K\rangle$ is zero. Equation (49b) can then be written in the form

$$\sum_{(rs)} \left(c_{pq,rs} + \sum_{k} \frac{b_{pq,K}\, b_{K,rs}}{E_0-E_K}\right) T_{rs} = -w_{pq}. \tag{50}$$

This equation forms the basis for the "two-step" NR method. It is obviously not a very practical equation to use in connection with large MC expansion, since we can in this case not easily obtain all the eigenvalues E_K. Attempts have therefore been made to truncate the summation in (50) to the lowest roots, with the inherent assumption that terms with small values in the denominator will give the largest contribution to the coupling term /7/. It turns out, however, that the convergence of the sum over the coupling terms is slow. It is therefore not very useful in practice to include the coupling between the orbital and CI rotations by means of equation (50), except in special cases with strong couplings between a few MC states, a situation which might obtain in calculations on excited states.

A commonly used approximation is to neglect the coupling term altogether. This leads to the unfolded two-step NR procedure, where the equations to be solved are

$$(\underline{\underline{H}} - E_0\underline{\underline{1}})\ \underline{C} = 0 \tag{51}$$

and

$$\underline{\underline{c}}\ \underline{T} = -\underline{w} \tag{52}$$

Even though this method is not quadratically convergent, it is frequently used in practical applications due to its much simpler form than the general second order approach. The secular problem (51) can be solved for very long MC expansions by means of efficient direct CI procedures /7/, especially for CAS-type wave functions. The orbital Hessian $\underline{\underline{c}}$ can be expressed in terms of first and second order reduced density matrix elements which are readily obtained from the MC wave function. Detaild expressions for $\underline{\underline{c}}$ have been given by Siegbahn et al. /7/. We shall not give them all here, but consider as a typical example $c_{ta,ub}$, where as usual t,u,v, and x labels active orbitals and a and b secondary orbitals,

$$c_{ta,ub} = 2 \sum_{v,x} \{P_{tuvx}(ab|vx) + (P_{txvu}+P_{txuv})(ax|bv)\} + $$
$$+ D_{tu}\, F^{I}_{ab} - \delta_{ab}\, F_{tu}, \tag{53}$$

where the interaction with the inactive subspace has been included in a Fock type matrix $\underline{\underline{F}}^{I}$ with elements

$$F^{I}_{pq} = h_{pq} + \sum_{i} [2(pq|ii)-(pi|qi)], \tag{54}$$

and a large contribution to the matrix element is collected into the MCSCF Fock operator

$$F_{pq} = \sum_{r} D_{pr}h_{qr} + 2 \sum_{r,s,t} P_{prst}\ (qr|st). \tag{55}$$

The calculation of (53) and the similar expressions for the other types of matrix elements appearing in $\underline{\underline{c}}$ is easily accomplished by one sequential read of the list of two-electron integrals, since the first and second order density matrix elements $\underline{\underline{D}}$ and $\underline{\underline{P}}$ can be held in the central memory of the computer. It is actually possible to set up an algorithm which computes the sum

$$Y_{ta} = \sum_{u,b} c_{ta,ub} T_{ub}/c_{ub,ub} \tag{56}$$

directly from the list of molecular integrals. Y_{ta} is the update vector needed to solve the linear equation system according to Pople's algorithm (cf. equations (44) and (45)). Such a direct NR algorithm for the orbital rotations has to my knowledge not yet been implemented. It has the advantage of making the program open endend with respect to the size of the orbital basis. Present implementations seem to be limited to about 60-80 basis functions, depending on the symmetry of the system.

One of the major computational problems in most applications is the need to compute the molecular two-electron integrals occurring in (53). In the NR procedure we need all integrals with two indices running over all orbitals and the remaining two indices restricted to the primary (inactive plus active) subspace. The number of operations needed to obtain these integrals in a transformation from the basic atomic integrals is proportional to $n_p \cdot n^4$, where n_p is the size of the primary subspace. Since n_p can be quite large for systems with many electrons, this transformation often becomes the most time-consuming part of an MCSCF iteration. An efficient algorithm for the transformation is therefore of importance. It is, for example, a great advantage to have the basic AO integrals computed over symmetry adapted orbitals /22/. The transformation can then be divided into symmetry blocks, resulting in considerable reductions in computer time (provided there is some symmetry in the system under study). Symmetry blocks in which there are no occupied orbitals are not needed and the corresponding primitive integrals can be sorted out before the transformation starts.

A method which can be used to further reduce the transformation time has recently been suggested by Olsen and Yeager /5,18/. It is not possible to describe this method in detail here, so the reader is referred to the references given above. In essence the simplification is achieved by introducing a transformation of the orbital rotations T_{pq} such that the MO coefficient matrix becomes diagonal in the inactive and secondary parts. The number of terms to be summed in the integral transformation is consequently reduced substantially. So far the method has not been tested in large scale applications, however.

5. THE SUPER-CI METHOD

A different approach to the MCSCF optimization problem was taken by Grein and coworkers /14/ in the early 70's. Their idea was to reach the optimal solution by annihilating the single excitations (the Brillouin states) in an iterative procedure. A slightly modified form of the method has been presented by Ruedenberg et al. /14/, and an approximate form, which has proven to be very effective in large scale applications, has been developed by the present author /7,23/. We shall now describe this method and illustrate its similarities with and deviations from the NR approach. The method will first be described in the unfolded two-step form. The extension to a folded one-step procedure will be indicated. We therefore assume that every MCSCF iteration starts by solving the secular problem (34) with the consequence that the MC reference state does not interact with the complementary space (1b). Now define a set of singly excited (SX) states according to

$$|pq\rangle = \hat{E}^{-}_{pq}|0\rangle \tag{57}$$

These states are called Brillouin states, since their interaction with the reference state gives the generalized Brillouin matrix elements (33b). A "Super-CI" wave function is defined as a linear combination of the SX states (57) and the reference state,

$$|SCI\rangle = |0\rangle + \sum_{p>q} t_{pq}|pq\rangle. \tag{58}$$

The super-CI method now implies solving the corresponding secular problem and using t_{pq} as the exponential parameters for the orbital rotations. Alternatively we can construct the first order density matrix corresponding to the wave function (58), diagonalize it, and use the natural orbitals as the new orbitals in $|0\rangle$. Both methods incorporate the effects of $|pq\rangle$ into $|0\rangle$ to second order in t_{pq}. We can therefore expect the magnitudes of t_{pq} to decrease in the next iteration. At convergence all t_{pq} vanish, which is equivalent to the condition

$$\langle 0|\hat{H}|pq\rangle = 0 \tag{59}$$

The super-CI method thus leads to the stationary point by a direct annihiliation of the single excitations (57).

We shall now study the secular equation in some detail in order to make a comparison between super-CI and the NR method in the augmented Hessian form. The first thing to note is the non-orthogonality between the SX states:

$$S_{pq,rs} = -\langle 0|\hat{E}^{-}_{pq}\hat{E}^{-}_{rs}|0\rangle \neq \delta_{pq,rs}. \tag{60}$$

They are not even normalized ($S_{pq,pq} \neq 1$), but they are orthogonal to the reference state, due to the trivial identity

$$\langle 0|\hat{E}^-_{pq}|0\rangle \equiv 0 \tag{61}$$

The matrix elements of $\hat{H}$ have the form

$$H_{0,pq} = \langle 0|\hat{H}|pq\rangle = \langle 0|\hat{H}\,\hat{E}^-_{pq}|0\rangle = w_{pq}, \tag{62a}$$

$$H_{pq,rs} = \langle pq|\hat{H}|rs\rangle = -\langle 0|\hat{E}^-_{pq}\hat{H}\,\hat{E}^-_{rs}|0\rangle = d_{pq,rs} \tag{62b}$$

where w_{pq} is a component of the gradient vector $\underline{w}$ in (36). The matrix elements $d_{pq,rs}$ are related to the orbital Hessian $\underline{\underline{c}}$ in (36) as will be demonstrated below. Using the matrix elements (60)-(62) the secular equation can now be written in the form

$$\begin{bmatrix} 0 & \underline{w}^\dagger \\ \underline{w} & \underline{\underline{d}}-E_0\underline{\underline{S}} \end{bmatrix} \begin{bmatrix} 1 \\ \underline{t} \end{bmatrix} = \varepsilon_{SCI} \begin{bmatrix} 1 & 0 \\ 0 & \underline{\underline{S}} \end{bmatrix} \begin{bmatrix} 1 \\ \underline{t} \end{bmatrix} \tag{63}$$

where $\varepsilon_{SCI} = E_{SCI} - E_0$, E_{SCI} being the super-CI energy:

$$E_{SCI} = \langle SCI|\hat{H}|SCI\rangle / \langle SCI|SCI\rangle \tag{64}$$

Equation (63) can be compared to the unfolded two-step version of the augmented Hessian method, which results in the secular equation

$$\begin{bmatrix} 0 & \underline{w}^\dagger \\ \underline{w} & \underline{\underline{c}} \end{bmatrix} \begin{bmatrix} 1 \\ \underline{T} \end{bmatrix} = \varepsilon_{AM} \begin{bmatrix} 1 \\ \underline{T} \end{bmatrix} \tag{65}$$

where $\underline{\underline{c}}$ is the orbital Hessian with elements (compare eq. (35b))

$$c_{pq,rs} = \tfrac{1}{2}\langle 0|\hat{E}^-_{pq}\,\hat{E}^-_{rs}\,\hat{H}|0\rangle + \tfrac{1}{2}\langle 0|\hat{H}\,\hat{E}^-_{pq}\,\hat{E}^-_{rs}|0\rangle$$

$$-\langle 0|\hat{E}^-_{pq}\,\hat{H}\,\hat{E}^-_{rs}|0\rangle \tag{66}$$

Comparing equations (63) and (65) we note two differences: the occurence of the overlap matrix in (63) and the replacement of $\underline{\underline{c}}$ by $\underline{\underline{d}}-E_0\underline{\underline{S}}$. Let us investigate the last difference in more detail:

$$d_{pq,rs} - E_0\,S_{pq,rs} = -\langle 0|\hat{E}^-_{pq}\hat{H}\,\hat{E}^-_{rs}|0\rangle + E_0\langle 0|\hat{E}^-_{pq}\hat{E}^-_{rs}|0\rangle \tag{67}$$

We can write $\hat{H}|0\rangle$ occurring in (66) as

$$\hat{H}|0\rangle = E_0|0\rangle + |\perp\rangle, \qquad (68)$$

where $|\perp\rangle$ represents the orthogonal complement to the CSF space $\{|0\rangle,|K\rangle\}$. In the CAS case $|\perp\rangle$ is a linear combination of configurations with one or two electrons in the secondary subspace and/or the number of inactive electrons decreased by one or two. For an incomplete reference space $|\perp\rangle$ also contains the difference between this space and the corresponding CAS space. Now, inserting (68) into (66) we obtain

$$c_{pq,rs} = d_{pq,rs} + \tfrac{1}{2}\langle 0|\hat{E}^-_{pq}\hat{E}^-_{rs} + \hat{E}^-_{rs}\hat{E}^-_{pq}|\perp\rangle. \qquad (69)$$

The difference between $\underline{\underline{c}}$ and $\underline{\underline{d}}$ thus involves the overlap between the orthogonal complement $|\perp\rangle$ and the double excitations

$$\hat{E}^-_{pq}\hat{E}^-_{rs}\;|0\rangle$$

The AM method is obtained from the super-CI equations by replacing $\underline{\underline{d}}-E_0\underline{\underline{S}}$ by $\underline{\underline{c}}$ and neglecting $\underline{\underline{S}}$ on the right hand side of (63), and it has therefore been regarded as an approximation to the super-CI method /24/. It is maybe more appropriate to regard the AM method as a mode-damped version of the NR approach, while super-CI is an approximation to the AM method. Both AM and super-CI, have the advantage of always converging to a local minimum (ε_{SCI} is always negative), in contrast to the NR method, which without sign and step size control can converge to a saddle point or a maximum. AM and super-CI therefore generally show better global convergence behaviour than NR and is often recommended to be used in the beginning of a calculation, until the local region is reached. In the local region, on the other hand the best convergence is obtained with the folded one-step NR procedure without any mode damping or step size control. The super-CI method does not exhibit true second order convergence, due to the approximations demonstrated above.

The super-CI method can also be given in a folded form, which includes the coupling between the CI and orbital rotations. This is done by including also the complementary space $|K\rangle$, (1b), into the super-CI secular problem. As in the NR approach it is more efficient to transform the equations back to the original CSF space, and thus work with a super-CI consisting of all CSF's plus the SX states. It is left to the reader as an excercise to construct the corresponding secular equation and compare it with the corresponding folded one-step NR equations (38).

Computationally the super-CI method is more complicated to work with than the NR approach. In the original formulation of

Grein et al., and also in the work by Ruedenberg et al. /14/, the matrix elements were constructed in two steps. The first step consisted in generating all matrix elements in the space containing the CSF's of the reference space plus all corresponding singly excited CSF's These matrix elements were, in the second step, contracted to the much smaller list of matrix elements between the SX states (note that the SX states are linear combinations of singly excited CSF's). This is a complicated procedure which cannot easily be extended to large reference spaces and many orbitals. The density matrix formulation, which works so well for the NR method, is complicated by the need of third and fourth order density matrices in the formulas for the matrix $\underline{\underline{d}}$. In the case of a CAS wave function only third order density matrices are needed however /7/, but even this introduces a complication, and a direct implementation of the super-CI method can therefore not compete with the NR method in computational efficiency. Since it is not expected to show any better global convergence than the AM method, it cannot be regarded as a favourable alternative.

An approximate form of the super-CI method has, however, been developed /7,23/, which leads to significant simplifications of the secular equation and also greatly reduces the transformation problem for the two-electron integrals. The idea is to approximate the matrix elements of $\underline{\underline{d}}$ by replacing the Hamiltonian by an effective one-electron operator constructed from the MCSCF operator (55). It has been demonstrated in a large number of applications both on ground and excited states, that this method preserves the good convergence of the exact super-CI method in most cases. It leads to a computationally very efficient procedure, where the most important feature is that only two-electron integrals with three indices in the active subspace are needed, resulting in a substantial reduction of the time spent in the transformation section. A very small fraction of the total iteration time is spent in the super-CI section of the program, due to the very simple structure of the secular equation. The solution of the MC secular equation therefore dominates the iteration time for large MC expansions, while the transformation time (still) dominates in the case of a large orbital space. CASSCF calculations using this method has been performed on systems described by more than one hundred basis functions and up to 30000 CSF's in the reference space.

6. SUMMARY AND CONCLUSIONS

This lecture note has discussed some of the basic features of the modern MCSCF techniques. Due to the limited space and time available the survey of the area is far from complete. A number of interesting aspects of the theory have been deleted or just mentioned in passing. Among these are a detailed discussion of the convergence behaviour for the different methods and methods used to improve convergence. The reader is referred to the extensive review by Olsen, Yeager, and Jörgensen for a more detailed treatment of this topic /5/. The integral transformation constitutes a major part of an MCSCF calculation and is therefore worthy of a detailed discussion in this context, with special emphasis on the problems connected with the partial transformation needed for MCSCF. The transformation problem will, however, be discussed in another lecture at this Institute and the reader is referred to the corresponding chapter of these notes for more details on the transformation.

The recent development of the MCSCF techniques has today reached a state, where large scale applications are possible and have to some extent become routine. We will therefore in the near future reach a stage where MCSCF will replace SCF as a routin tool for obtaining zeroth order (and in some cases more than that) approximations to the wave function for many chemically interesting molecular processes. It should, however, be pointed out, that MCSCF is much less of a black box than SCF. The problems connected with the choice of a proper active orbital subspace and an MC reference space are almost never trivial, and require understanding of the general features of molecular wave functions. We still have much to learn in this area, but on gaining this knowledge we will also gain a deeper understanding of what the electrons really are doing in the molecules.

REFERENCES

1. Frenkel, J., "Wave Mechanics, Volume 2: Advanced General Theory" (Clarendon, Oxford, 1934), pp. 460-462.

2. Hartree, D.R., Hartree, W. and Swirles, B., Phil.Trans.Roy.Soc. (London) A238, 229 (1939).

3. "Recent Developments and Applications of Multiconfiguration Hartree-Fock Methods" (NRCC Proceedings No. 10), Report LBL-12151, Lawrence Berkeley Laboratory, University of California, Berkeley (1981).

4. Levy, B., Thesis CNRS No. A05271, Paris (1971); Levy, B., Int.J. Quantum Chem. 4, 297 (1970); Kendrick, J. and Hillier, I,. Chem.Phys. Letters 41, 283 (1976); Yeager, D.L. and Jörgensen, P., J.Chem.Phys. 71, 755 (1979); Roothaan, C.C.J., Detrich, J. and Hopper, D.G., Int.J. Quantum Chem. S13, 93 (1979); Werner, H.-J. and Meyer, W., J.Chem.Phys. 73, 2342 (1980); Lengsfield, B.H., J.Chem.Phys. 73, 382 (1980).

5. For a review of the second order optimization techniques see: Olsen, J. and Yeager, D.L., Adv. in Chem.Phys. (in press).

6. Roos, B., Chem.Phys. Letters 15, 153 (1972); Roos, B.O. and Siegbahn, P., in: "Methods of Electronic Structure Theory" (ed. H.F. Schaefer III), Plenum Press (New York, 1977).

7. Roos, B.O., Taylor, P.R. and Siegbahn, P.E.M., Chem.Phys. 48, 157 (1980); Siegbahn, P.E.M., Heiberg, A., Roos, B.O. and Levy, B., Phys.Scr. 21, 323 (1980); Siegbahn, P.E.M., Almlöf, J., Heiberg, A. and Roos, B., J.Chem.Phys. 74, 2384 (1981).

8. A recent MCSCF study of Cr_2 yielded a wave function (for the $^1\Sigma_g^+$ ground state) comprising 3088 terms, where all CI coefficients were in the range -0.09 to +0.09: Bagus, P.S., Nelin C., Roos, B.O. and Siegbahn, P., to be published.

9. M. Moshinsky, "Group Theory and the Many-Body Problem"(Gordon and Breach, New York, 1968).

10. Paldus, J. in: "Theoretical Chemistry: Advances and Perspectives" (eds. H. Eyring and D. Henderson), Vol. 2, (Academic Press, New York, 1976) p. 131.

11. Shavitt, I., Int.J. Quantum Chem. S11, 131 (1977) and S12, 5 (1978).

12. Wahl, A.C. and Das, G. in: "Methods of Electronic Structure Theory" (ed. H.F. Schaefer III), Plenum Press (New York, 1977); Hinze, J., J.Chem.Phys. 59, 6424 (1973); Hinze, J., Int.J. Quantum Chem. S15, 69 (1981).

13. Levy, B. and Berthier, G., Int.J. Quantum Chem. 2, 307 (1968).

14. Grein, F. and Chang, T.C., Chem.Phys. Letters 12, 44 (1977); Grein, F. and Banerjee, T.C., Int.J.Quantum Chem. S9, 147 (1975); J.Chem.Phys. 66, 1054 (1977); Cheung, L.M., Elbert, S.T. and Ruedenberg, K., Int.J. Quantum Chem. 14, 1069 (1979).

15. Lengsfield III, B.H., J.Chem.Phys. (in press); Lengsfield III, B.H. and Liu, B., J.Chem.Phys. 75, 478 (1981).

16. Yaffe, L. and Goddard III, W.A., Phys.Rev. A13, 1682 (1976).

17. Dalgaard, E. and Jörgensen, P., J.Chem.Phys. 69, 3833 (1978).

18. Olsen, J. and Yeager, D.L., preprint.

19. Pople, J.A., Krishnan, R., Schlegel, H.B. and Binkley, J.S., Int.J. Quantum Chem. S13, 225 (1979).

20. Yarkony, D.R., Chem.Phys. Letters 77, 634 (1981).

21. MacDonald, J.K.L., Phys.Rev. 43, 830 (1933).

22. AO integrals over symmetry orbitals are for example produced by the MOLECULE program system: J. Almlöf, Report 74-29 (December 1974), University of Stockholm, Institute of Physics, Stockholm, Sweden.

23. Roos, B., Int.J. Quantum Chem. S14, 175 (1980).

24. Shepard, R., Shavitt, I. and Simons, J., J.Chem.Phys. 76, 543 (1982).

THE DIRECT CI METHOD

Per E.M. Siegbahn

Institute of Theoretical Physics
University of Stockholm
Vanadisvägen 9
S–113 46 Stockholm
Sweden

ABSTRACT

The general direct CI method for a configuration expansion consisting of all single and double replacements from a leading set of reference states is described. The strategy involves the initial construction of a formula tape for the internal space with a subsequent direct CI treatment of the external space, where the internal formulas are processed parallel to the molecular integrals. The recent development of this method is reviewed including a matrix formulation, particularly efficient for array-processing, and the internal spin driven scheme. Within the same general direct CI framework some approximate methods are also described. These methods include the internally and externally contracted CI methods. The externally contracted CI method is shown to correspond to the work of one half of an ordinary CI iteration in the large basis set limit where the overhead is negligible.

G. H. F. Diercksen and S. Wilson (eds.), Methods in Computational Molecular Physics, 189–207.

I. INTRODUCTION

The total energy for an atomic or molecular system is usually divided into two parts. One part is the SCF energy which is obtained from a wave function consisting of a single configuration and the other part is the correlation energy. In many situations it is further convenient to subdivide the correlation energy into two parts with different physical origins. For chemical reactions where bonds are broken and formed, and for most excited states, the major part of the correlation energy can be obtained by adding only a few extra configurations besides the Hartree-Fock configuration. This part of the correlation energy is due to near degeneracy between different configurations and has its origin quite often in artefacts of the Hartree-Fock approximation. The physical origin of the second part of the correlation energy is the dynamical correlation of the motion of the electrons and is therefore sometimes called the dynamical correlation energy. Since the Hamiltonian operator contains only one- and two-particle operators this part of the correlation energy can be very well described by single and double replacements from the leading, near degenerate, reference configurations. It is this type of general CI-expansion which will be discussed in the present chapter.

The conventional CI-method is described in a different chapter of this book and the main characteristica of this method will therefore be only briefly described here. As a starting point the integrals over the basis functions are computed and transformed to a chosen molecular orbital basis. The Hamiltonian matrix elements between the selected configurations then have to be constructed. Since the configurations are generally spin adapted, which means that each configuration is a linear combination of several determinants with spin coupling coefficients, these matrix elements can be quite complicated. For this reason the formulas for the matrix elements are usually constructed once and for all and stored on peripheral storage. This data set contains the sequence numbers of the configurations, the indices of the integrals and the spin coupling coefficients, and is usually termed the symbolic formula tape. From the symbolic formula tape and the molecular integrals, numerical values for the Hamiltonian matrix elements can now be obtained and the matrix diagonalized by the use of iterative techniques. The number of Hamiltonian matrix elements can easily be very large since it grows as the square of the number of configurations. With 10^4 configurations and with only 10 % of the matrix elements different from zero, the number of matrix elements if of the order of 10^7, which starts to become a storage and retrieval problem even on large computers. The idea behind the direct CI-method is to avoid the construction of the Hamiltonian matrix and thereby remove the major bottle neck for very large CI-expansions. The term "direct" comes from the fact that wave-functions are constructed directly from the molecular integrals without constructing an intermediate Hamiltonian matrix. This is obviously possible in principle since each Hamiltonian matrix element is just a linear combination of molecular integrals. The problem is to do it efficiently for general CI-expansions, which is required to solve realistic chemical problems.

The direct CI-method was introduced for the first time in 1972 by Roos [1] for a configuration expansion consisting of all single and double replacements from a closed shell configuration. All configurations were properly spin-adapted. In the following years the method was extended to treat also the case of all single and double replacements from an unrestricted Hartree-Fock determinant and complete CI-expansions for three electrons [2]. For a review of the earlier work on the direct CI-method, see Ref. 3. A few years later a generalization to an arbitrary set of closed shell reference configurations was made [4]. Following the same lines as the work of Lucchese and Schaefer for triplets [5], Handy et al. developed a direct CI-method applicable to CI-expansions where the reference configuration can be a high spin determinant or an open shell two determinant function [6]. Despite this important progress the complete generalization of the direct CI-method to any number and type of reference configurations seemed like an almost impossible problem. The solution of this problem was the realization that a CI-calculation can be divided into two independent parts. In the first part the small internal orbital space, which defines the reference configurations is treated. This is usually a short step and results in a formula tape, which is used in the second step where the large external orbital space is treated. The structure of the latter step is entirely independent of the number and type of reference configurations and is therefore very similar to the original direct CI-method for one closed shell reference configuration. The main difference is the use of the short symbolic formula tape. For the formula generation conventional CI-methods can be used and the resulting method is therefore a hybrid between conventional and direct CI-methods. This is the essential features behind the method which was first presented in Ref. 7. In that paper the powerful unitary group language as developed particularly by Paldus [8] and Shavitt [9] was used. The insight into the complicated structure of the general CI-problem, as provided by the graphical unitary group approach, was important for the earlier development of the method but is actually not necessary in principle. Once the method is outlined it is therefore possible, and sometimes even advantageous [10], to go back to the more familiar formulation in terms of determinants and spin- and space-coupling coefficients. In the present chapter the unitary group language will be used however, and some of the most essential features of this language is therefore briefly described in the next section.

Since the method in Ref. 7 was presented a large number of papers have been devoted to methods which are very similar in general structure. Further generalizations and important improvements have therefore appeared. Recently, for example, Saunders and van Lenthe [11] presented a detailed analysis of the general direct CI-method. This method is probably the most efficient implementation of the direct CI-method to this date, particularly for vector processors, but also for purely scalar machines. To obtain maximum vectorization they combined the general strategy of Ref. 7 with the elegant matrix formulation of the SCEP-method as developed particularly by Meyer [12] and Ahlrichs and Driessler [13]. Other new features are also presented in Ref. 11 and these will be described in the coming sections. The strategy of the method implemented by Liu and Yoshimine [10] is also very similar to the method in Ref. 7 but their method is more general, particularly in the use of point group symmetry. Linear and even atomic symmetry can be fully used. Also the choice of spin-coupling of the configurations is completely general, which could be important for the optimal use of the interacting space concept [14]. Further details of this method will not be discussed here since it is not yet published. Other recent papers with important contributions to the general direct CI-method with the same strategy as in Ref. 7, have been presented by Lischka et al. [15], Duch and Karwowski [16] and Ruttink and van Schaik [18].

Another area which will be briefly discussed in the present chapter is approximate CI-methods. Even though configuration expansions of the order of 10^6 can be treated by the general direct CI-method the calculations will be quite time consuming, particularly on smaller machines like mini-computers. The so called contracted CI-methods are approximate methods with a strategy which is very similar to the above described general direct CI-strategy. There are two types of contracted methods. In the internally contracted CI-method the number of variational parameters is reduced by a contraction in the internal orbital space. This method has been discussed by Meyer [19], by the present author [20] and by Werner and Reinsch [21]. When the number of variational parameters is reduced by a contraction in the external orbital space the method is termed externally contracted and is presented in detail in Ref. 22. Approximate CI-methods based on configuration selection have been very successful in connection with the conventional CI-method [23]. Possibilities to incorporate configuration selection schemes into the general direct CI-strategy will therefore also finally be discussed in this chapter.

II. GENERAL FORMALISM

In this section there will be a brief description of the necessary concepts of the unitary group formalism. It should be emphasized again that the unitary group formalism is not necessary for the method, but very convenient in many respects for giving proper definitions and for showing the structure of the complex general CI-problem in a compact form. The description given here will by necessity be very short and the reader is referred to the articles by Paldus [8] and by Shavitt [9] for details.

Since the idea in the direct CI-method is to avoid all references to explicit Hamiltonian matrix elements and instead refer directly to the integrals, the Hamiltonian operator is best written in second quantized form, which after spin summation becomes

$$\hat{H} = \sum_{pq} (p|\hat{h}|q)\,\hat{E}_{pq} + \frac{1}{2}\sum_{pqrs} (pq|rs)\,(\hat{E}_{pq}\hat{E}_{rs} - \delta_{qr}\hat{E}_{ps})$$

where p, q, r, s are space orbitals. The generators of the unitary group are consequently defined as

$$\hat{E}_{pq} = \sum_{\sigma} \hat{X}^{+}_{p\sigma}\hat{X}_{q\sigma}$$

where $p\sigma$, $q\sigma$ are spin orbitals and $\hat{X}$ are the usual annihilation- and creation-operators. From the commutation relations defined for $\hat{X}$ the following commutation relation can be derived for $\hat{E}$,

$$\hat{E}_{pq}\hat{E}_{rs} - \hat{E}_{rs}\hat{E}_{pq} = \delta_{qr}\hat{E}_{ps} - \delta_{ps}\hat{E}_{rq}$$

which defines the algebra for these operators. In the CI-problem finally, the Hamiltonian matrix is diagonalized which is done either by an iterative method or by perturbation theory. In both these cases the main computational step in each iteration is the construction of a vector σ as,

$$\sigma_{\mu} = \sum_{\nu} H_{\mu\nu} C_{\nu}$$

where μ, ν are labels for the chosen configuration basis and C is the wave-function from the previous iteration. Combining the above form of the Hamiltonian operator with the expression for σ the main working equation in the direct CI-method is reached. This equation can consequently be written

$$\sigma_{\mu} = \sum_{\nu}\sum_{pq} A^{\mu\nu}_{pq}\,(p|\hat{h}|q)C_{\nu} + \sum_{\nu}\sum_{pqrs} A^{\mu\nu}_{pqrs}\,(pq|rs)\,C_{\nu} \tag{1}$$

where

$$A^{\mu\nu}_{pq} = \langle \mu | \hat{E}_{pq} | \nu \rangle$$
$$A^{\mu\nu}_{pqrs} = \frac{1}{2} \langle \mu | \hat{E}_{pq} \hat{E}_{rs} - \delta_{qr} \hat{E}_{ps} | \nu \rangle \tag{2}$$

are the so called direct CI-coupling coefficients. In order to evaluate these coupling coefficients a spin-coupling scheme for the N-electron spin-adapted basis Φ_μ must be defined. For the type of wave function which will be discussed here, namely single and double replacements from a set of reference functions, the simplest form of the coupling coefficients A is obtained if the external orbitals are spin-coupled first. For configurations containing only one external orbital such a spin-adapted configuration can be written simply as (see also Ref. 17 and 19)

$$\Phi_\mu = \Phi^a_\varkappa = [\Phi_\varkappa(N-1)\,\varphi_a] \tag{3}$$

and with two occupied external orbitals as

$$\Phi_\mu = \Phi^{ab}_\varkappa = [\Phi_\varkappa(N-2)\cdot(2+2\delta_{ab})^{-1/2}(\varphi_a\bar{\varphi}_b + p\varphi_b\bar{\varphi}_a)] \tag{4}$$

where p is +1 for singlet coupling and −1 for triplet coupling of a and b, and where the square bracket indicates proper total spin-coupling. If the internal functions $\Phi_\varkappa$, which contain no external orbitals, are assumed orthonormal and the external orbital set is orthonormalized the spin-adapted configuration basis will also be orthonormal.

A very convenient way of constructing a spin-adapted configuration basis which fulfils the above requirements is to use the Shavitt graph [9], and place the external orbitals at the bottom of the graph. The characteristica of these types of graphs will be briefly described here. The general multi-reference graph of the external space is given in Fig. 1, and the graph of a special internal space is given in Fig. 2. A path from the bottom of the external graph to the top of the internal graph describes completely a single spin-adapted configuration, a so-called Gelfand state. The vertical position of each vertex represents the value of the orbital index i and each vertex is further characterized by the values of a and b, where 2a+b is the number of electrons placed in the orbitals with sequence number ≤ i. There are four different types of arcs connecting the vertices. These arcs are given case numbers s which characterize the occupation of orbital i and the spin-coupling to the electrons in the orbitals < i. s = 0, which is a vertical line and does not change a or b, is given for unoccupied orbitals. s = 3, which has the least slope and changes only

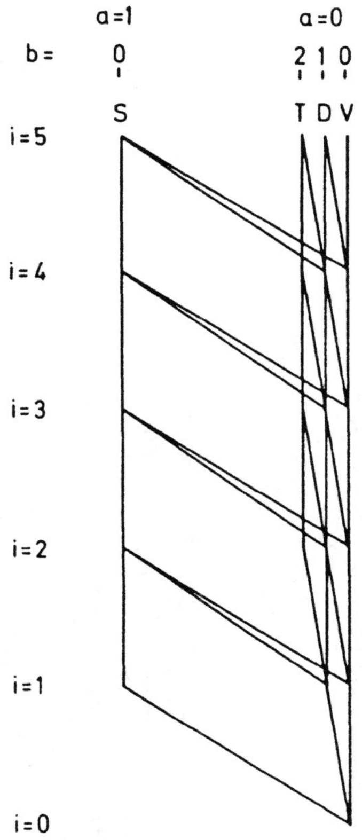

Figure 1. Graph of the external space for a general multi-reference CI-expansion in five external orbitals.

a by +1, is given for doubly occupied orbitals. None of these arcs obviously change the spin, which at each orbital level is equal to 2 b. For the singly occupied orbitals there must be two different types of arcs since the total spin can go up or down by 1/2. $s=1$ gives $\Delta S=+1/2$ and therefore changes b by +1 and does not change a. $s=2$ gives $\Delta S=-1/2$ and changes b by -1 and a by +1. As can be seen from the figures the external orbital space has a much simpler structure than the internal space, and is further completely general, independent of the type and number of reference states. These properties are both exploited in the general direct CI-method. The letters S, T, D and V in Fig. 1-2 characterize the type of configurations which passes through these vertices. Configurations passing through V, valence configurations, have no occupied external orbitals. Those passing through S and T have two electrons in the external space. S means singlet-coupling of these electrons, $p=+1$ in (4), and T means triplet coupling, $p=-1$ in

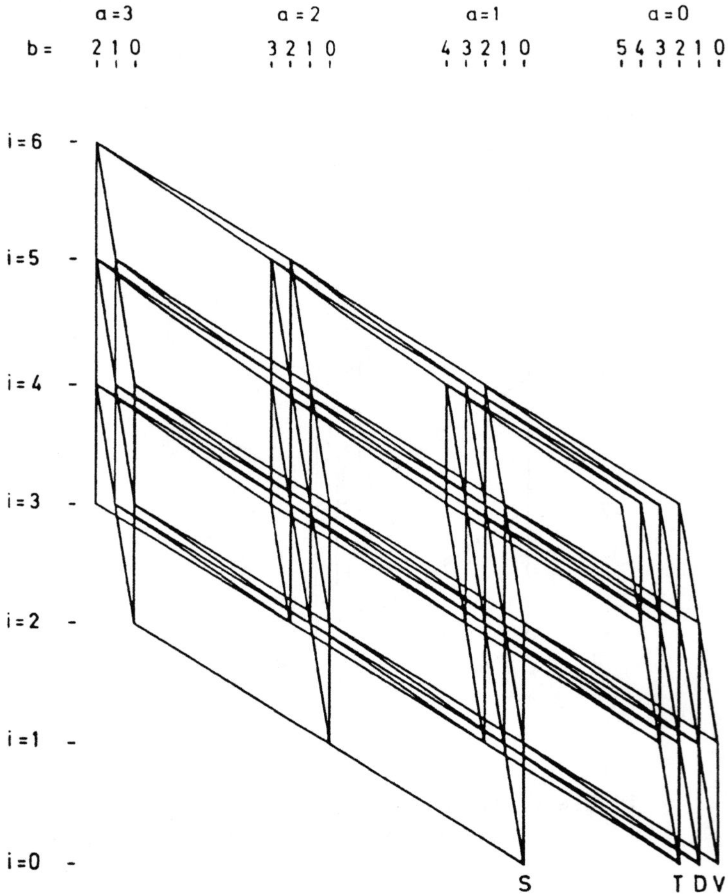

Figure 2. Graph of the internal space for eight electrons in six orbitals in a triplet spin state.

(4). D finally has a single electron in the external space which therefore leads to a spin doublet. It should be pointed out that the number of electrons in the external orbitals should not be confused with the number of orbital replacements of the reference states, since some of the replacements can of course be made with other internal orbitals.

Detailed formulas for evaluating the coupling coefficients (2) over Gelfand states have been given by Shavitt [9], and by Paldus and Boyle [24], for all possible cases. These formulas can be summarized as

$$A^{\mu\nu}_{pq} = \prod_{t=p}^{q} W(T_t, b_t) \qquad (p<q)$$

$$A^{\mu\nu}_{pqrs} = \{ \prod_{t \in S_1} W(T_t, b_t)\} \sum_{x=0,1} \{ \prod_{t \in S_2} W_x(T_t, b_t)\} \tag{5}$$

where S_1 and S_2 are the sets of orbital levels within just one or both, respectively, of the ranges (p, q) and (r, s), and x indicates the type of intermediate coupling. The $W(T_t, b_t)$ are segment values characterized by the shape and position of the segment in the graph, see further details in Ref. 9. The only important part of the formulas (5) in the present context is that the one-electron coupling coefficients are given as a simple product of contributions from each orbital, and that something similar holds also for the two-electron coupling coefficients. This fact will be used in the next section.

With the general equation (1) and the formulas (5) one can in principle formulate a general direct CI-method. By storing the vectors σ and C in core storage and reading the integrals sequently from peripheral storage the update of σ could be made by calculating the coupling coefficients A as they are needed. The formulas for A turn out to be much too complicated to be practically useful, however, except in the simplest cases with a single reference state. Storing the coefficients is not possible either, other then in small calculations, which can anyway be treated by conventional methods. The solution of this problem is described in the next section.

III. THE DIRECT CI-METHOD

The three most important facts from the preceding section, which can be used to formulate the general strategy for the direct CI-method, are the following. First, the structure of the external part of the Shavitt graph, Fig. 1, is very much simpler than the internal part, Fig. 2. Second, the external part of the graph is independent of the choice of reference states. It is even independent of the total spin. Third, the direct CI-coupling coefficients are factorized into contributions from each orbital according to the formulas (5). From the structure of the one electron coupling coefficients it is immediately clear from (5) that we can write

$$A^{\mu\nu}_{pq} = B^{\mu\nu}_{pq} \cdot D^{\mu\nu}_{pq} \tag{6}$$

where $B^{\mu\nu}_{pq}$ is a product of contributions from orbitals in the internal space and $D^{\mu\nu}_{pq}$ can be entirely evaluated in the simple external space. The $B^{\mu\nu}_{pq}$ are therefore called internal and $D^{\mu\nu}_{pq}$ external coupling coefficients. It is more complicated to show that the same type of factorization holds also for the two electron coefficients, so that

$$A^{\mu\nu}_{pqrs} = B^{\mu\nu}_{pqrs} \cdot D^{\mu\nu}_{pqrs} \tag{7}$$

This is shown in detail in Ref. 7, following the formalism developed by Shavitt [9]. Due to the usually small number of orbitals in the internal space the number of different B-coefficients is generally also small and can be calculated once and for all in the beginning of the calculation and stored on peripheral storage. Due to the simple and general structure of the external part of the graph the external coupling coefficients can be evaluated very easily and have the simple values ± 1, $\pm\sqrt{2}$ and $+2$. The values further depend only on the type of interaction, which is classified according to which boundary vertex V, D, S or T the interacting configurations passes through. All possible external coupling coefficients in the general multireference CI case are derived and given in Ref. 7.

For most of the interactions the external coupling coefficients are products of sign-factors, appearing when one of the interacting configurations has a triplet coupling of the external orbitals, and $\sqrt{2}$, appearing for singlet coupling when $a=b$. The appearance of these factors can be understood from the definition of Φ^{ab}_{μ} according to (4). Triplet coupling of a and b in Φ^{ab}_{μ} leads to anti-symmetry with respect to a permutation of a and b, which in turn leads to the sign-factor in the matrix elements. For the case of singlet coupling and $a=b$ the normalization factor introduces a factor of $1/\sqrt{2}$. With $a=b$ the coulomb and exchange integral in the matrix-elements further becomes the same integral, which leads to a factor of 2 for this integral and all together a factor of $\sqrt{2}$ for the coupling coefficient. Following the strategy of the SCEP-method the external coupling coefficients can be entirely removed from the formalism [12, 17]. This is done by introducing a matrix notation where the coefficient matrix $C^{ab}_{\mu\nu}$, with a and b singlet coupled, is defined as a symmetric matrix in a and b and where the diagonal elements are multiplied by $\sqrt{2}$, and when a and b are triplet coupled the matrix is defined as anti-symmetric. This leads to a formulation where all interactions are defined as simple matrix multiplications without any reference to external coupling coefficients.

Since the external coupling coefficients are equal to $+1$ for the interactions between the functions (3) - (4) provided $a > b$ in (4), the internal coupling coefficients $B^{\mu\nu}_{pqrs}$ could also be defined by identifying them with the total coupling coefficients $A^{\mu\nu}_{pqrs}$ in the interaction between such configurations. In this way the B-coefficients do not need to be evaluated as the internal segment product in the Shavitt graph, but can equally well be calculated with conventional techniques using ordinary spin-adapted configurations of the type (3) - (4).

The strategy of the direct CI-method is best illustrated by taking a detailed example. This example will here be the treatment of integrals with two external indices, which is in most cases the dominating part of the CI-calculation. For the internal i, j-pair block, equation (1) can in this case, with the factorization of the coupling coefficients, be written with $\Phi_\mu = \Phi_\varkappa^{ac}$ and $\Phi_\nu = \Phi_\lambda^{bc}$,

$$\Delta\sigma_\varkappa^{ac}(i,j) = \sum_\lambda \sum_b \{ \delta_{\varkappa\lambda} D_{ab}^{\mu\nu} (a|\hat{h}|b) + \quad (8)$$

$$+ B_{ijab}^{\mu\nu} \cdot D_{ijab}^{\mu\nu} (ij|ab) + B_{iajb}^{\mu\nu} \cdot D_{iajb}^{\mu\nu} (ia|jb)\} C_\lambda^{bc}$$

By introducing the matrix-notation described above, which eliminates the external coupling coefficients, this equation simplifies to

$$\Delta\sigma_\varkappa^{ac}(i,j) = \sum_\lambda \sum_b \{ \delta_{\varkappa\lambda} (a|\hat{h}|b) + B_{ijab}^{\mu\nu} (ij|ab) + \quad (9)$$

$$+ B_{iajb}^{\mu\nu} (ia|jb)\} \cdot C_\lambda^{bc}$$

Defining the integral matrix with matrix-elements

$$I_{ij}^{ab} = \delta_{\varkappa\lambda} (a|\hat{h}|b) + B_{ijab}^{\mu\nu} (ij|ab) + B_{iajb}^{\mu\nu} (ia|jb) \quad (10)$$

equation (9) can be written as a matrix-product

$$\Delta\underline{\underline{\sigma}}_\varkappa = \underline{\underline{I}}_{ij} \times \underline{\underline{C}}_\lambda \quad (11)$$

For symmetry reasons the opposite interaction will also occur

$$\Delta\underline{\underline{\sigma}}_\lambda = \underline{\underline{I}}_{ij} \times \underline{\underline{C}}_\varkappa \quad (12)$$

Spatial symmetry is easily taking care of in these interactions by symmetry blocking the corresponding matrices. The main advantage of using (11) - (12) is otherwise that they are very well vectorizable, which makes the method very efficient on the modern array-processors.

In the paper by Saunders and van Lenthe [11] a further modification of the direct CI-equations has introduced improved efficiency. This improvement is easiest seen directly in equation (1), where the two-electron term can be rewritten

$$\sigma_{\mu}^{(2)} = \sum_{\nu} \sum_{pqrs} A_{pqrs}^{\mu\nu} (pq|rs)\, C_{\nu} = \sum_{pqrs} (pq|rs) \sum_{\nu} A_{pqrs}^{\mu\nu} C_{\nu} = \sum_{pqrs} (pq|rs)\, A'^{\mu}_{pqrs} \qquad (13)$$

The longer the summation ν is for the intermediate quantities A' the more efficient is the reformulation. It is easily realized that in the summation over ν all configurations Φ_{ν} have the same orbital occupation so that the summation is in practice over the different possible internal spin-couplings. The use of (13) is therefore called the internal spin-driven scheme. In this formulation the combination of coulomb and exchange integrals as in (10) can on the other hand not be made, so that more than two different internal spin-couplings must be possible before (13) wins over (9) - (12). For the integrals with two external orbitals this condition is practically always satisfied. It should be added that with the Saunders-van Lenthe reformulation, the spin-recoupling for obtaining the smallest possible interacting space [14] is usually not worth while.

In the preceding formulas an orthonormal external orbital set has been assumed. It turns out that essentially the same formulas are obtained if the non-orthogonal atomic basis set is used instead. Arranging the external molecular orbital coefficients into a rectangular matrix $\underline{\underline{d}}$ with each row given by one external orbital the orthonormality yields

$$\underline{\underline{d}} \times \underline{\underline{d}}^{+} = \underline{\underline{1}}$$

Inserting this into (11) and multiplying from the left by $\underline{\underline{d}}^{+}$ and to the right by $\underline{\underline{d}}$ gives

$$(\underline{\underline{d}}^{+} \times \Delta \underline{\underline{\sigma}}_{\varkappa} \times \underline{\underline{d}}) = (\underline{\underline{d}}^{+} \times \underline{\underline{I}}_{ij} \times \underline{\underline{d}}) \times (\underline{\underline{d}}^{+} \times \underline{\underline{c}}_{\lambda} \times \underline{\underline{d}})$$

which can be rewritten

$$\Delta \underline{\underline{\sigma}}'_{\varkappa} = \underline{\underline{I}}'_{ij} \times \underline{\underline{C}}'_{\lambda}$$

where the matrix elements of $\underline{\underline{I}}'_{ij}$ are given from (10) by changing the external orbital indices a and b to the atomic basis function indices p and q. The matrix products involving integrals with three [17] and four external indices can be rewritten similarly. In this way the full integral transformation can be avoided [11-13, 17]. Since the dimension of the matrix multiplications (11) - (12) is increased, the gain in the transformation part is usually lost in the CI part, however. The advantages and disadvantages of this approach is discussed in Ref. 11.

In the actual implementation of the direct CI-method there are many more details than given here, which can all be found in the references given. The more interested readers are particularly recommended to study the paper by Saunders and van Lenthe [11], whose detailed formulations must be very close to what can optimally be achieved.

IV. APPROXIMATE METHODS

Since the solution of the CI-problem with a long CI-expansion is time consuming there is often a need for faster approximate methods. In this section three different ways of approximating the CI-problem will be discussed still using the general direct CI-strategy described in the preceding section.

a. Externally contracted CI

The CI wave function which should be approximated is of the general form

$$\Psi = \sum_{\substack{\varkappa \\ (N)}} C_\varkappa \Phi_\varkappa + \sum_{\substack{\varkappa \\ (N-1)}} \sum_a C_\varkappa^a \Phi_\varkappa^a + \sum_{\substack{\varkappa \\ (N-2)}} \sum_{ab} C_\varkappa^{ab} \Phi_\varkappa^{ab} \tag{14}$$

where the spin-adapted configurations are the ones described by (3) - (4), with $\varkappa$ denoting the internal parts of these configurations. The approximate wave-function suggested in the externally contracted CI-method is then

$$\Psi = \sum_{\substack{\varkappa \\ (N)}} C_\varkappa \Phi_\varkappa + \sum_{\substack{\varkappa \\ (N-1)}} C_\varkappa \sum_a \tilde{C}_\varkappa^a \Phi_\varkappa^a + \sum_{\substack{\varkappa \\ (N-2)}} C_\varkappa \sum_{ab} \tilde{C}_\varkappa^{ab} \Phi_\varkappa^{ab} = \sum_\varkappa C_\varkappa \varphi_\varkappa \tag{15}$$

where the $C_\varkappa$ are obtained variationally but the $\tilde{C}_\varkappa^a$ and $\tilde{C}_\varkappa^{ab}$ are obtained by perturbation theory as

$$\tilde{C}_{\varkappa}^{ab} = \frac{\langle 0|\hat{H}|\Phi_{\varkappa}^{ab}\rangle}{E_0 - \langle \Phi_{\varkappa}^{ab}|\hat{H}|\Phi_{\varkappa}^{ab}\rangle} \qquad (16)$$

$|0\rangle$ is formed by diagonalizing the matrix containing all valence configurations $\Phi_\varkappa$, and E_0 is the corresponding variational energy. The number of variational parameters in (14) is therefore drastically reduced with the price that each function $\varphi_\varkappa$ is much more complicated than the original $\Phi_\varkappa^{ab}$. A typical reduction is from 10^5 to 10^3 parameters. In practice the method is technically very close to the first two iterations in the direct CI-method described in Section III, where the first iteration is very fast and corresponds to (16). The calculation of the matrix elements between the configurations $\varphi_\varkappa$ represents the by far dominating part of the calculation. A contribution to a matrix element from integrals with two external idices is

$$\Delta H_{\varkappa\lambda}(i,j) = \sum_{ab} \{ \delta_{\varkappa\lambda} (a|\hat{h}|b) + B_{ijab}^{\varkappa\lambda} (ij|ab) + B_{iajb}^{\varkappa\lambda} (ia|jb)\} \sum_c \tilde{C}_{\varkappa}^{ac} \cdot \tilde{C}_{\lambda}^{bc} \qquad (17)$$

or in matrix notation

$$\Delta H_{\varkappa\lambda}(i,j) = \underline{\underline{I}}_{ij} \cdot (\underline{\underline{C}}_\varkappa \times \underline{\underline{C}}_\lambda^+) \qquad (18)$$

where the second product is an ordinary n^3 matrix product and the first product is an n^2 scalar product. (17) and (18) can now be compared with (9) - (12), and it is seen that when the ordinary matrix product dominates, the formation of the contracted CI-matrix should take half the time of a direct CI-iteration, since there is only one rather than two matrix multiplications. When the contracted matrix is formed it can be easily diagonalized using conventional methods.

A drawback of the present contraction scheme is that the efficient internal spin-driven strategy of Ref. 11 can not be used. One way out of this problem is to reduce the number of variational parameters further by grouping together all configurations with the same internal orbital occupation. Another more complicated way, is to diagonalize the internal coupling coefficient matrix for each spin block interaction. When the matrix multiplications (18) have been performed in this new basis the contributions to the contracted matrix have to be backtransformed to the original spin-coupling. If this procedure is worth-while remains to be shown.

The loss of accuracy by the external contraction is usually around 1.5 % of the correlation energy defined as the difference between the final energy and the reference space energy. For further details of this method, see Ref. 22.

b. Internally contracted CI

In the simplest formulation of the internally contracted CI-method there is just a single reference state composed of a linear combination of reference configurations

$$\Psi_0 = |0\rangle = \sum_{\varkappa}^{(N)} C_\varkappa \Phi_\varkappa \tag{19}$$

where the $C_\varkappa$ and the $\Phi_\varkappa$ are ideally taken from an MC-SCF calculation but could in principle be obtained in a different way. The single and double replacement states are defined by applying the generators of the unitary group,

$$\Psi_i^a = \hat{E}_{ai} |0\rangle = \sum_{\varkappa}^{(N-1)} d_\varkappa (N-1) \Phi_\varkappa^a \tag{20}$$

$$\Psi_{ij}^{ab} = (\hat{E}_{ai} \hat{E}_{bj} + p \hat{E}_{aj} \hat{E}_{bi}) |0\rangle = \sum_{\varkappa}^{(N-2)} d_\varkappa (N-2) \Phi_\varkappa^{ab} \tag{21}$$

where p is +1 for singlet coupling and -1 for triplet coupling of a and b as in (4). The expressions for $d_\varkappa$ are easily obtained [20] as,

$$\begin{aligned} d_\varkappa (N-1) &= \sum_{\lambda}^{(N)} B_{ai}^{\varkappa\lambda} C_\lambda \\ d_\varkappa (N-2) &= \sum_{\lambda}^{(N)} B_{aibj}^{\varkappa\lambda} C_\lambda \end{aligned} \tag{22}$$

where the C-coefficients are the same as in (19) and the internal coupling coefficients B are defined as in Section III and are thus independent of the external indices. In general the state functions (19) - (21) have to be orthonormalized and linear dependencies have to be removed. Since the state functions (19) - (21) are linear combinations of configurations with coefficients entirely determined in the internal space the method is called internally contracted.

The matrix elements between the state functions (19) - (21) can be expressed using the same formalism as in the preceding sections. A contribution from integrals with two external indices can thus be written,

$$\Delta \sigma_{\varkappa}^{ac} = \sum_{\lambda} \sum_{b} \{ \delta_{\varkappa\lambda} (a|\hat{h}|b) + \sum_{ij} [B_{ijab}^{\varkappa\lambda} (ij|ab) + B_{iajb}^{\varkappa\lambda} (ia|jb)] C_{\lambda}^{bc} \tag{23}$$

Introducing intermediate quantities,

$$F_{\varkappa\lambda}^{ab} = \delta_{\varkappa\lambda} (a|\hat{h}|b) + \sum_{ij} [B_{ijab}^{\varkappa\lambda} (ij|ab) + B_{iajb}^{\varkappa\lambda} (ia|jb)] \tag{24}$$

(23) can be rewritten in matrix form as,

$$\Delta \underline{\underline{\sigma}}_{\varkappa} = \underline{\underline{F}}_{\varkappa\lambda} \times \underline{\underline{C}}_{\lambda} \tag{25}$$

The expressions (23) - (25) differs from the corresponding expressions (9) - (11) only in the intermediate summations over internal orbitals i and j. This summation could have been introduced in (9) - (11) but is not believed to be very efficient in the ordinary multireference case. The reason is that the summation in (24) usually would include very few terms due to the sparsity of the internal coupling coefficient matrix in the basis of ordinary configurations. With the more complex state functions (19) - (21) this is not generally true anymore and the formation of the $\underline{\underline{F}}$-matrices should be more efficient at least in the limit of many terms in (19). The formation of (24) can, however, be an organization problem since the generated list of F-elements can be quite long, sometimes longer than the two-electron integral file, and is formed out of two potentially long lists of internal coupling coefficients B and integrals (pq|rs).

The main advantage in the internal contraction scheme is that the CI expansion length only depends on the number of internal orbitals and not on the number of reference configurations in (19). The method should therefore have its main applicability for cases with many reference configurations. For these cases there are, however, some problems with the semi-internal excitations as has been pointed out in Ref. 20. These problems are of course not worse than in the exact case, following the procedure in Section III, but many of the advantages of the internal contraction seems to be lost. For further details of the method the reader is referred to Ref. 19-21. In Ref. 21 the method has been programmed and tested and shows a high degree of accuracy and efficiency.

c. Configuration selection

Individual configuration selection as used in the conventional CI-method [23] can not be used in the general direct CI-method as described in Section III, without destroying the simple matrix structure. The full advantage of the reduction in configuration space can therefore not be expected. If the coefficients for the deleted configurations are simply set to zero, however, the increased sparsity of the $\underline{\underline{C}}$ matrix can be utilized almost fully if the matrix multiplications (11) - (12) are organized properly [25]. The simultaneous sparsity of $\underline{\underline{C}}$ and $\Delta\,\underline{\underline{\sigma}}$ can however not be used, which therefore leads to many unnecessary operations if the configuration list is drastically reduced. The improvement in efficiency by setting small CI-coefficients equal to zero can also be used in an exact treatment, by scanning through the CI-vector in each iteration [26]. Coefficients deleted in one iteration may then get contributions in a later iteration and no accuracy is lost.

Full advantage of configuration selection can be achieved if groups of configurations with identical internal parts are deleted. This just leads to a reduction in the number of internal coupling coefficients which have to be processed in the CI-part. This way of deleting configurations follows the same principles as the reference state selection [7]. With this restriction in the way configurations are allowed to be deleted it is clear that more configurations are needed for the same accuracy as compared to an entirely general selection. How much accuracy is lost and how high the efficiency will be with this procedure is a matter of experience. Investigations along these lines are in progress [27].

V. SUMMARY

The general formalism and the structure of the multireference CI problem has been described. The emphasis was on the simplicity and generality of the external space, the complexity of the internal space and on the factorization of the direct CI-coupling coefficients. With this background the strategy of the general direct CI-method was outlined. Detailed formulas for the most complicated case of the treatment of integrals with two external indices were given. These formulas were shown to have a simple matrix structure particularly convenient for array processing. The matrices could be blocked to take care of space symmetry and also transformed to an atomic orbital basis for avoiding the integral transformation, which is not recommended in general, however. The recent development on the direct CI-method was reviewed with particular emphasis on the work by Saunders and van Lenthe [11]. Their internal spin-driven scheme was described.

Approximate methods within the general direct CI-strategy were finally described. Formulas used for matrix-element evaluation in the internally and externally contracted CI-schemes were given. The externally contracted CI-method was shown to take the time of one half of a CI-iteration in the limit of a large basis set. The internally contracted CI-method was shown to be potentially well suited for treating cases with very many reference states. Different ways of implementing configuration selection within the direct CI-method were also discussed.

REFERENCES

1. B.O. Roos, Chem. Phys. Lett. 15, 153 (1972)
2. P.E.M. Siegbahn in: Proceedings of SRC Atlas Symposium No 4, 'Quantum Chemistry – the State of the Art', V.R. Saunders and J. Brown, eds., Atlas Computer Laboratory, Chilton Didcot, Oxfordshire (1975).
3. B.O. Roos and P.E.M. Siegbahn in 'Modern Theoretical Chemistry', Vol. 3, H.F. Schaefer, ed., Plenum Press, New York, N.Y., (1977), Chap. 7.
4. B.O. Roos and P.E.M. Siegbahn, Int. J. Quantum Chem. 16, 485 (1980).
5. R.R. Lucchese and H.F. Schaefer, J. Chem. Phys. 68, 769 (1978).
6. N.C. Handy, J.D. Goddard and H.F. Schaefer, J. Chem. Phys. 71, 426 (1979).
7. P.E.M. Siegbahn, J. Chem. Phys. 72, 1467 (1980).
8. J. Paldus, in: Theoretical Chemistry: Advances and Perspectives, eds. H. Eyring and D.J. Henderson, Vol. 2 (Academic Press, New York, 1976) p. 131.
9. I. Shavitt, Int. J. Quantum Chem. Symp. 12, 5 (1978); and in Lecture Notes in Chemistry: The Unitary Group, ed. J. Hinze (Springer, Berlin, 1981) p. 51.
10. B. Liu and M. Yoshimine, J. Chem. Phys. 74, 612 (1981).
11. V.R. Saunders and J.H. van Lenthe, to be published.
12. W. Meyer, J. Chem. Phys. 64, 2901 (1976).
13. R. Ahlrichs and F. Driessler, Theoret. Chim. Acta 36, 275 (1975).
14. A.D. McLean and B. Liu, J. Chem. Phys. 58, 1066 (1973).

15. H. Lischka, R. Shepard, F.B. Brown and I. Shavitt, Int. J. Quantum Chem. Symp. 15, 91 (1981).
16. W. Duch and J. Karwowski, Int. J. Quantum Chem. (in press).
17. R. Ahlrichs, in'Proceedings of the 5th Seminar on Computational Methods in Quantum Chemistry, Groningen, September 1981', P.T. van Duijnen and W. C. Nieuwpoort, eds., (Max Planck Institut, Garching, München).
18. P.J.A. Ruttink and M.M.M. van Schaik, same as Ref. 17.
19. W. Meyer, same as Ref. 3, Chap. 11.
20. P.E.M. Siegbahn, Int. J. Quantum Chem. 18, 1229 (1980).
21. H.J. Werner and E.A. Reinsch, same as Ref. 17.
22. P.E.M. Siegbahn, same as Ref. 17.
23. R.J. Buenker and S.D. Peyerimhoff, Theor. Chim. Acta 35, 33 (1974).
24. J. Paldus and M.J. Boyle, Physica Scripta 21, 295 (1980).
25. V.R. Saunders and M.F. Guest, Comput. Phys. Comm., to be published.
26. V.R. Saunders, private communication.
27. I. Shavitt, private communication.

PAIR CORRELATION THEORIES.

R. Ahlrichs

Institut f. Phys. Chem. u. Elektrochemie
Lehrstuhl f. Theoretische Chemie
University Karlsruhe
D 7500 Karlsruhe, Germany (FRG)

In this review we first recall in chapter 2 the special features of two-electron wave functions. Chapter 3 deals with the concepts of pairs in n-electron wave functions, the structure of Hamiltonian matrix elements and residual vectors, and the characterization of pair methods in connection with CI(SD) techniques. Chapter 4 is devoted to the discussion of cluster corrections and size consistency problems in order to give an introduction to coupled pair methods.

1. INTRODUCTION

Pair theories build upon the simple structure of wave functions for two-electron systems in order to derive more or less approximate theories of the n-electron case. Since the Hamiltonian contains only two particle terms and since the Pauli principle prevents three or more electrons to occupy the same point in space it may, furthermore, be expected at least intuitively that pair correlation effects account for the most important features of electron correlation. The term pair theories is not defined in a precise way but the collection of methods carrying a P (for pair) or G (for geminal) yields a useful characterization: the APSG method of Hurley et al [1], Cizek's CP-MET [2], the methods of Sinanoglu [3] and Nesbet [4] now referred to as IEPA, and the IEPA-PNO variant [5], Meyer's CEPA and PNO-CI [6,7], Meyer's SCEP [8] and

G. H. F. Diercksen and S. Wilson (eds.), Methods in Computational Molecular Physics, 209-226.

related procedures developped by Ahlrichs et al [9] and Dykstra [10], Taylor's procedure which could be abbreviated MR-PNO-CI [11], the MR-SCEP of Werner and Reinsch [12], and finally Siegbahn's pair contracted MR-CI [13].

Whereas earlier pair methods were of rather approximate nature - the APSG could account only for certain parts of total correlation and the IEPA involved rather crude approximations, as we now know - recent developments show a pronounced convergence of methods. SCEP and MR-SCEP could be considered as technical variants of direct CI schemes [14], and recent direct CI schemes take advantage of certain features of pair approaches such as the organization of integrals or CI vectors in matrices [15,16]. An important feature of most pair techniques is the attempt to account for cluster corrections - or higher excitations - at least in an approximate way [17], but even here one observes a convergence as is obvious from the popularity of the Davidson correction [3,18].

In the present paper we first consider briefly two-electron systems, then the concept of electron pairs in the n-electron case on the CI(SD) level, and finally discuss cluster corrections.

2. TWO-ELECTRON SYSTEMS

Let $|a\rangle, |b\rangle$ denote MOs and $|x\rangle$, $|y\rangle$ AOs

$$|a\rangle = \sum_x V(x,a)\, |x\rangle \, , \quad \langle a|b\rangle = \delta(a,b) \tag{1}$$

The straightforward CI expansion for the wave function ψ in terms of CSFs is

$$\psi = \sum_{a\leq b} \tilde{C}(ab)\,\tilde{\psi}(p,ab)\sigma(p) \, , \quad \tilde{\psi}(p,ab) = N(ab+pba) \tag{2}$$

where $N=(2+2\delta(ab))**(-1/2)$, and $\sigma(p)$ denotes the appropriate spin function for singlet, p=1, and triplet states, p=-1. The very fact that ψ factorises into space and spin parts permits an alternative formulation which provides additional insight and will turn out to be useful later on. If

$$\psi(p,ab) = 1/2(ab+pba) = 1/(2N)\,\tilde{\psi}(p,ab) \tag{3}$$

$$C(ab) = pC(ba) = (2-\delta(ab))**(-1/2)\tilde{C}(ab), \tag{4}$$

one can write the space part U of ψ, $\psi = U\sigma(p)$ as

$$U = \sum_{ab} C(ab)\psi(p,ab) = \sum_{ab} C(ab)ab = \sum_{xy} D(xy)xy \qquad (5)$$

where C(ab) and D(xy) can be considered as matrices $\underline{\underline{C}}$ and $\underline{\underline{D}}$

$$\underline{\underline{C}}^{\dagger} = p\underline{\underline{C}} \quad , \quad \underline{\underline{D}} = \underline{\underline{V}}\,\underline{\underline{C}}\,\underline{\underline{V}}^{\dagger} \qquad (6)$$

The representation (5) of U together with (6) suggests immediately to perform a unitary transformation of MOs $|a\rangle \rightarrow |d\rangle$ such that $\underline{\underline{C}}$ becomes diagonal [19]

$$p=1 \qquad U = \sum_{d} C(d)\ dd \qquad (7a)$$

$$p=-1 \qquad U = \sum_{d} C(d)\ (dd' - d'd) \qquad (7b)$$

which is the expansion of U in terms of its NOs $|d\rangle$. This has the best convergence properties of all possible CI expansions [1,19] and a relatively small number of NOs are usually sufficient to exhaust a basis set. The determiantion of NOs requires the knowledge of the pair function or the solution of MC-SCF eqs [20] . However, efficient methods have been developped to get quite accurate approximations to the PNOs - e.g. a loss of 1-2% in correlation energy - which have made PNO techniques a useful tool in electronic structure theory [6,7,11,20,21] .

Let us now turn to the integral driven direct CI treatment for the two-electron case. The eqs to be solved read

$$R(ab) = \langle ab|\ \hat{H} - E\ |\ U\rangle = 0 \qquad (8)$$

where R(ab) denotes the residual "vector" or matrix $\underline{\underline{R}}$. A straightforward computation yields

$$\underline{\underline{R}} = \underline{\underline{C}}\,\underline{\underline{h}} + \underline{\underline{h}}\,\underline{\underline{C}} + \underline{\underline{K}}[\underline{\underline{C}}] - E\,\underline{\underline{C}} \qquad (9)$$

$$= T(p)\ \{ 2\,\underline{\underline{C}}\,\underline{\underline{h}} + \underline{\underline{K}}[\underline{\underline{C}}] - E\,\underline{\underline{C}} \}$$

where

$$T(p)\ \underline{\underline{A}} = 1/2\ (\underline{\underline{A}} + p\underline{\underline{A}}^{\dagger}) \qquad (10)$$

$$(\underline{\underline{K}}[\underline{\underline{C}}])_{ab} = \sum_{cd} C(cd)\ (ac|bd) \qquad (11)$$

i.e. $\underline{\underline{K}}[\underline{\underline{C}}]$ is the usual exchange operator obtained with $\underline{\underline{C}}$ as input matrix. Eq. (9) for $\underline{\underline{R}}$ has a very simple structure which is well suited for all kinds of computers. With the aid of (9) it is then an easy matter to solve the CI eqs by means of any of the standard iterative procedures.

The evalutation of $\underline{\underline{R}}$, eqs (9), (11), does in fact not require an integral transformation [8,20]. Recalling the definition (1) of MOs it is easily verified that [9]

$$\underline{\underline{K}}\,[\underline{\underline{C}}] = \underline{\underline{V}}^{\dagger}\ \tilde{\underline{\underline{K}}}\,[\underline{\underline{V}}\ \underline{\underline{C}}\ \underline{\underline{V}}^{\dagger}]\ \ \underline{\underline{V}} \tag{12}$$

where $\tilde{\underline{\underline{K}}}$ denotes the exchange operator in the AO basis obtained from an input matrix also given in the AO basis. By virtue of (12) one avoids the integral transformation (~N**5) at the cost of two matrix transformations (~N**3) in each iteration. The solution of the full CI then requires roughly the same time as an SCF computation [20].

3. PAIRS IN N-ELECTRON WAVE FUNCTIONS

In order to take advantage of the simple structure of pair functions one has to identify electron pairs in the n-electron case which is actually possible for a large variety of methods. Let us first consider CI techniques and discuss cluster corrections and size consistency problems in the next chapter.

For a general definition of pair functions one divides the MOs into two disjoint sets, the internal (labelled i,j,k,l) and the external (a,b,c,d) orbitals. This partitioning is closely connected with the idea of single or multiple reference CI(SD) methods where internal MOs are occupied in at least one reference CSF and external MOs are unoccupied in all references. The possibility to identify pairs in a MR-CI(SD) depends only on the CSF basis chosen, and arises in a natural way if the CSFs are constructed as properly antisymmetrized and spin coupled products of internal and external CSFs with 0 or 1 or 2 external MOs. Spin coupling to the desired n-electron CSF is then uniquely defined by the spins of internal and external parts. Since the external CSFs factorize into space and spin parts (just 1 or 2 electrons), it is convenient to use the following notation. P (or Q) is a combined label which specifies the internal (n-2)-electron CSF $\tilde{\Phi}(P)$ and the parity p (and, hence, the spin) of the external space part. We will write $\phi(P)=\tilde{\phi}(P)\sigma(p)$, which is the product of an (n-2)-electron CSF times the external spin function $\sigma(p)$ for the remaining electrons. The label s (or t) similarly specifies the internal (n-1)-electron CSF, and $\phi(s)$ includes again the spin for the n'th electron. The $\phi(P)$ and $\phi(s)$ are assumed to form orthonormal sets. One then gets the following form for the singles and doubles CSFs and their orthogonality relationships.

$$\psi(s,a)=[\phi(s)a],\langle\psi(s,a)|\psi(t,b)\rangle=\delta(st)\delta(ab) \quad (13)$$

$$\psi(P,ab)=[\phi(P)\ 0.5\ (ab+pba)] \quad (14)$$

$$\langle\psi(P,ab)|\psi(Q,cd)\rangle=\delta(PQ)T(P,ab)T(Q,cd)\langle a|c\rangle\langle b|d\rangle \quad (15)$$

where [ ] indicates proper antisymmetrization and spin coupling, T(P,ab) the (anti)symmetrizer analogous to (10), and $\psi(P,ab)$ has been chosen in close analogy to (3).

If $\psi(R)$ denotes the all internal CSFs one has the following form of the total MR-CI(SD) wave function

$$\psi=\sum_R C(R)\psi(R)+\sum_{sa} C(s,a)\psi(s,a)+\sum_{Pab} C(P,ab)\psi(P,ab) \quad (16)$$

This suggests immediately the definition of spinless external (effective) one-electron functions u(s) and pair functions U(P) together with the corresponding contributions $\psi(s)$ and $\psi(P)$ to the total wave function.

$$u(s)=\sum_a C(s,a)a, \qquad \psi(s)=[\phi(s)\ u(s)] \quad (17)$$

$$U(P)=\sum_{ab} C(P,ab)ab, \qquad \psi(P)=[\phi(P)\ U(P)] \quad (18)$$

$$\psi=\sum_R C(R)\psi(R)+\sum_S \psi(s)+\sum_P \psi(P) \quad (16a)$$

One could, of course, have started directly from (16a) and the definitions (17) and (18) for $\psi(s)$ and $\psi(P)$, where u(s) and U(P) are arbitrary provided U(P) has parity p and the following (strong) orthogonality conditions are met.

$$\langle u|i\rangle=0,\ \langle U|i\rangle=0,\ \langle\phi(s)|a\rangle=0,\langle\phi(P)|a\rangle=0 \quad (19)$$

Using the product basis (13), (14) implies no loss of generality and appears as a natural choice rather than an oddity. It is not only a common feature of pair methods since existing MR-CI(SD) techniques based on the GUGA [22-24] also couple external MOs first which are then coupled to internal walks which is just (13), (14).

Let us now consider the structure of Hamiltonian matrix elements and treat the case $\langle\psi(P)|\hat{H}|\psi(Q)\rangle$, see eq. (18), in detail, where we follow basically Meyer [21]. Since one has the same number of internal and external MOs on either side of $\hat{H}$, only the terms $(ij|kl)$, $\langle i|\hat{h}|j\rangle$, $\langle a|\hat{h}|b\rangle$, $(ab|cd)$,

$$(ij|ab)=\langle a|\hat{J}(ij)|b\rangle,\quad (ia|jb)=\langle a|\hat{K}(ij)|b\rangle \qquad (20)$$

enter as a consequence of (19) and the fact that $\hat{H}$ includes only one- and two-particle operators. Performing the integration over internal MOs and over all spins one gets

$$\langle\psi(P)|\hat{H}|\psi(Q)\rangle=\langle U(P)|H(PQ)+\hat{h}(PQ,1)+pq\hat{h}(PQ,2)+\delta(PQ)\ 1/r_{12}\,|U(Q)\rangle \qquad (21)$$

$$H(PQ)=\langle\phi(P)|\hat{H}(n-2)|\phi(Q)\rangle \qquad (22)$$

$$\hat{h}(PQ)=\delta(PQ)\hat{h}+\sum_{ij} BJ(PQ,ij)\hat{J}(ij)+BK(PQ,ij)\hat{K}(ij) \qquad (23)$$

where BJ and BK denote the corresponding coefficients of coulomb and exchange type operators. The derivation of eqs (21)-(22) is straightforward. The l.h.s. of (21) is a bilinear form in U(P) and U(Q), compare (17) and (18), which can be represented by an operator which must have parity pq with respect to an interchange of particles. The r.h.s. of (21) constitutes the most general form for such an operator and the explicit expressions (22), (23) are obtained by collecting the all internal, all external and the coupling terms (ij|ab) and (ia|jb). Inserting the explicit expansion (18) for U(P) one obtaines immediately the following expression (21a) which has to be evaluated in pair contracted CI calculations [13]

$$\langle\psi(P)|\hat{H}|\psi(Q)\rangle= \mathrm{tr}\{\underline{\underline{C}}^{+}(P)[H(PQ)+2\underline{\underline{h}}(PQ)]\underline{\underline{C}}(Q)\} + \delta(PQ)\ \mathrm{tr}\{\underline{\underline{C}}^{+}(P)\ \underline{\underline{K}}[\underline{\underline{C}}(P)]\} \qquad (21a)$$

The remaining cases lead to completely analogous formulae from which one gets the corresponding

Schrödinger equations for u(s) and U(P) (which constitute a coupled set of equations) by means of the variation principle.

We follow instead a more conventional but equivalent route and evaluate the following matrix element [11,12,15,16,25]

$$\langle \Psi(P,ab) | \hat{H} | \Psi(Q,cd) \rangle = T(P,ab)T(Q,cd) \quad (24)$$
$$\{ \delta(PQ)(ac|bd) + \langle a|c\rangle\langle b|d\rangle H(PQ) +$$
$$2 \langle b|d\rangle\langle a|\hat{h}(PQ)|c\rangle \}$$

which follows immediately from (14), (20)-(23) for the "primitive" expressions U(P)=0.5(ap+pba)=T(P,ab)ab, and U(Q)=T(Q,cd)cd. The corresponding remaining cases as well as the explicit expressions for the residual vector are collected in the appendix [16].

Eq. (24) shows some features which are worth to be mentioned
(i) Only internal coupling coefficients occur in (24). This is a consequence of the special choice (15).
(ii) In the derivation of (24) it has nowhere been assumed that external MOs occuring on the left (a,b) and right (c,d) of $\hat{H}$ belong to the same orthonormal set, compare eq. (21).
(iii) Since the proper symmetry with respect to p and q is guaranteed by means of T(P,ab)T(Q,cd), one gets the shortest possible list of B-coefficients in (24).
(iv) Eq. (24) leads immediately to matrix formulated expressions for the residualvector [12,16], see appendix, and the corresponding derivation is actually facilitated by the "T(P,ab)T(Q,cd)-formulation".
(v) The coupling coefficients depend, of course, on the actual choice of ϕ(s), ϕ(P); they can be obtained from any symbolic matrix element procedure by simple comparison.
(vi) The formulation given above is well suited for an integral driven procedure. A complete integral transformation can again be avoided as in the two-electron case [12,16].

The present treatment has been kept as general as possible which facilitates the characterization and comparison of various techniques. In the single reference case one usually evaluates the coupling coefficients beforehand on paper. These then occur in the programs as fixed numbers 1, $\sqrt{2}$, $\sqrt{3}$, etc [8,9,14].

This is clearly not possible for the multiple reference case, where general applicability requires the computation of a "symbolic matrix tape" and processing similar to the one sketched here [16,22-24].

In SCEP techniques one usually [9b] performes a partial integral transformation to get (ij|kl), (ij|ka), (ij|ab), and (ia|jb), but avoids a complete integral transformation by means of (12). One thereby avoids an N**5 step; and SCEP procedures appear to be preferable for sufficiently large basis sets or if a large fraction of two-electron AO-integrals "vanishes" for numerical reasons such as in the treatment of intermolecular interactions [26] (and the evaluation of (12) is fast).

Since the total number of CSFs become rapidly prohibitive in MR-CI(SD) treatments various proposals have been made to cope with this problem, besides the selection procedures [27]. A natural way appears to be a PNO-CI procedure, especially since sufficiently accurate "pair natural orbitals" can be obtained by a diagonalization of C(P,ab) as obtained in first order of perturbation theory [6,7,11,20,21]. Within the PNO scheme one represents each pair function U(P), eq. (18), in its PNO expansion , eq. (7). PNOs of different pairs are then non-orthogonal but this causes no serious problems as discussed in connection with eqs (21) and (24). However, PNO-CI procedures are not necessarily faster than direct CI or SCEP techniques [9,11].

A drastic reduction in computational effort is achieved by Siegbahn's pair contracted CI, which optimizes just a single variational parameter per s and P [13]

$$\Psi = \sum_R C(R)\Psi(R) + \sum_s C(s)\Psi(s) + \sum_P C(P)\Psi(P) \quad (25)$$

$\Psi(s)$ and $\Psi(P)$ are obtained according to (17) and (18) in using C(s,a) and C(P,ab) as obtained from first order perturbation theory. This method requires only the effort of roughly a single CI iteration and implies a loss of 2% of the correlation energy (as compared to uncontracted CI).

An alternative to the pair or external contraction is the internal contraction proposed by Meyer [21], Werner and Reinsch [12] (in an MR-SCEP formalism). Here one starts from an MC-SCF function as zeroth order approximation

$$\Psi_0 = \sum_R C(R)\, \Psi(R) \tag{26}$$

and then constructs a basis of CSFs spanning the first order interacting space with respect to Ψ_0. For this purpose one first annihilates an internal pair from Ψ_0 to get a set of two-hole states $\phi(P)$ to which the external, semiinternal and all internal parts are coupled, for more details the reader is referred to the literature [12]. This procedure may be considered as a straightforward extension of the single reference CI(SD) or SCEP to the multiple reference case. As compared to the ordinary MR-CI(SD) one achieves a substantial reduction of the CSF space dimension - and of computational effort - especially for large sets of references included in (26). Applications published so far indicate only minor losses in energy as resulting from the internal contraction - e.g. 0.2% of correlation energy for the saddlepoint of $CH_2(T)+H_2 \rightarrow CH_3+H$ [28] - but further comparisons should be made to establish the merits of internal and external contraction schemes.

4. CLUSTER CORRECTIONS AND SIZE CONSISTENCY

CI(SD) treatments are not size consistent or size extensive - i.e. the correlation energy does not scale properly with molecular size. This is most easily demonstrated for a system of N He-atoms at sufficiently large interatomic distances (to facilitate derivations). Writing the isolated - exact or CI(SD) - two electron function as

$$\Psi = a\,\Psi(o) + b\,U \quad , \langle \Psi(o) | U \rangle = 0 \tag{27}$$

the total 2N-electron function is a simple product since antisymmetrization is immaterial in this case. One then gets immediately the weight W(i) of all terms in the exact wave function linear in U**i

$$W(i) = a^{**}(2N-2i)^{*}b^{**}(2i)^{*}\binom{N}{i} \tag{28}$$

which shows the increasing importance of higher substitutions if N increases.

One can even solve "CI(SD)" equations, i.e. a CI with Ψ(SCF) - again the product of SCF wavefunctions - and all terms linear in U. The result is

$$E=E(SCF)+A/2-(A**2/4+N*W**2)**0.5 \tag{29}$$

$$A=-\langle\psi(o)|\hat{H}|\psi(o)\rangle+\langle U|\hat{H}|U\rangle\ ,\quad W=\langle\psi(o)|\hat{H}|U\rangle \tag{30}$$

With realistic values, A=1au, W=0.2au, one verifies that eq. (29) recovers for N=5 and 10, respectively, only 90% and 80% of the correlation energy. Eq. (29) furthermore shows the wrong asymptotic behaviour for large N

$$CI(SD):\qquad E\propto W*N**0.5, \tag{31}$$

which is already useful for N > 10, i.e. more than 20 electrons. The wrong asymptotic behaviour persists even if the Davidson correction is added.

For a further analysis and possible remedies of these deficiencies we consider a single reference CI within the UHF formalism. To simplify matters further let us consider only double, quadruple,..., excitations. Only the doubles enter in second and third order of perturbation theory for the energy; singles, triples and quadruples enter in 4th order but the much larger number of quadruples indicates their relatively large importance (in an analysis based on CI formalism, this may be different in MBPT-type reasonings). Using the intermediate normalization we then have if $\psi(o)$ denotes the SCF function and $\psi(ij,ab)$ etc the double substitutions in an obvious nomenclature

$$\psi=\psi(o)+\sum C(ij,ab)\psi(ij,ab)+\sum C(ijkl,abcd)*$$
$$*\psi(ijkl,abcd)+\ldots \tag{32}$$
$$=\psi(o)+\psi(D)+\psi(Q)+\ldots$$

This leads to the CI eqs

$$\langle\psi(o)|\hat{H}-E|\psi\rangle=0\rightsquigarrow E(corr)=\langle\psi(o)|\hat{H}|\psi(D)\rangle \tag{33}$$

$$\langle\psi(ij,ab)|\hat{H}-E|\psi(o)+\psi(D)\rangle=-\langle\psi(ij,ab)|\hat{H}|\psi(Q)\rangle \tag{34}$$

and so on. Eq. (34) clearly shows the reason for the problems of a CI(D): the neglect of the r.h.s.which couples the equations for doubles to the quadruples.

In order to remedy this deficiency one has to include the r.h.s. of (34) in a CI(SD)-type treatment which requires an approximation for $\psi(Q)$ [9,17,29]. Let us apply perturbation theory to get an insight into the structure of $\psi(Q)$. In the lowest orders of Moeller-Plesset perturbation theory - i.e. with the Fock operator as unperturbed operator - one gets

$$C^{(1)}(ij,ab)=\langle\psi(ij,ab)|\hat{R}\hat{H}|\psi(o)\rangle=\langle ij||ab\rangle D(ij,ab) \quad (35)$$

$$C^{(2)}(ijkl,abcd)=\langle\psi(ijkl,abcd)|\hat{R}\hat{H}\hat{R}\hat{H}|\psi(o)\rangle \quad (36)$$

where $\hat{R}$ denotes the reduced resolvent, a diagonal operator in the basis of Slater determinants from Fock-MOs,

$$\langle\psi(i\ldots,a\ldots)|\hat{R}|\psi(i\ldots,a\ldots)\rangle=D(i\ldots,a\ldots)=$$

$$(\varepsilon(i)+\ldots-\varepsilon(a)-\ldots)**-1 \quad (37)$$

where $\varepsilon(i)$, $\varepsilon(a)$ denote SCF orbital energies and

$$\langle ij||ab\rangle = (ia|jb) - (ib|ja) \quad (38)$$

The quadruple excitation performed by $\hat{R}\hat{H}\hat{R}\hat{H}$ in (36) can be achieved in 36 diffferent ways, depending on which excitation, e.g. $ij\rightarrow ab$, is performed first (6 possibilities each to pick the first pair of occupied and virtual orbitals). Combining the two partitionings which differ only in the order of pair replacements one gets a contribution, e.g. for $ij\rightarrow ab$ and $kl\rightarrow cd$

$$D(ijkl,abcd)\langle ij||ab\rangle\langle kl||cd\rangle(D(ij,ab)+D(kl,cd))=$$

$$D(ij,ab)\langle ij||ab\rangle\ D(kl,cd)\langle kl||cd\rangle, \quad (39)$$

which is just a product of double replacements, compare (35). The explicit expression for (36) is then a sum of 18 terms (39) corresponding to the 18 (=36/2) different partitionings of $ijkl\rightarrow abcd$ into pair replacements which can be written as

$$C^{(2)}(ijkl,abcd) = 1/32 \sum_{PP'}(-1)^{P+P'}$$

$$C^{(1)}(ij,ab)C^{(1)}(kl,cd) \quad (40)$$

where P and P' denote permutations of (ijkl) and (abcd) respectively. Because of the usual convergence problems of perturbation expansions one does not use (40) directly, but the structure displayed by (40) is carried over in putting

$$C(ijkl,abcd) \approx \tilde{C}(ijkl,abcd) =$$

$$= 1/32 \sum_{PP'} (-1)^{P+P'} C(ij,ab)\ C(kl,cd)$$

$$= C(ij,ab)\ C(kl,cd) - C(ik,ab)\ C(jl,cd) + \ldots \qquad (41)$$

We note that 18 terms contribute to $\tilde{C}$ which has, of course, the required antisymmetry. Inserting (41) into (34) we get the CP-MET equations of Cizek [2]

$$\langle \Psi(ij,ab) | \hat{H}-E | \Psi(o) + \Psi(D) \rangle = -\sum_{k<l} \sum_{c<d} \langle kl || cd \rangle \tilde{C}(ijkl,abcd) \qquad (42)$$

which together with (33) and (41) determine C(ij,ab) and E.

It is an important aspect of eq (40) that quadruples occur in the lowest order of perturbation theory as products of doubles or cluster contributions, and that genuine quadruples enter in higher order only. Similar relationships hold for other types of excitations and can be formulated in a systematic way, see e.g. [17]. Let us define the excitation operators

$$\hat{S}(1) = \sum_{ia} C(i,a)\ \hat{a}^{+}\ \hat{i} \qquad (43)$$

$$\hat{S}(2) = \sum_{i<j} \sum_{a<b} C(ij,ab)\ \hat{a}^{+}\ \hat{i}\ \hat{b}^{+}\ \hat{j} \qquad (43a)$$

(and $\hat{S}(3)$ etc correspondingly), where $\hat{i}$, $\hat{a}^{+}$ denote annihilation (of $|i\rangle$) and creation (of $|a\rangle$) operators. One is then lead to the Coester-Kämmel ansatz [30]

$$\Psi = \exp\ (\hat{S}(1) + \hat{S}(2) + \ldots)\ \Psi(o) \qquad (44)$$

The parameters C(i,a), C(ij,ab) are determined from the CC(SD) eqs.

$$\langle \Psi(i,a) | \hat{H}-E |\ \exp(\hat{S}(1) + \hat{S}(2))\ \Psi(o) \rangle = 0 \qquad (45)$$

$$\langle \Psi(ij,ab) | \hat{H}-E |\ \exp(\hat{S}(1) + \hat{S}(2))\ \Psi(o) \rangle = 0 \qquad (46)$$

(together with (33)) if we restrict the treatment to singles and doubles. The eqs (45) and (46) do not lead to an upper bound for the energy, but a strictly variational treatment, i.e. optimization of $\langle \Psi | \hat{H} | \Psi \rangle$ with Ψ from (44), appears to be too complicated.

Coupled cluster techniques have recently been implemented by Taylor et al [31], Koch and Kutzelnigg [32] (both with the PNO expansion for the pair functions), by Pople et al [33], Dykstra et al [10] (an SCEP version), and by Bartlett and Purvis [34], and we refer the reader to these articles for more details. Since full CI computations have recently become available [35] it was possible to compare these energies with those from the CC(SD) method. For a small basis set(of DZ type) water calculation the CC(SD) yield a correlation energy of -0.14624 au, which has to be compared with the CI(SD) result, -0.14018 au, and the full CI, -0.14803 au [34]. The CC(SD) thus accounts for 77% of the correlation energy which in a CI would be attributed to higher than double substitutions.

CEPA methods simplify the coupled cluster eqs (42) by further approximations of the r.h.s., which results in a less strongly coupled set of equations and thereby achieve a reduction of computational expense by about 30%. The CEPA equations have the form

$$\langle \Psi(ij,ab) | \hat{H} - E(SCF) - A(ij) | \Psi(o) + \Psi(D) \rangle = 0 \quad (47)$$

if we neglect singles for the sake of simplicity, for more details the reader is again referred to the literature [6-10,14,29,32]. The parameters A(ij) have different values in different CEPA versions. Let

$$e(ij) = \sum_{a<b} C(ij,ab) \langle \Psi(ij,ab) | \hat{H} | \Psi(o) \rangle \quad (48)$$

$$E(corr) = E - E(SCF) = \sum_{i<j} e(ij) \quad (49)$$

then

$$A(ij) = 0 \qquad \text{CEPA(0)} \quad (50)$$

$$A(ij) = e(ij) \qquad \text{CEPA(2)} \quad (51)$$

$$A(ij) = 0.5 \sum_{k} (e(ik) + e(jk)) \qquad \text{CEPA(1)} \quad (52)$$

$$A(ij) = \sum_{k} (e(ik) + e(jk)) - e(ij) \qquad \text{CEPA(3)} \quad (53)$$

where CEPA(0) is identical to the linear version of CP-MET [2].

These choices may be justified in the following way. If one includes only the first term of the r.h.s. of (41) it is possible to sum the r.h.s. of (42) which just gives the correlation energy and thus cancels the latter in the l.h.s. of (42). Proceeding this way one has in fact summed terms such as C(ij,ab) C(ij,cd), which violate the Pauli principle and are cancelled by the remaining terms in (41). If one just drops the excitations from the same pair in (42), i.e. the terms (kl)=(ij), one gets CEPA(2), if all Pauli principle violations with respect to occupied MOs are omitted [36] one gets CEPA(3). CEPA(1) can then be considered as an average of CEPA(1) and CEPA(3). The CEPA versions 1,2,and 3 depend on the actual choice of occupied MOs whereas the CP-MET (42) and CEPA(0) are unitarily invariant. However, CEPA(1) is unitarily invariant for the important special case of separated electron pairs, e.g. N He atoms, and is generally close to unitary invariance [9]. The CEPA equations are usually not solved in a spin-orbit basis but rather in using spin irreducible pairs, as described in sections 2 and 3. This requires slight changes in the definitions of A(ij) parameters, due to the necessary spin averaging, which is treated in detail by Hurley [29].

As far as comparisons of CEPA(SD) with CC(SD) or complete CI calculations are possible, it appears that CEPA(1) is usually very close to CC(SD), and that CEPA(1) and CEPA(2) usually bracket the full CI: CEPA(0) $<$ CEPA(2) $<$ CI(full) $<$ CEPA(1) $<$ CEPA(3). The CEPA equations (47)-(53) involve only slight changes with respect to the CI(D) or CI(SD) and lead to a marked improvement in potential curves and other properties, as has been demonstrated in a number of applications of which we mention especially those of Meyer et al [37] as an example.

Let us finally treat the r.h.s. of (42) by means of perturbation theory on top of a CI(SD), i.e. one first neglects this term and then evaluates the correction in using the CI coefficients from the CI(SD). Considering again only the first term for C(ijkl,abcd) of (41) in eq. (42), we get by multiplication with C(ij,ab) and summation over ij and ab a correction ΔE to E in lowest order of perturbation theory

$$\Delta E=(E(SD)-E(SCF))\sum_{i<j}\sum_{a<b}|C(ij,ab)|^{2} \qquad (54)$$

which is a Davidson type correction [3,18] in the intermediate normalization. For a normalized CI(SD) wavefunction this yields

$$E=(E(SD)-E(SCF))(1-C(o)^{2})/C(o)^{2} \qquad (55)$$

This correction seems to work remarkably well. The CI(SD) certainly yields too small CI coefficients C(ij,ab), as discussed in section 3, but the CEPA(0) overestimates the cluster corrections which leads to a cancellation of errors, at least in part.

5. CONCLUSIONS

The present author has found it difficult to write this review of pair theories since recent developments have resulted in a large overlap between these and standard CI methods, as discussed in section 3. The concept of electron pairs has rather become an aspect of CI techniques and the most recent pair methods, e.g. SCEP [8-10], are perhaps better described as matrix formulated direct CI procedures [9].

With the development of MR-CI(SD) programs it could be hoped that there was no urgent need for cluster corrections as a consequence of markedly increased accuracy. However, recent experience indicates [38,39] the importance of cluster corrections even for these cases. Although CEPA(0) and Davidson's correction formula are easily extended to this MR-CI(SD) one certainly needs more carefull investigations to settle this problem.

Acknowledgements. The author is indebted to Dr. C. Votava for valuable discussions and help in preparing the manuscript.

APPENDIX: STRUCTURE OF CI MATRIX ELEMENTS AND RESIDUAL VECTORS.

For the sake of completeness we report the structure of CI matrix elements for the CSFs defined in eqs (13) and (14) [12,15,16]. The nomenclature is the same as in section 3, see especially eqs (21)-(25)

$$\langle\Psi(R)|\hat{H}|\Psi(s,a)\rangle = H(Rs,a) = \sum_{i}\Big[A(Rs,i)h(ia) + \sum_{jk} B(Rs,ijk)(ai|jk)\Big] \tag{A1}$$

$$\langle\Psi(s,a)|\hat{H}|\Psi(t,b)\rangle = H(st,ab) = \langle a|b\rangle\langle\Phi(s)|\hat{H}|\Phi(t)\rangle \tag{A2}$$

$$+\delta(st)h(ab) + \sum_{ij} BJ(st,ij)(ij|ab) + BK(st,ij)(ia|jb)$$

$$\langle \Psi(R) | \hat{H} | \Psi(P,ab) \rangle = H(RP,ab) = \qquad (A3)$$

$$T(P,ab) \sum_{ij} B(RP,ij)(ia|jb)$$

$$\langle \Psi(sa) | \hat{H} | \Psi(P,cd) \rangle = T(P,cd) \Big[\langle a|c \rangle H(sP,d) + \qquad (A4)$$

$$\sum_{i} B(sP,i)(ic|ad) \Big]$$

$$H(sP,d) = \sum_{i} \Big[A(sP,i)h(id) + \sum_{jk} B(sP,ijk)(di|jk) \Big] \qquad (A5)$$

$$\langle \Psi(P,ab) | \hat{H} | \Psi(Q,cd) \rangle = T(P,ab)T(Q,cd) \Big[\delta(PQ)(ac|bd)$$

$$+ 2\langle b|d \rangle h(PQ,ac) + \langle a|c \rangle \langle b|d \rangle \; H(PQ) \Big] \qquad (A6)$$

$$h(PQ,ac) = \langle a | \hat{h}(PQ) | c \rangle \qquad (A7)$$

In these equations we have deleted the explicit electron numbers for $\hat{H}$ - e.g. n, n-1 or n-2 - since these are obvious from the context, and we have further introduced for convenience the auxiliary quantities H(st,ab) etc. We note again that these equations even hold if the external MOs occuring on the left and right of $\hat{H}$ do not belong to the same orthonormal set which is the basis of PNO methods. However, the explicit expression for

$$G(I) = \langle \Psi(I) | \hat{H} | \Psi \rangle, \qquad (A8)$$

where I denotes any of the CSFs and Ψ is as in eq. (16), is simplified if one has just a single orthonormal set of external MOs. One then gets immediately from (A1)-(A7):

$$G(R) = \sum_{S} \langle \Psi(R) | \hat{H} | \Psi(S) \rangle C(S) + \sum_{sa} H(Rs,a)C(s,a)$$

$$+ \sum_{Pab} H(RP,ab)C(P,ab) \qquad (A9)$$

$$G(s,a) = \sum_{R} H(Rs,a)C(R) + \sum_{tb} H(st,ab)C(t,b) + \qquad (A10)$$

$$+ \sum_{Pd} H(sP,d)C(P,ad) + \sum_{cdi} \Big[\sum_{P} B(sP,i)C(P,cd) \Big] (ic|ad)$$

$$G(P,cd)=\sum_{R} H(RP,cd)C(R)+T(P,cd)\Big\{\sum_{s} H(sP,d)C(s,c)$$

$$+\sum_{ia}\Big[\sum_{s} C(s,a)B(sP,i)\Big](ic|ad)+\sum_{Qb} 2h(PQ,cb)C(Q,bd)\Big\}$$

$$+\sum_{Q} H(PQ)\ C(Q,cd)+\sum_{db} C(P,ab)(ac|bd) \qquad (A11)$$

These equations involve mainly matrix algebra (for the most time consuming processing of two particle - two hole integrals) and the construction of general exchange operators, see eq. (11), which can be performed by an AO integral driven procedure, eq. (12), i.e. one has a typical SCEP formulation. As for the absorption of (ic|ad) integrals into the processing of the (ab|cd) integrals the reader is referred to the literature [12,16].

REFERENCES

1. Hurley, A.C., Lennard-Jones, J., and Pople, J.A.: 1953, Proc. Roy. Soc. (London) A220, 446.
2. Cizek, J.: 1966, J. Chem. Phys. 45, 4256; 1969, Adv. Chem. Phys. 14, 35.
3. Sinanoglu, O.: 1962, J. Chem. Phys. 36, 706; 1969, Adv. Chem. Phys. 14, 237.
4. Nesbet, R.K.: 1958, Phys. Rev. 109, 1632; 1969, Adv. Chem. Phys. 9, 311.
5. Jungen, M., and Ahlrichs, R.: 1970, Theoret. Chim. Acta (Berlin) 17, 339.
6. Meyer, W.: 1971, Int. J. Quant. Chem. S5, 341; 1973, J. Chem. Phys. 58, 1017.
7. Ahlrichs, R., Lischka, H., Staemmler, V., and Kutzelnigg, W.: 1975, J. Chem. Phys. 62, 1225.
8. Meyer, W.: 1976, J. Chem. Phys. 64, 2901.
9. Ahlrichs, R.: 1979, Comp. Phys. Comm. 17, 31; Zirz, C., and Ahlrichs, R., in: 1979, Electron correlation: Proceedings of the Daresbury study weekend, Guest, H.F., and Wilson, S., eds., Daresbury, England: Sciences Research Council.
10. Chiles, R.A., and Dykstra, C.E.: 1981, J. Chem. Phys. 74, 4544.
11. Taylor, P.R.: 1981, J. Chem. Phys. 74, 1256.
12. Werner, H.-J., and Reinsch, F.-A.: 1982, J. Chem. Phys. 76, 3144.
13. Siegbahn, P.E.M.: 1977, Chem. Phys. 25, 197; 1981, to be published.
14. Roos, B.O.: 1972, Chem. Phys. Letters 15, 153; Roos, B.O., and Siegbahn, P.E.M., in: 1977, Modern Theoretical Chemistry 3, Schaefer, H.F., ed., Plenum,

New York.
15. Siegbahn, P.E.M.: 1980, Int. J. Quant. Chem. 18,1229.
16. Ahlrichs, R., in: Proceedings of the 5th seminar on computational methods in quantum chemistry, Groningen 1981, Max Planck Institut fär Physik und Astrophysik, München.
17. Kutzelnigg, W., in: Modern Theoretical Chemistry 3, see ref 14b.
18. Davidson, E.R., in: 1974, The world of quantum chemistry, Daudel, R., and Pullman, B., eds., Reidel, Dordrecht; Langhoff, S.R., and Davidson, E.R.: Int. J. Quant. Chem. S9, 183.
19. Löwdin, P.-O., and Shull, H.: 1956, Phys. Rev. 101, 1730; see also ref 1.
20. Ahlrichs, R., and Driessler, F.: 1975, Theoret. Chim. Acta (Berlin), 36, 275.
21. Meyer, W., in: Modern Theoretical Chemistry 3, see ref 14b.
22. Siegbahn, P.E.M.: 1980, J. Chem. Phys. 72, 1647.
23. Brooks, B.R., Laidig, W.D., Saxe, P., Handy, N.C., and Schaefer, H.F.: 1980, Physica Scripta 21, 312.
24. Lischka, H., Shepard, R., Brown, F.B., and Shavitt, I.: 1981, Int. J. Quant. Chem. S15, 91.
25. Liu, B., and Yoshimine, M.: 1981, J. Chem. Phys. 74, 612.
26. Saunders, V.R., and van Lenthe, J.H.: 1982, Mol. Phys. (submitted).
27. Buenker, R.J., Peyerimhoff, S.D., and Butscher, W.: 1978, Mol. Phys. 35, 771.
28. Werner, H.-J.: 1982, Habilitationsschrift, Fachbereich Chemie, Universität Frankfurt.
29. Hurley, A.C.: 1976, Electron correlation in small molecules, Academic Press, London.
30. Coester, F., and Kümmel, H.: 1960, Nucl. Phys. 17, 477.
31. Taylor, P.R., Bacskay, G.B., Hush, N.S., and Hurley, A.C.: 1976, Chem. Phys. Letters 41, 444; 1978, J. Chem. Phys. 69, 4669.
32. Koch, S., and Kutzelnigg, W.: 1981, Theoret. Chim. Acta (Berlin) 59, 387.
33. Pople, J.A., Krishnan, R., Schlegel, H.B., and Binkley, J.S.: 1978, Int. J. Quant. Chem. 14, 545.
34. Purvis, G.D., and Bartlett, R.J.: 1982, J. Chem. Phys. 76, 1910.
35. Saxe, P., Schaefer, H.F., and Handy, N.: 1981, Chem. Phys. Letters 79, 202.
36. Kelly, H.P.: 1964, Phys. Rev. 134A, 1450.
37. Meyer, W., and Rosmus, P.: 1975, J. Chem. Phys. 63, 2356.
38. Siegbahn, P.E.M.: private communication.
39. Werner, H.-J.: private communication.

ON A GREEN'S FUNCTION METHOD FOR THE CALCULATION OF IONIZATION SPECTRA IN THE OUTER AND INNER VALENCE REGION

W. von Niessen

Institut für Physikalische und
Theoretische Chemie
Technische Universität Braunschweig
D-33 Braunschweig, W.-Germany

J. Schirmer and L. S. Cederbaum

Institut für Physikalische Chemie
Universität Heidelberg
D-69 Heidelberg, W.-Germany

I. Introduction

The one-particle Green's function contains all information about ionization spectra and electron affinities, in particular the information on the energies and relative intensities of the lines. The vibrational structure as well as vibronic coupling phenomena can also be obtained from the Green's function. It turns out that the structure of the ionization spectrum is very different in general in the outer and inner valence region. In the inner valence region the molecular orbital model of ionization may break down completely. The intensity then becomes distributed over numerous lines and a main line ceases to exist. In this energy region due to the high density of states a computational method should be used which correctly gives the general structure of the spectra - such a method is the two-particle-hole Tamm-Dancoff Green's function approximation (2ph-TDA), whereas in

G. H. F. Diercksen and S. Wilson (eds.), Methods in Computational Molecular Physics, 227–248.

the outer valence region high accu-racy is required. This is supplied by an extended 2ph-TDA method (e-2ph-TDA) or a method especially adapted to this energy range, the outer valence Green's function or OVGF method. The latter method is based on a finite perturbation expansion with renormalization of the interaction, whereas the former one is an infinite order approximation leading to systems of equations. In section II we are going to introduce the concept of the Green's function and present the essentials of the OVGF and e-2ph-TDA approximation schemes. Sections III and IV are devoted to the numerical implimentation of the OVGF and the e-2ph-TDA methods and section V will give an introduction to the applicability of the methods.

II. The one-particle Green's function and its OVGF and e-2ph-TDA approximation schemes.

The one-particle Green's function is defined in time, state space as the expectation value with respect to the exact ground state wave function of a time-ordered product of annihilation and creation operators for electrons in one-particle states:

$$G_{k\ell}(t,t') = -i \left\langle \Psi_0^N \middle| T\{a_k(t) a_\ell^{+}(t')\} \middle| \Psi_0^N \right\rangle . \quad (1)$$

For a useful reference on Green's functions see ref.1 $a_k(t)$, $a_k^{+}(t)$ are operators in the Heisenberg representation with H the full Hamiltonian of the system:

$$a_k^{(+)}(t) = e^{iHt} a_k^{(+)} e^{-iHt} . \quad (2)$$

They annihilate (create) electrons in one-particle states $|k\rangle$. The operators fulfill the anticommutation relations $[a_k, a_\ell^{+}]_+ = \delta_{k\ell}$ with all other anticommutators vanishing. The Hamiltonian is

$$H = \sum_i \varepsilon_i a_i^{+} a_i + \frac{1}{2} \sum_{i,j,k,\ell} V_{ijk\ell}\, a_i^{+} a_j^{+} a_\ell a_k \quad (3)$$

$$- \sum_{i,j} \sum_\ell V_{i\ell[j\ell]}\, a_i^{+} a_j n_\ell ,$$

where

$$V_{ijkl} = \iint \varphi_i^*(1)\varphi_j^*(2)\frac{1}{r_{12}}\varphi_k(1)\varphi_l(2)\,d\tau_1 d\tau_2 \tag{4}$$

$$V_{ij[kl]} = V_{ijkl} - V_{ijlk} \quad . \tag{5}$$

The ε_i are the orbital energies and $n_l = 1$ for l an occupied orbital and $n_l = 0$ for l a virtual orbital. This form of the Hamiltonian is the appropriate one if one starts from a Hartree-Fock solution. T is Wick's time ordering operator which orders the operators so that time increases from right to left. A permutation of the operators from the original ordering by the action of T is accompanied by a change of sign. With the help of the Fourier transformation

$$G_{kl}(\omega) = \int_{-\infty}^{\infty} G_{kl}(t,t')e^{i\omega(t-t')}d(t-t') \tag{6}$$

and inserting the decomposition of unity one arrives at the spectral representation of the Green's function

$$G_{kl}(\omega) = \lim_{\eta\to+0}\left\{\sum_n \frac{\langle\Psi_0^N|a_k|\Psi_n^{N+1}\rangle\langle\Psi_n^{N+1}|a_l^+|\Psi_0^N\rangle}{\omega + A_n + i\eta}\right. \tag{7}$$

$$\left. + \sum_m \frac{\langle\Psi_0^N|a_l^+|\Psi_m^{N-1}\rangle\langle\Psi_m^{N-1}|a_k|\Psi_0^N\rangle}{\omega + I_m - i\eta}\right\}$$

with $A_n = E_0^N - E_n^{N+1}$ the vertical electron affinity and $I_m = E_m^{N-1} - E_0^N$ the vertical ionization energy. By calculating the poles of the Green's function one thus obtains directly the ionization energies and electron affinities. The amplitudes of the Green's function

$$x_p^{(n)} = \langle\Psi_n^{N-1}|a_p|\Psi_0^N\rangle \tag{8}$$

$$Y_p^{(n)} = \langle \Psi_n^{N+1} | a_p^+ | \Psi_0^N \rangle \tag{9}$$

are related to the relative spectral intensities for electron ejection and capture, respectively. In the limit of high kinetic energy of the ejected electron the intensity of line (n) in the ionization spectrum is given by

$$I^{(n)} = \sum_{\varepsilon} \left| \sum_p \tau_{\varepsilon p} X_p^{(n)} \right|^2 , \tag{10}$$

where $\tau_{\varepsilon p}$ is the dipole matrix element for ionization out of orbital p; ε denotes (energy conserving) continuum states. In general, for a particular ionic state n, only one of the amplitudes $x_p^{(n)}$ is appreciable. In this case equation (10) simplifies to

$$I^{(n)} = |X_p^{(n)}|^2 \sum_{\varepsilon} |\tau_{\varepsilon p}|^2 . \tag{11}$$

The relative intensity or pole strength is given by

$$P_p^{(n)} = |X_p^{(n)}|^2 . \tag{12}$$

The ionization energies and electron affinities and their pole strengths are best calculated from the Dyson equation

$$G(\omega) = G^0(\omega) + G^0(\omega) \Sigma(\omega) G(\omega) , \tag{13}$$

which connects the Green's function with the Hartree-Fock Green's function $G^0_{ke} = \delta_{ke}/(\omega - \varepsilon_k)$ and the self-energy $\Sigma(\omega)$. $\Sigma(\omega)$ is the exact potential seen by an electron due to the interaction with its surroundings. The Green's function has an inverse:

$$G^{-1} = (G^0)^{-1} - \Sigma(\omega) . \tag{14}$$

Ionization energies and electron affinities are thus calculated by obtaining the zeros of G^{-1}. In a diagonal approximation this takes the form:

$$G^{-1}_{kk} = 0 = \omega - \varepsilon_k - \Sigma_{kk}(\omega) \tag{15}$$

The energies fulfilling this equation are the ionization energies and electron affinities. They can be calculated by obtaining the intersection points of the straight line $y = \omega - \varepsilon_k$ with $\Sigma_{kk}(\omega)$. This and the structure of Σ is represented schematically in fig. 1.

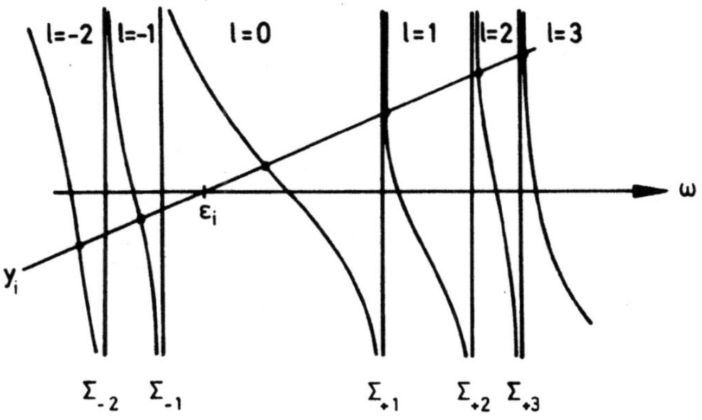

Fig.1 Schematic plot of $\Sigma_{kk}(\omega)$ and of the solution of the Dyson equation.

It has to be noted that this simple picture applies only in case of a diagonal approximation to Σ. If Σ is nondiagonal the Green's function is nondiagonal and the picture is quite different. There may be 0,1 or many intersections of the straight line $y = \omega$ with $\varepsilon + \Sigma(\omega)$. Σ itself has poles and is a monotonically decreasing function of ω between the poles. For closed shell systems Σ has always a large interval free of poles. In this interval the outer valence ionization energies are situated and for their calculation high accuracy is required. In this region far away from the poles a perturbation expansion of Σ in the electron-electron-interaction is justified. This has been done by Cederbaum[2,3]. All terms up to and including the third order terms are included. Higher order terms are taken into account by a renormalization procedure

$$\Sigma = \Sigma^{(2)} + \Sigma^{(3)} + \Sigma^{(R)} \quad . \qquad (16)$$

The first order term vanishes if one starts from the Hartree-Fock solution. The expansion of Σ is done with the help of diagrammatic techniques. The second and third order diagrams in the Abrikosov notation are given in fig. 2 and fig.3

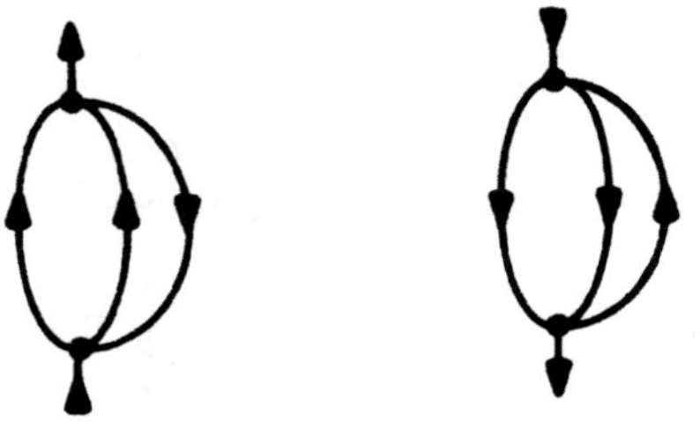

Fig. 2 The time-ordered self-energy diagrams of second order.

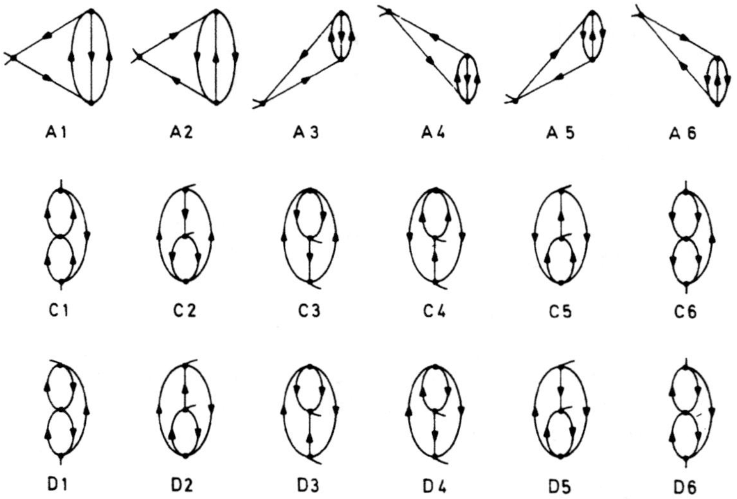

Fig. 3 The notation of the time-ordered self-energy diagrams of third order.

In these diagrams the dots represent antisymmetrized Vijkl matrix elements and the lines are Green's function lines. In the renormalization procedure where geometric series are summed, which are derived based on a physical model of the interaction, no new diagrams are required. In second and third order the essential structure of the formulae obtained by evaluating the diagrams is the following (omitting any indices and details).

$$\Sigma^{(2)} = \sum \frac{V \cdot V}{\omega - \Delta \varepsilon} \tag{17}$$

$$\Sigma^{(3)} = \begin{cases} \sum \dfrac{V \cdot V \cdot V}{\Delta\varepsilon \cdot \Delta\varepsilon} & \text{for A-type diagrams} \quad (18) \\[2ex] \sum \dfrac{V \cdot V \cdot V}{(\omega - \Delta\varepsilon) \cdot \Delta\varepsilon} & \text{for C2 - C5 and D2 to D5 diagrams} \quad (19) \\[2ex] \sum \dfrac{V \cdot V \cdot V}{(\omega - \Delta\varepsilon)(\omega - \Delta\varepsilon)} & \text{for C1, C6, D1, D6 diagrams} \quad (20) \end{cases}$$

i. e. in second order we have only simple poles (which holds also for the exact self energy), whereas in third order we can have energy independent diagrams, simple poles and quadratic poles. At the latter poles the self-energy necessarily behaves incorrectly. In the OVGF method the pole strength is calculated from the following formula

$$P_p^{(n)} = \left(1 - \frac{\partial \Sigma_{pp}(\omega)}{\partial \omega}\right)^{-1} \qquad (21)$$

which applies if the Green's function is diagonal and the main ionic state n derives from the simple electron ejection from orbital p. These pole strengths are necessary as an essential part of the final results and also for the iterative solution of the Dyson equation. Thus in addition to the diagrams their derivatives must be calculated, but this can be done at very little extra cost.

The finite perturbation expansion including renormalization can only be used for the calculation of the outer valence ionization energies which are far from the poles of the self-energy. For the calculation of the ionization energies in the inner valence region, for the calculation of the satellite lines and for the calculation of the ionization energies which are still situated in the main interval of the self-energy but close to the poles a method is required which correctly accounts for the pole structure of the self-energy. Quite new phenomena may occur in this energy range. In the inner valence region the one-particle picture of ionization may break down completely. The method used in this energy region is the 2ph-TDA method and its extended version[4,5].

The exact self-energy has a constant and an energy dependent term which we denote by M (ω).

$$\Sigma(\omega) = \Sigma(\infty) + M(\omega) \quad . \tag{22}$$

$\Sigma(\infty)$ can be obtained from M (ω) and the Green's function which in turn is obtained from the solution of the Dyson equation. As an algebraic equation we obtain

$$\Sigma_{pq}(\infty) = -\sum_i V_{pi[qi]}\, n_i + \sum_{i,j} V_{pi[qj]} \frac{1}{2\pi i} \oint G_{ji}(\omega)\, d\omega \tag{23}$$

which together with the Dyson equation (13) indicates the iterative process. We now investigate the term M (ω) further. The lowest order term of M (ω) has the correct analytical structure but supplies only a poor approximation. Any higher finite order approximation will introduce higher order poles and is not appropriate. We thus have to construct infinite summations which lead to the correct pole structure. This can be achieved in the following way. (We will omit in general any details and indices). We separate M (ω) into its external vertices and the remainder

$$M = V^T \cdot \Gamma \cdot V \quad , \tag{24}$$

where the product includes a number of summations. Γ, the kernel of the self-energy, has itself a perturbation expansion. For Γ the following equation is obtained

$$\Gamma(\omega) = (\omega - K - C)^{-1} \tag{25}$$

Where

$$K_{jkl,j'k'l'} = \delta_{jj'}\,\delta_{kk'}\,\delta_{ll'}\,(\varepsilon_k + \varepsilon_l - \varepsilon_j) \tag{26}$$

and

$$C_{jkl,j'k'l'} = -\delta_{jj'}\, V_{kl[k'l']} + \delta_{kk'}\, V_{jl'[j'l]} + \delta_{ll'}\, V_{jk'[j'k]} \tag{27}$$

The index space jkl is restricted to be either the two-particle-one-hole space (2p1h) (this is called the A block later on and will supply the affinity poles of the self-energy finally) or the two-hole-one-particle space (2h1p) (this is called the B block later on and will supply the ionization poles of the self-energy). Thus we have effectively summed all diagrams which contain either two hole and one particle line or one hole and two particle lines between any two interaction points. Equation (25) is solved by diagonalization:

$$\mathcal{U}^{+}(\omega - K^{A,B} - C^{A,B})\,\mathcal{U} = \omega - \omega^{A,B} \tag{28}$$

$$(\omega - K^{A,B} - C^{A,B})^{-1} = \sum_n \frac{1}{\omega - \omega_n^{A,B}}\, \mathcal{U}^{(n)}\mathcal{U}^{(n)+} \quad . \tag{29}$$

The self energy is then

$$\Sigma_{pq}(\omega) = A_{pq} + \sum_n \frac{z^A_{pn} z^{A*}_{qn}}{\omega - \omega_n{}^A} + \sum_m \frac{z^B_{pm} z^{B*}_{qm}}{\omega - \omega_m{}^B} \tag{30}$$

with

$$z = V \cdot \mathcal{U} \tag{31}$$

where again a summation is implied. The details are given in ref. 4-6. Having obtained the poles and amplitudes of the residues of the self-energy one can solve the Dyson equation for the ionization energies and electron affinities either by a pole search procedure [7] or by diagonalization of the matrix.

$$Y = \begin{pmatrix} \omega - \varepsilon - A & z^A & z^B \\ z^{A+} & \omega - \omega^A & 0 \\ z^{B+} & 0 & \omega - \omega^B \end{pmatrix} \tag{32}$$

That this is equival-ent to solving the Dyson equation can be shown by using partitioning. The formulas given above define the 2ph-TDA scheme. The e-2ph-TDA differs in the calculation of the amplitudes of the self-energy. z is then schematically given by

$$z = V \cdot \mathcal{U} + \frac{V \cdot V}{\Delta \varepsilon} \cdot \mathcal{U} \tag{33}$$

i. e. a series of terms which are products of V matrix elements are added. The 2ph-TDA method is exact to second order, whereas the e-2ph-TDA method is exact to third order, i. e. it represents a much richer summation which in particular gives better results than the 2ph-TDA method for the main lines. Not only diagrams which have two hole and one particle lines or one hole and two particle lines between any two interaction points are summed but also the diagrams C2 - C5 and D2 - D5 of fig. 3 and higher order diagrams of these types. Originally the e-2ph-TDA scheme has been derived by a careful analysis of the third order diagrams but now it can be derived by a rigorous procedure called algebraic diagrammatic construction.[8] Also approximation schemes which are exact to fourth or to fifth order have been derived.[8]

III. Computational procedure for the OVGF method

It has been mentioned above that the OVGF method can only be applied to states far from the poles of the self-energy. These are in general the first ionization energies for each symmetry. From the OVGF method it cannot be determined whether this method can be applied, a 2ph-TDA calculation must first be performed to find this out. For larger molecules already the first ionization process may give rise to appreciable satellite structure and there may be molecules where not a single ionization energy is far away from the poles (e. g. CuF and CuCl). These limitations are inherent in a finite order perturbational approach. We are going to restrict ourselves now to the case that the Green's function is diagonal, this restriction is, however, not necessary. We have to consider two problems here, the evaluation of the diagrams and the solution of the Dyson equation.

Evaluation of diagrams

The diagrams consist in second order of triple summations over double products of Vijkl matrix-elements (and combinations thereof) in third order of quintuple summations over triple products. The problem lies in the number of such matrix elements and the addressing of these. Symmetry is very essential in dealing with this problem. There are two types of symmetry which are compatible and both can be made full use of: the permutational symmetry and the point group symmetry of the nuclear framework. Only integrals nonzero by point group symmetry and unique by permutational symmetry are kept. At present no use is made of symmetry groups with degenerate representations. Abelian subgroups are used instead. Three different methods are available for the evaluation of the diagrams. In the first one we keep all unique and nonzero integrals in fast core as a linear array. The addressing is done by a set of three linear index arrays. Second and third order diagrams can be treated by similar procedures.

Sometimes - this is true for the constant diagrams - can the diagrams be separated and a part be calculated ahead of time. In the second method which has been described by Silver and Wilson[9] not all of the integrals are kept in core so that some have to be read in from disk. The diagrams to be evaluated contain integrals of the type Vijkl,Vijka,Vijab,Viabc and Vabcd, where i,j, k,l refer to occupied orbitals and a,b,c,d, to virtual orbitals. In general the number of virtual orbitals is considerably larger than the number of occupied ones. Thus the first blocks at least for some range of index pairs can be kept in core, but the block Vabcd, e. g. not. The blocks in core are stored in individual linear arrays. The others are read into core when needed.

If we can keep only a small fraction of the integrals in core then a method devised by Diercksen[10] can be used to evaluate the diagrams. The method is similar to the algorithm of Yoshimine[11] for the four-index transformation of the integrals. The two-electron integrals Vijkl are subdivided into coreloads. A coreload is the number of integrals which will be kept in core in the process of evaluating the diagrams. Instead of the four indices defining the integral we will use only two (n,m) the first one (n) is the coreload the integral is going to appear in and the second one (m) is the sequence number in the coreload. The diagrams consist of terms of the structure.

$$(p,p)\,C\,\underbrace{(abcd)}_{I(n_1,m_1)}\underbrace{(efgh)}_{II(n_2,m_2)}\underbrace{(ijkl)}_{III(n_3,m_3)} \qquad (34)$$

where (p,p) denotes the matrix element of the self-energy (diagonal here), C is a numerical constant and the three sets of indices, I,II and III, denote the two-electron integrals. Some of the indices a,b,c,... can be equal. The terms are thus created as symbolic elements consisting of one real word and 8 integers. We will step by step evaluate these terms starting with the index group III, which we will replace by the numerical value of the integral. As the integrals are not kept in core we have to order the terms with respect to the indices of III such that all terms in a given group can be filled once a coreload of integrals has been read into core. To each coreload n we assign a box n. The terms will be put into these boxes such that all terms will be put into box n if the integral defined by III will appear in coreload n. The sequence number m of this integral will be assigned to the term in box n. The boxes are subdivided into subparts and a subpart is written onto a direct access file once it is filled. We then start with the next subpart of this box. (The size of these subparts is calculated iteratively once the available core storage is given). The address on the direct access file is put into an identification section which is the first part of the next subpart. Once all terms are ordered the addresses of all the last subparts are noted for each box. In reading the boxes one rolls up the process from the end. We then start reading the integrals by coreloads. For each coreload we read in the corresponding box and replace the index set III by the numerical value of the integral which we combine with the constant. The resulting terms take then the form

$$(p,p)\,C_1^{\prime}\ \mathrm{I}\ \underline{\mathrm{II}} \qquad (35)$$

which are written on disk. This procedure ist continued for index sets II and then I. At the end we arrive at a list of terms. $(p,p)T_1, (p,p)T_2, \ldots. (q,q)T_k$. All terms having the same index pair (p,p) are combined. Their sum gives the matrix element $\Sigma(\omega)$.

Solution of the Dyson equation

The solution of the Dyson equation (13) requires an iterative process as the self-energy itself depends on the quantity ω. We have a case of Brillouin-Wigner perturbation theory. The solutions of the Dyson equation correspond to finding the intersection of the straight line $y = \omega - \varepsilon_p$ with $\Sigma_{pp}(\omega)$. The graphical representation is given in fig. 1. The iteration formula for the solution ω_p

$$\omega_p = \varepsilon_p + \Sigma_{pp}(\omega_p) \equiv \varphi_p(\omega_p) \qquad (36)$$

is based on a Taylor expansion of ω_p about the result of the previous step. One obtains

$$\omega_p^{(i)} = P_p^{(i-1)}\left[\varphi_p(\omega_p^{(i-1)}) - \omega_p^{(i-1)}\varphi_p^{\prime}(\omega_p^{(i-1)})\right] \qquad (37)$$

with P_p the pole strength for this solution (eq.21). Thus in addition to the diagrams we need their derivatives which are evaluated together with the diagrams at little extra cost. The starting values for the iteration process are the results of the next lower order approximation. The process converges extremely quickly (2 to

3 steps) except if one is close to poles of the self-energy where the method should not be used.

In this context a few other convergence problems should be shortly addressed from the point of view of experience alone. The perturbation expansion itself is in general an alternating series with respect to the final "converged" result. Thus its convergence behaviour should be excellent. This is certainly not sufficient for a reliable extrapolation of the series but it gives a good chance for success. The renormalization which has proved so successful is based on a physical model, on the information on the effective interaction which can be deduced from the second and third order results when the terms are separated into the two different time orderings. This is discussed in ref. 2 and 3. Within each order of the perturbation series the diagrams are of opposite sign and cancel to a fair degree. This has already been mentioned in ref. 2 and in part led to the development of a renormalization procedure. It also has another consequence which has proved to be important in practice. The convergence of the individual diagrams has been investigated both with respect to increasing the number of virtual orbitals for a fixed atomic basis set and with respect to enlarging the basis set. In the first case the convergence is monotonic and asymptotic for each diagram. As they are of opposite sign the sum of the diagrams tends to converge quicker than the individual diagrams. It appears now sensible that an extension of the basis influences positive and negative diagrams to a nearly equal extent. This would explain the relative basis set insensitivity of the results. This behaviour has been observed in fact in a fairly large number of cases. Both exhaustion of the molecular orbital basis set as enlargement of the atomic basis set tends to have a similar effect in magnitude on the positive and negative diagrams. The effect thus sums effectively to zero if one does not look for fine details and highly accurate results. This expected basis set insensitivity of the ionization energies in turn explains the good results obtained when using double-zeta basis sets. There is certainly no garantee that this cancellation effect works. Sure enough exceptions have been found. The larger errors practically always disappeared when using polarizations functions in the basis set and exhausting it (and if the OVGF method is applicable). The convergence behaviour turned out to be unproblematic in molecular orbital basis sets obtained without the use of polarization functions. If many polarization functions are included in the bases one should attempt to exhaust a given basis set as orbitals with a large contribution from polarization functions tend to be high in orbital energy. These orbitals thus tend to give important contributions to the diagrams which may not cancel out entirely in summing the diagrams. This basis set insensitivity has been made use of in calculating the ionization energies of a fairly large number of molecules in order to assign their photoelectron spectra. Only basis sets of double-zeta quality have been used for this purpose in general. It should be pointed out that the largest basis set is necessarily the best one and a given basis set should also be exhausted as much as possible. Except for the first investigations this has always been the aim. To obtain high accuracy large basis sets with several sets of polarization functions have been used.

The renormalization methods in use have been described in detail in ref. 2 and 3. There are three variants. Method A involves a single geometric series and the constant diagrams are renormalized as well. This is in the outer valence region a high energy renormalization. Method B involves the summation of two geometric series, where the terms are separated according to the time ordering of the diagrams. The constant diagrams are not renormalized in this case. It is in the outer valence region a low energy renormalization. The results of both methods tend to approach each other upon enlarging the atomic basis set. Method C is a rather special case and applies only if the ionization energy in second order is dominated by the affinity poles of the self-energy, i. e. when $\Sigma^{(2)}$ is negative or positive but small.

IV. Numerical implementation of the extended 2ph-TDA method

In the extended 2ph-TDA method we have to solve the Dyson equation (13) with the self-energy of the form of eq (30). The solution of the Dyson equation with this self-energy is equivalent to diagonalizing the matrix y of eq (32). Thus we have two means of solving the Dyson equation for the ionization energies and electron affinities and the corresponding pole strengths. Both the pole search method and the diagonalization method are shortly described below. We first turn to the calculation of the quantities $\omega^{A,B}$ and $z^{A,B}$. These quantities are obtained by solving eq (25) via the diagonalization (eq.'s (28) and (29)).

The diagonalization is done separately to obtain the affinity poles and the ionization poles. In the e-2ph-TDA method the matrices consist of combinations of orbital energies and Vijkl matrix elements and should be diagonalized completely. In the present version of the program the maximum dimension is set each to 350 to 400. It would be desirable to considerably increase this dimens-ion but the computation time increases dramatically. Another possibility would be to construct matrices of dimension 2 - 5000 and obtain the lowest 200 - 300 roots. This has to be addressed in the future. The A and B block matrices could be constructed similarly to CI-type matrices by making use of symbolic matrix elements. This is, however, not done and it does not appear to be necessary for some time. The problem associated with the e-2ph-TDA scheme lies in the dimension of the matrices and not in the number of two-electron integrals if proper use is made of symmetry. Thus the procedure of storing and addressing the integrals as described above is used here. The eigenvalues of the matrices give directly the poles. The amplitudes are obtained from the eigenvectors and the Vijkl matrix elements schematically given in eq. (31) and eq. (33).

Given the quantities ε_i, Aij (see below), $\omega^{A,B}$, and $z^{A,B}$ we can construct the matrix y of eq. (32) and obtain the ionization energies and electron affinities as the eigenvalues of this matrix. From the eigenvectors the residues of the Green's function, i.e. the pole strengths, can be obtained but since not the residues but the amplitudes are calculated directly one also obtains the sign which is important in the case the intensity of a given transition is borrowed from several simple transitions and interference effects can be found.

The pole search method for solving the Dyson equation is faster than the diagonalization method but is numerically more delicate. An important advantage is that more poles can be taken into account. The details are given in ref. 7. We can give here only a very sketchy outline. We consider pole n_o at which the self-energy goes to infinity and isolate this pole

$$\omega - [\underbrace{\{\varepsilon_{ij}\} + \{\Sigma'_{ij}\}}_{\{D_{ij}\}} + \{\Sigma_{ij}^{n_0}\}] = G^{-1} \quad . \qquad (38)$$

The eigenvalues of G^{-1} at $\omega = \omega^{n_o}$ are to be calculated, of which one diverges and is separated off. For this purpose the eigenvalues and eigenvectors of $\{\Sigma_{ij}^{n_o}\}$ are obtained. The only non-zero eigenvalue is $\lambda_N = \sum_{i=1}^{N} |z_{in}|^2$. Let U be the matrix of eigenvectors. We evaluate the matrix D at $\omega = \omega^{n_o}$, i.e. D^{n_o} form

$$\mathcal{U}^{+} D^{n_o} \mathcal{U} = F \tag{39}$$

and delete the column and row referring to the eigenvalue λ_N to obtain F' which is diagonalized. The resulting eigenvalues μ are the eigenvalues of G^{-1} at pole n_o which do not diverge. Next we investigate the intervall I between pole n_o and n_{o+1} to find out how the straight line $y = \omega$ intersects one of the eigenvalues. The intersection points are calculated iteratively via a Newton-Raphson procedure. The derivatives μ' can be obtained from the eigenvectors S of G^{-1}

$$\mu' = S^{+} G^{-1\prime} S \tag{40}$$

where

$$S^{+} G^{-1} S = \mu \quad . \tag{41}$$

If the result of a Newton-Raphson step leaves the interval one turns to the procedure of interval bisecting. The convergence should not be judged by the difference $\omega^{(i-1)} - \omega^{(i)}$ but by $\mu(\omega^{(i)}) - \omega^{(i)}$. The calculation of the eigenvectors of the Dyson matrix is done from the following equations:

a) component with respect to the one-particle block

$$R_{ip} = S_{ip} \left[1 - \mu'_p\right]^{-1/2} \tag{42}$$

b) component with respect to the 2h1p and 2p1h-block (jkl = n)

$$R_{np} = \Big(-\sum_{i=1}^{N} z_{ni} \cdot R_{ip}\Big) \Big/ (\mu_p - \omega_n) \tag{43}$$

The pole strength of solution p with respect to orbital i is then R_{ip}^2 for i an occupied orbital.

The calculation of the energy-independent diagrams follows from eq. (23) where the Green's function is obtained from the solution of the Dyson equation (eq. (13)). One notes that an iterative procedure must be used to calculate $\Sigma(\infty)$. The Green's function is calculated blockwise by symmetry. For any given symmetry contributions to $\Sigma(\infty)$ are obtained from all symmetry components of G. After all these different symmetry contributions are obtained and put together the next iteration step in the calculation of the Green's function can be started. It should be noted that not only the contributions from the symmetry species of the orbitals occupied in the ground state are required but also from the symmetry species of low lying virtual orbitals. For the calculation of $\Sigma(\infty)$ in general a starting value is required. If $\Sigma(\infty)$ is set equal to zero one may encounter convergence problems and in this case it is advisable to multiply $\Sigma^{i}(\infty)$ for the first three steps by numerical factors less than unity and approaching unity. A good starting point is obtained by using the Hartree-Fock Green's function. The values thus obtained differ by less than 10 % in general from the values obtained with the full Green's function. The calculation of the constant diagrams is an expensive process because of the many diagonalization which are required. Their inclusion gives in most cases a considerable improvement in the values for the main ionization energies. There is little effect on the values of the satellite lines as the pole structure is not affected by the constant diagrams.

The use of symmetry is essentially done as in the OVGF method. The ionization energies and the Green's function is calculated separately for each irreducible representation and even in the calculation of the constant diagrams a separate contribution is obtained from any given symmetry. The only true mixing of symmetries occurs in the calculation of $\Sigma(\infty)$ with the Hartree-Fock Green's function because of a required inversion. If a molecular symmetry group contains degenerate representations a further simplification can be achieved in the solution of the Dyson equation which reduces the dimension of the matrices. This does not involve the use of explicit vector coupling schemes for molecular point groups. The only thing which is needed are correlation tables. The A and B block matrices are constructed as usual. No simplification is achieved here. But the eigenvectors of these matrices transform as irreducible representations of the full group containing degenerate representations. Since the amplitudes of the residues of the self-energy are calculated from these eigenvectors they also transform as irreducible representations of the full group. This is taken advantage of in the separation of the components. The Dyson equation for each symmetry species of the subgroup separates into a number of components, i. e. we obtain a block diagonal form. This is made use of. Details are given in ref. 12.

Numerical Approximations

If one wishes to use the e-2ph-TDA method also for moderately large molecules or for small molecules with large basis sets additional numerical approximations need to be introduced. The main problem is the large dimension of the matrices which have to be diagonalized completely. We will first deal with the approximations used in the calculation of the poles of the self-energy to turn later to the approximations used for solving the Dyson equation.

To reduce the dimension of the A and B-block matrices it is simplest to discard as many virtual orbitals as necessary. This is not quite satisfactory as too few virtual orbitals may remain. The method used at present consists in discarding some of the high-lying virtual orbitals and using configuration selection together with a diagonal approximation for the terms neglected in the selection procedure. At present about 40 to 50 orbitals are taken into account. This may still result in dimensions for the A- and B-block matrices of 500 to many thousand. For the configuration selection (which is done separately in the A and B blocks) two schemes are used. The simplest one has proved to be most successful so far. All diagonal matrix elements and the "diagonal" amplitudes of the self-energy are calculated which arise in the given orbital space. They are separated into the two possible spin couplings leading to a doublet state. Those configurations are selected which have the lowest diagonal energies in either of the two spin coupling situations. Both spin couplings are either selected or neglected. About 350 to 400 configurations are selected for each symmetry in this way. In this configuration space the A- or B-block matrix is constructed and diagonalized. The configurations which are not selected are not neglected but taken into account in a diagonal approximation in the Dyson equation, i. e. only the diagonal matrix elements and the corresponding amplitudes of the self-energy are used for them. The other configuration selection method which is still under development takes into account the coupling strength of a given pole (i. e. the diagonal residue of the self-energy) besides the diagonal matrix element.

The A block matrix is constructed from 2p1h configurations, i. e. configurations of the negative ion. It may be thought that these terms should be much less important than the 2h1p configurations. It should be kept in mind, however, that this mixing of N-1 and N+1 electron states in the Dyson equation is used in the Green's function method to introduce ground state correlation. The size of the A block matrix should be as large as possible. It was only found in the 2ph-TDA method that a relatively small dimension of 100 to 150 would suffice to give nearly converged results. With the e-2ph-TDA method which includes more ground state correlation this is no longer the case.

If one uses the pole search method for the solution of the Dyson equation no further numerical approximations need to be introduced and all poles of the self-energy can be taken into account. If, on the other hand, the diagonalization of the matrix Y of eq. (32) is done additional approximations cannot be avoided. If one has taken eg 350 poles in the A and B block matrices into account and if there are 20 orbitals of the given symmetry then the matrix to be diagonalized in the Dyson equation is of dimension 720 not taking into account the diagonal approximation to the poles mentioned above. One has to reduce the dimension again to about 350 to 400. This is done in two ways. Weak poles both of the affinity and the ionization block are neglected. No further poles of the B block are neglected. The ionization poles neglected in the configuration selection procedure are taken into account in a diagonal approximation by forming effective poles and effective amplitudes of the residues according to

$$\omega_p^{C} = \sum_{n>M} \omega_n z_{pn}^2 \,/\left(z_p^{C}\right)^2 \tag{44}$$

$$z_p^{C} = \sqrt{\sum_{n>M} z_{pn}^2} \; , \tag{45}$$

where M is the number of poles taken into account exactly in the Dyson equation. For each orbital p of the given symmetry there will be one centroid of energy $\omega_p^{C(B)}$ and amplitude $z_p^{C(B)}$. This increases the dimension of the matrix again. The significant reduction comes from the treatment of the affinity poles. Only 10 to 20 affinity poles are taken into account exactly. The remaining poles and amplitudes which result from the diagonalization are combined with the diagonally approximated poles to form effective poles as above. This looks like a very severe approximation in particular with respect to what has been said above about the size of the A-block matrix. For the calculation of ionization energies, however, it has been found that these effective poles work extremely well. Increasing the number of poles which are taken into account exactly in the Dyson equation from 10 to 20 has an effect of only 10^{-3}eV on the lowest ionization energies, whereas increasing the dimension of the A block matrix from 200 to 300 may have an effect as large as 0.1-0.2 eV for the same system (10 affinity poles were taken into account in both calculations). Not enough experience has been gained, however, whether this restriction may have an effect on the values of the constant diagrams.

V. Applicability of the OVGF and e-2ph-TDA methods

This section serves to summarize very shortly the limits of applicability of these approximations. First of all it should be pointed out that all approximations discussed here apply only if the ground state of the molecule is a closed shell state. The 2ph-TDA and the e-2ph-TDA method are universally applicable in the valence region and in principle also in the core region. But ionic main states are treated more accurately than satellite lines in this scheme. To arrive at an equivalent accuracy for the satellite lines one would have to add 3p2h and 3h2p terms. This approximation scheme is being worked out[8]. It will be exact to fourth and fifth order in the electron-electron interaction, respectively, depending on the calculation of the residues of the self-energy. If the coupling between ionic main states and satellite states is strong these higher excitations also are crucial for an accurate description of the main state. This occurs in the valence region if intense satellite lines accompany the main line and in particular in the core region and the missing of such terms is the reason for the poorer performance of the method in the calculation of core hole states. The (e)-2ph-TDA method is applicable if the Green's function is nondiagonal. The OVGF method on the other hand, is applicable only far away from the poles of the self-energy, if there are no intense satellite lines (i. e. if the pole strength of the main line as calculated with the 2ph-TDA method is larger than approximately 0.9) and if the Green's function is diagonal (due to the restriction made above, which in this instance is not necessary). The poles of the self-energy as they occur in this method are "Hartree-Fock" type poles, i. e. differences of orbital energies, whereas in the 2ph-TDA method they are renormalized poles. In addition quadratic poles appear in the OVGF method. In some instances these shifts of the poles (and the change of character from quadratic to simple poles) due to this renormalization are so large that the OVGF method may become inapplicable. This has been found e.g. for CuF, CuCl. The OVGF method is inapplicable to any ionization process of these molecules not because of the appearance of intense satellite lines but because of important renormalization of the poles. Several of the ionization energies of CuF and CuCl have pole strengths larger than 0.9 in the e-2ph-TDA method, but when the OVGF method is used pole strengths larger than unity are encountered indicating the close presence of quadratic poles. The pure perturbation expansion breaks down in spite of the applicability of the one-particle model of ionization. A similar situation arises in the case of SnO and similar molecules. For SnO, however, intense satellite lines do appear limiting the applicability of the OVGF method also from this side.

For small molecules and for medium sized basis sets both methods are in the meantime of the same accuracy. One could certainly ask why worry about the OVGF method at all. The reason for still using it is mainly technical. Much larger molecules can be treated and for a given molecule larger basis sets can be emploied with the OVGF method. With the 2ph-TDA method we may be in a hopeless situation due to the large dimension of the matrices. Also the OVGF method is much faster than the e-2ph-TDA method if one wishes the ionization energies of the main ionic states.

Acknowledgment

This work has been partially supported by the Fonds der Deutschen Chemischen Industrie.

References

1) A. L. Fetter and J. D. Walecka, Quantum Theory of Many-Particle Systems, McGraw Hill, New York 1971

2) L. S. Cederbaum, Theor. Chim. Acta 31, 239 (1973), J. Phys. B 8, 290 (1975)

3) L. S. Cederbaum and W. Domcke, Adv. Chem. Phys. 36, 205 (1977)

4) J. Schirmer and L. S. Cederbaum, J. Phys. B 11, 1889 (1978)

5) O. Walter and J. Schirmer, J. Phys. B 14, 3805 (1981)

6) W. von Niessen, L. S. Cederbaum and W. Domcke "Excited States in Quantum Chemistry" Proceedings of the NATO Advanced Study Institute, Kos, Greece, 1978. ed. C. A. Nicolaides and D. R. Beck, D. Reidel Publishing Company, Dordrecht 1978, p. 183

7) O. Walter and L. S. Cederbaum, in preparation

8) J. Schirmer, Phys. Rev A in press and in preparation

9) D. M Silver Computer Phys. Comm. 14, 71, 81 (1978), S. Wilson ibid. 14, 91 (1978)

10) G. H. F. Diercksen in preparation

11) M Yoshimine IBM Corp. Technical Report RJ 555 (1969), San Jose; G. H. F. Diercksen, Theor. Chim. Acta 33, 1 (1974)

INTRODUCTORY POLARIZATION PROPAGATOR THEORY

Jens Oddershede

Department of Chemistry
Odense University
DK-5230 Odense M
Denmark

ABSTRACT An overview over the current polarization propagator methods is presented. These methods include the random phase approximation, the multi-configurational random phase approximation, the selfconsistent polarization propagator approximation, and the second order polarization propagator approximation. The paper concentrates on the common properties of the polarization propagator approaches rather than describing the details of the individual methods. It is thus intended as an introduction for the reader who is unfamiliar with polarization propagator methods.

1. SCOPE OF THE PRESENTATION

Molecular and atomic applications of random-phase like approximations are less than 20 years old (1,2) even though many of the techniques that we are now using have been known in nuclear (3) and solid state (4) physics much longer. Several different formulations of these methods have emerged in the quantum chemical literature (5-11) all of which are essentially equivalent. Many of these methods have been reviewed previously but, even though the extensions of the random phase approximation (RPA) is treated in some recent textbooks (7,12), it is the impression of the present author that there is a need for a review of the equations which form the basis for RPA like methods. Therefore, the scope of the lectures will be to introduce RPA like methods rather than discussing the more technical nature of the various approaches, including how well they reproduce actual physical observables. The latter aspect has been reviewed elsewhere (13-15). This means that in order to have a coherent presentation it will be necessary to include some wellknown textbook material in the present

G. H. F. Diercksen and S. Wilson (eds.), Methods in Computational Molecular Physics, 249–271.

article.

We will use the propagator (7) or Green's function formalism but we will also make frequent comparison with another extensively used formulation of RPA like method, namely the equation-of-motion (EOM) method (16). The material is presented such that we try answer three common questions (1) what is a propagator?, (2) why do we use propagators?, and (3) how do we perform propagator calculations?, that is, we first define the concept of a propagator, then derive an equation motion for it and then finally discuss the current computational polarization propagator methods. We will, among other things, also show why the RPA in the "natural reference" approximation in polarization propagator approaches.

2. DEFINING THE POLARIZATION PROPAGATOR

Let us first address the question: what is a propagator? The phrase 'propagator' was first introduced by Feynman and it is just another word for a double-time Green's function (17), $G(t,t')$. The reason for using this name is that $G(t,t')$ describes how a disturbance propagates in an interacting system from time t to time t'. Depending on the choice of "disturbance" we can define different propagators. If the disturbance consists of addition of an electron to the system at time t and removal from the system at time t' ($t >$ or $< t'$) we arrive at the so-called electron propagator (7). This is the one-electron propagator that describes ionization processes and Öhrn and Born (18) have recently reviewed the present state of electron propagator calculations. The electron propagator approach is basically the same as the Green's function method (19) of Cederbaum, von Niessen and coworkers and a description of this method is presented by von Niessen at this Study Institute.

The polarization propagator, which is the topic of these notes, describes how a density disturbance or polarization propagates through an interacting system. The "polarization" $P(t)$ may be any number conserving one electron disturbance of the system given in the general form

$$P(t) = \sum_{rs} P_{rs} a_r^\dagger(t) a_s(t) \qquad (1)$$

where $a_r(t)$ and $a_s^\dagger(t)$ are annihilation and creation operators, respectively, obeying the usual anticommutation relations

$$[a_r(t), a_s(t)]_+ = a_r(t)a_s(t) + a_s(t)a_r(t) = 0$$
$$[a_s(t), a_r^\dagger(t)]_+ = \delta_{rs}$$
$$[a_r^\dagger(t), a_s^\dagger(t)]_+ = 0 \qquad (2)$$

Equation (1) is the second quantized expression for an arbitrary one-electron operator (12) and we can thus define polarization propagators for a range of phenomena depending on the choice of P. If we wish to consider the polarization propagator pertaining to electronic excitations P should be chosen as $\vec{r}$ or $\vec{p}$, the dipole operators, as we shall demonstrate in sec. 3. In most applications this is the choice that is made. Another possibility that has been used in actual calculation is to choose $P = H_F$, the Fermi-contact Hamiltonian, in which case we can determine the indirect nuclear spin-spin coupling constants (20). It is, however, important, to notice that polarization propagators can be defined for any operator P(t). Also, there is no restriction, e.g. with respect to being particle or hole operators, on the creation and annihilation operators in Eq. (1).

Mathematically the polarization propagator $\langle\langle P(t);Q(t')\rangle\rangle$ is defined as (17) ($\hbar = 1$)

$$\langle\langle P(t);Q(t')\rangle\rangle = -i\theta(t-t')\langle 0|P(t)Q(t')|0\rangle - i\theta(t'-t)\langle 0|Q(t')P(t)|0\rangle \qquad (3)$$

where $\theta(t)$ is the Heaviside step function

$$\theta(t) = \begin{cases} 1 & t \geq 0 \\ 0 & t < 0 \end{cases} \qquad (4)$$

The causal Green's function in Eq. (3) is defined for both positive and negative time intervals t-t'. It is a sum of the advanced and the retarded Green's functions [the two terms in Eq. (3)]. The propagator is defined for a particular state $|0\rangle$. This is an exact eigenstate for the system and is often called the <u>reference state</u> since the polarization propagator refer to that particular state. In most applications $|0\rangle$ is the ground-state. Equation (3) clearly exhibits the physical interpretation that was mentioned earlier; $\langle\langle Q(t);P(t)\rangle\rangle$ is the probability amplitude for the process: a polarization of state $|0\rangle$, $P(t)|0\rangle$, that was created at time t has become a polarization $Q(t')|0\rangle$ at time t'. In the polarization propagator method this probability amplitude replaces the wavefunction of conventional Schrödinger quantum mechanics as the fundamental dynamical variable. Within this transformation lies the key to the successes (and shortcomings) of the polarization propagator method. Since we have a different variable than in standard wave mechanics we can, by treating the propagator at various levels of approximation, obtain approximate schemes that often cannot be derived in a statefunction description. Also, the direct output from a polarization propagator calculation refer to two states rather than just a single state as in the wavefunction approach. On the other hand, the fact that we take an average value in Eq. (3) means that we loose some of the information that might have been in $|0\rangle$. For instance it is straightforward to obtain a two-particle density matrix from the polarization propagator (7,21) whereas it is not possible to extract the wavefunction itself.

In order to obtain the propagator equation of motion which replaces the time-dependent Schrödinger equation in the state picture we use that any time dependent operator obeys the Heisenberg equation of motion, i.e.

$$i\frac{dP(t)}{dt} = [P(t),H] \tag{5}$$

where H is the Hamiltonian ($\hbar = 1$). If we apply d/dt to Eq. (3) and use Eq. (5) and the relation $d\theta(t-t')/dt = \delta(t-t')$ we obtain

$$i\frac{d}{dt}\langle\langle P(t);Q(t')\rangle\rangle = \delta(t-t')\langle 0|[P(t),Q(t)]|0\rangle + \langle\langle [P(t),H];Q(t')\rangle\rangle \tag{6}$$

as the equation of motion for the propagator. It expresses the time variation of one propagator in terms of another even more complicated propagator and can of course not be solved exactly for any system of physical interest. However, it provides a starting point for approximate polarization propagator calculations. The two most common approximations are (1) a decoupling of Eq. (6), i.e. expressing $\langle\langle [P(t),H];Q(t')\rangle\rangle$ in terms of other $\langle\langle R(t);S(t')\rangle\rangle$ propagators (22), and (2) an iterative solution to Eq. (6) whereby we obtain higher and higher order commutation of the form $\langle 0|[[\ldots[P,H],H\ldots],Q]|0\rangle$ until a certain level of accuracy is achieved. Neither of those methods correspond to any schemes that have been used in state function methods. We shall elaborate more on methods of the second kind in the sec. 3.

However, before proceeding to a discussion of the physical content of the polarization propagator we wish to show that $\langle\langle P(t);Q(t')\rangle\rangle$ is only a function of the time interval t-t'. To this end we utilize that the time dependency of any Heisenberg operator can be represented as (23)

$$P(t) = \exp(iHt)P\exp(-iHt) \tag{7}$$

where P is P(0). Inserting Eq. (7) in Eq. (3) we obtain

$$\langle\langle P(t);Q(t')\rangle\rangle = -i\theta(t-t')\sum_{n\neq o}\langle 0|P|n\rangle\langle n|Q|0\rangle\exp[i(t-t')(E_o-E_n)] -i\theta(t'-t)\sum_{n\neq o}\langle 0|Q|n\rangle\langle n|P|0\rangle\exp[i(t-t')(E_n-E_o)] \tag{8}$$

which demonstrates the dependence on t-t'. In deriving Eq. (8) we have inserted the resolution of the identity, $|n\rangle\langle n| = 1$, where $\{|n\rangle\}$ is the set of exact eigenstates for H. Strictly speaking (15) we should have considered $P(t) - \langle 0|P(t)|0\rangle$ rather than just P(t) in Eq. (3). This makes the term n=o disappear as indicated in the sums in Eq. (8).

3. PHYSICAL CONTENT OF THE PROPAGATOR

Let us now see how we can extract physical information from the polarization propagator. In almost (24) all applications of propagator methods it turns out to be convenient to consider the propagator in terms of the conjugated variable, the energy E, in very much the same way as it sometimes is of interest to consider $\Psi(\vec{p})$ rather than $\psi(\vec{r})$. Like the $\vec{r}\rightarrow\vec{p}$ transformation the $\langle\langle P;Q\rangle\rangle_E$ propagator is obtained by taking the Fourier transform

$$\langle\langle P;Q\rangle\rangle_E = \int_{-\infty}^{\infty} d(t-t')\langle\langle P(t);Q(t')\rangle\rangle \exp(iE(t-t')) \tag{9}$$

where we have used the result of Eq. (8), that is, $\langle\langle P(t);Q(t')\rangle\rangle$ is only a function of t-t'. The factor in front of the integral is conventionally (7,17,25) taken to be unity which means that we must have a factor $(2\pi)^{-1}$ in the inverse transformation. If we insert Eq. (8) into Eq. (9) we obtain the <u>spectral representation</u> of $\langle\langle P,Q\rangle\rangle_E$

$$\langle\langle P;Q\rangle\rangle_E = \lim_{\eta\rightarrow 0^+} \sum_{n\neq o} \left[\frac{\langle 0|P|n\rangle\langle n|Q|0\rangle}{E-E_n+E_o+i\eta} - \frac{\langle 0|Q|n\rangle\langle n|P|0\rangle}{E-E_o+E_n-i\eta}\right] \tag{10}$$

The positive infinitesimal η in Eq. (10) assure that the improper integrals that we otherwise would obtain when Eq. (8) is inserted in Eq. (9) actually exist. Consider for example the integral arising from the first term in Eq. (8) when we include the convergence factor

$$I_1 = \lim_{\eta\rightarrow 0^+} \int_0^{\infty} d(t-t')\exp[i(t-t')(E_o-E_n+E)]\exp(-\eta(t-t')) \tag{11}$$

As long as $\eta>o$, symbolized by $\eta\rightarrow 0^+$, this integral is convergent and its value is

$$I_1 = -\frac{1}{i(E-E_n+E_o+i\eta)} \tag{12}$$

and we get the first term in Eq. (10). The last term is obtained similarly.

Equation (10) shows that the real values of the poles of the polarization propagator are

$$E = \pm(E_n - E_o) \tag{13}$$

which are the excitation energies of the system if $|0\rangle$ is the ground-state. Furthermore, for the $\langle\langle\vec{r};\vec{r}\rangle\rangle_E$ polarization propagator the residues of the propagator are $\langle 0|\vec{r}|n\rangle$, the transition dipole matrix elements. For other choices of P and Q we may compute other transition quantities as residues. However, the poles will be the same which shows that the 'kernel' of the polarization propagator which we shall call the principal propagator (26)) remains the same when P and Q changes (provided we use the same reference state). As will be demonstrated later, the main computational effort in a polarization propagator calculation is spent in the construction of the principal propagator and it is thus possible to determine several transition quantities at little extra cost once we have computed the principal propagator. From the excitation energies and the transition moments (oscillator strengths) we may also compute second order response properties by using the sum-over-states experssion. However, in methods which goes beyond RPA it is often very tedious to compute all excitations within a given basis set and it becomes computationally favourable to obtain the response properties directly from the propagator (27). To illustrate how this may be done let us consider the real part of Eq. (10) for the $\langle\langle\vec{r};\vec{r}\rangle\rangle_E$ polarization propagator in the limit $\eta\to 0$:

$$\mathrm{Re}\langle\langle\vec{r};\vec{r}\rangle\rangle_E = -2\sum_{n\neq o}\frac{\langle 0|\vec{r}|n\rangle\langle n|\vec{r}|0\rangle}{(E_n-E_o)^2-E^2}(E_n-E_o) \tag{14}$$

The right hand side Eq. (14) is readily recognized as the standard expression for the dynamic dipole polarizability tensor at the frequency $\omega = E/\hbar$. Thus, the $\langle\langle\vec{r};\vec{r}\rangle\rangle_E$ propagator <u>is</u> the dynamic polarizability at any frequency, also when E is <u>not</u> equal to an excitation energy. Similarly, we may obtain expression for other second order properties by simply replacing $\vec{r}$ by the appropriate polarization operators (20,27). Relations like Eq. (14) show that the polarization propagator determines the linear response of a system to an external perturbation. In the case of Eq. (14) the external perturbation is the electric field. Actually the double-time Green's functions were introduced by Zubarev (17,23), as a mean of describing linear response phenomena. A more detailed discussion of the relation between linear response theory and the calculation of dynamic polarizability from the causal Green's function is given elsewhere (28).

4. EQUATION OF MOTION

Knowing state functions and eigenvalues it is in principle possible to compute $<<P;Q>>_E$ from Eq. (10). However, it is the philosophy of the propagator method to compute $<<P;Q>>_E$ directly by some other equation and then utilize Eq. (10) to identify poles and residues of the propagator. The equation of motion in energy space, or rather some approximate form of it, is the equation used for that purpose. From the inverse relation of Eq. (9), i.e. the Fourier transform of $<<P;Q>>_E$, we see that

$$i\frac{d}{dt} <<P(t);Q(t')>> = \frac{1}{2\pi}\int_{-\infty}^{\infty} Ee^{-iE(t-t')}<<P;Q>>_E \, dE \tag{15}$$

and it thus follows from Eq. (6) that the equation of motion in the energy representation is

$$E<<P;Q>>_E = <0|[P,Q]|0> + <<[P,H];Q>>_E \tag{16}$$

where we have utilized that

$$\delta(t-t') = \frac{1}{2\pi}\int_{-\infty}^{\infty} e^{-iE(t-t')}dE \tag{17}$$

Iterating on Eq. (16) as discussed after Eq. (6) we obtain

$$<<P;Q>>_E = E^{-1}<0|[P,Q]|0> + E^{-2}<0|[[P,H],Q]|0> + E^{-3}<0|[[[P,H],H],Q]|0> + \ldots . \tag{18}$$

which is referred to as the moment expansion (7) of the polarization propagator. This equation can be expressed in a more compact form if we use the so-called superoperator formalism (29). The operators P and Q are construed as elements in a superoperator space with a binary product (30) defined as

$$(P|Q) = <0|[P^+,Q]|0> \tag{19}$$

The effect of the superoperator Hamiltonian, $\hat{H}$, and of the superoperator identity, $\hat{I}$, on an element of the superoperator space is defined to be

$$\hat{H}P = -[P,H] \quad \text{and} \quad \hat{I}P = P \tag{20}$$

Using the superoperator formalism Eq. (18) may be rewritten as

$$<<P;Q>>_E = (P^+|E^{-1}\hat{I}|Q) + (P^+|E^{-2}\hat{H}|Q) + (P^+|E^{-3}\hat{H}^2|Q) + \ldots \tag{21}$$

or

$$\langle\langle P;Q\rangle\rangle_E = (P^\dagger|(E\hat{I}-\hat{H})^{-1}|Q) \tag{22}$$

where the inverse operator in Eq. (22), called the superoperator resolvent, is defined though the series expansion

$$(E\hat{I}-\hat{H})^{-1} = \frac{1}{E}[\hat{I}+\sum_{n=1}^{\infty}(\frac{\hat{H}}{E})^n] \tag{23}$$

As the last step in reformulating the equation of motion we will introduce a matrix representation of the superoperator resolvent in Eq. (22). Let h be a complete operator manifold which spans all possible excitations when h operates on $|0\rangle$. Within this operator space the resolution of the identity [see Eq. (20)] is

$$\hat{I} = |h)(h|h)^{-1}(h| \tag{24}$$

which inserted in (22) gives

$$\langle\langle P;Q\rangle\rangle_E = (P^\dagger|h)(h|h)^{-1}(h|(E\hat{I}-\hat{H})^{-1}|h)(h|h)^{-1}(h|Q) \tag{25}$$

Using an identity first proposed by Simons (31)

$$(h|h) = (h|(E\hat{I}-\hat{H})(E\hat{I}-\hat{H})^{-1}|h) \tag{26}$$

which by means of Eq. (24) becomes

$$(h|h) = (h|(E\hat{I}-\hat{H}|h)(h|h)^{-1}(h|(E\hat{I}-\hat{H})^{-1}|h) \tag{27}$$

we find that

$$(h|(E\hat{I}-\hat{H})^{-1}|h) = (h|h)(h|E\hat{I}-\hat{H}|h)^{-1}(h|h) \tag{28}$$

which inserted in Eq. (25) yields the final form for the equation of motion

$$\langle\langle P;Q\rangle\rangle_E = (P^\dagger|h)(h|E\hat{I}-\hat{H}|h)^{-1}(h|Q) \tag{29}$$

In polarization propagator calculations the complete (30) excitation manifold is

$$h = \{h_2, h_4, h_6, \ldots\} \tag{30}$$

where

$$h_2 \equiv \{q^\dagger, q\} \equiv \{a_m^\dagger a_\alpha, a_\alpha^\dagger a_m\} \tag{31}$$

$$h_4 \equiv \{q^\dagger q^\dagger, qq\} \equiv \{a_m^\dagger a_\alpha a_n^\dagger a_\beta, a_\beta^\dagger a_n a_\alpha^\dagger a_m\} \quad m>n,\ \alpha>\beta \tag{32}$$

The manifolds $h, h_2, h_4, \ldots$ and $q^\dagger, q, q^\dagger q^\dagger, qq, \ldots$ are row vectors with elements given in the curly brackets in Eq. (31) and (32). Roman indices refer to 'particle' operators, that is, electron addition ($a_m^\dagger$) or electron removal (a_m) from an unoccupied Hartree-Fock (HF) orbital while Greek indices refer to hole operators (occupied HF orbitals). Operators with unspecified occupation have indices i,j,k,l. Thus, when operating on the HF ground-state, $|HF\rangle$, $q^\dagger$ generate singly excited states, $q^\dagger q^\dagger$ generate doubly excited states, etc. The adjoint operators (q,qq etc.) give zero when working on $|HF\rangle$. Thus, it is obvious that h produces all possible excited states when $|0\rangle = |HF\rangle$. However, it is not a trivial matter to convince oneself that it is also sufficient to include in h only the 'HF supported' operators in Eqs. (31) and (32) when $|0\rangle$ is a more general (correlated) state. The proof for that has been given by Dalgaard (30) provided $\langle HF|0\rangle \neq 0$.

Equation (29) is exact provided h is the complete manifold and $|0\rangle$ is the exact reference state. Note that $|0\rangle$ appears in Eq. (29) through the three superoperator binary products (cf. Eq. (19)). Approximation to the polarization propagator is naturally obtained by using approximate reference states and/or truncating the projection manifold. In the next sections we will discuss some of the approximate polarization propagator methods obtained in this manner.

5. RANDOM PHASE APPROXIMATION

The simplest possible approximation to Eq. (29) is obtained when $|0\rangle = |HF\rangle$ and $h = h_2$ in which case the polarization propagator becomes

$$\langle\langle P;Q\rangle\rangle_E = \{(P^\dagger|q^\dagger),(P^\dagger|q)\}\, F(E) \begin{Bmatrix} (\tilde{q}^\dagger|Q) \\ (\tilde{q}|Q) \end{Bmatrix} \quad (33)$$

where F(E) is the <u>principal propagator</u> (26)

$$F(E) = \begin{Bmatrix} \langle\langle \tilde{q};q^\dagger\rangle\rangle_E & \langle\langle \tilde{q};q\rangle\rangle_E \\ \langle\langle \tilde{q}^\dagger;q^\dagger\rangle\rangle_E & \langle\langle \tilde{q}^\dagger;q\rangle\rangle_E \end{Bmatrix}$$

$$= \begin{Bmatrix} E - A & -B \\ -B^* & -E - A^* \end{Bmatrix}^{-1} \quad (34)$$

The upper left block of F(E), $\langle\langle\tilde{q};q^{\dagger}\rangle\rangle_E$, is the particle-hole propagator which is a matrix of dimensions (ph) x (ph) while the whole F(E) is of dimension 2(ph) x 2(ph) where (ph) are the number of particle-hole excitation in the vector $q^{\dagger}$. A ~ indicates a transpose vector, that is, $\tilde{q}$ is a column vector. It follows from Eqs. (29)-(31) and (33)-(34) that the elements of the matrices A and B, both of dimension (ph) x (ph), are defined as

$$A_{m\alpha,n\beta} = (q^{\dagger}_{m\alpha}|\hat{H}|q^{\dagger}_{n\beta}) = -\langle HF|[a^{\dagger}_{\alpha}a_m,[a^{\dagger}_n a_{\beta},H]]|HF\rangle \quad (35)$$

$$B_{m\alpha,n\beta} = (q^{\dagger}_{m\alpha}|\hat{H}|q_{n\beta}) = -\langle HF|[a^{\dagger}_{\alpha}a_m,[a^{\dagger}_{\beta}a_n,H]]|HF\rangle \quad (36)$$

In deriving Eq. (34) we have also used that

$$\langle HF|[\tilde{q},q^{\dagger}]|HF\rangle = -\langle HF|[\tilde{q}^{\dagger},q]|HF\rangle = 1 \quad (37)$$

It is often useful to analyze the equation of motion for the polarization propagator in terms of a perturbation expansion. As zeroth order one normally choose an SCF calculation for the reference state $|0\rangle$. The total Hamiltonian is therefore written in the form

$$H = F + V \quad (38)$$

where

$$F = \sum_k \varepsilon_k a^{\dagger}_k a_k \quad (39)$$

and V is the so-called fluctuation potential

$$V = \frac{1}{4}\sum_{\substack{ij\\kl}} (ik||jl)a^{\dagger}_i a^{\dagger}_j a_l a_k - \sum_{\alpha ij} (\alpha\alpha||ij)a^{\dagger}_i a_j \quad (40)$$

The electron repulsion integrals are defined as

$$(ij|kl) = \int u^*_i(1)u_j r^{-1}_{12}u^*_k(2)u_l(2)d\tau_1 d\tau_2 \quad (41)$$

and

$$(ij||kl) = (ij|kl) - (il|kj) \quad (42)$$

Inserting Eqs. (38)-(40) in Eqs. (35)-(36) we obtain

$$A_{m\alpha,n\beta} = \delta_{mn}\delta_{\alpha\beta}(\varepsilon_m-\varepsilon_\alpha) + (m\alpha|\beta n) - (mn|\beta\alpha) \quad (43)$$

$$B_{m\alpha,n\beta} = (m\beta|n\alpha) - (m\alpha|n\beta) \tag{44}$$

The A matrix is thus of order zero and one in V while B is of order one. The transition matrix elements in Eq. (33) are (see Eq. (1))

$$(P^{\dagger}|q^{\dagger}_{m\alpha}) = P_{\alpha m} \tag{45}$$

$$(P^{\dagger}|q_{m\alpha}) = -P_{m\alpha} \tag{46}$$

and the only part of $\langle\langle P;Q\rangle\rangle_E$ that carries an order in perturbation theory is thus the principal propagator, F(E). Furthermore, we have included all possible first order terms in the propagator since extension of the excitation manifold to include h_4 (see Eqs. (30)-(32) yields terms in second (and higher) order of perturbation theory (32). Also, expansion of the reference state in orders of V using Rayleigh-Schrödinger perturbation theory gives

$$|0\rangle = |HF\rangle + |0\rangle^{(1)} + |0\rangle^{(2)} + \ldots. \tag{47}$$

where $|0\rangle^{(1)}$ are doubly excited states (12). If $|0\rangle^{(1)}$ replaces one of the $|HF\rangle$ states in Eqs. (35) and (36) the A and B matrix vanish when we compute the H=F part (thus, no first order terms) and the only nonvanishing terms (H=V) appear in second order (see sec. 8). Thus, we obtain a consistent first order approximation to the polarization propagator by choosing $|0\rangle = |HF\rangle$ and $h = h_2$.

From sec. 3 we know that the excitation energies are the poles of $\langle\langle P;Q\rangle\rangle_E$, that is, according to Eqs. (33)-(34) and (45)-(46), the eigenvalues of the following eigenvalue problem

$$\begin{Bmatrix} A & B \\ B^* & A^* \end{Bmatrix} \begin{Bmatrix} Z & Y^* \\ Y & Z^* \end{Bmatrix} = \begin{Bmatrix} E & 0 \\ 0 & -E \end{Bmatrix} \begin{Bmatrix} Z & Y^* \\ Y & Z^* \end{Bmatrix} \tag{48}$$

Equation (48) is readily recognized as the standard RPA eigenvalue problem (13,14,16) and we have thus demonstrated that the first order polarization propagator approximation is identical to RPA. In the present context (calculation of response properties) RPA is also the same as the time-dependent HF (TDHF) approximation (1,3). The latter name may be somewhat misleading: the 'time-dependent' refer to the derivation of TDHF and does not mean that there is any time-dependency in the resulting equations. One should probably therefore rather use the name RPA than TDHF for the first order polarization propagator approximation. We may compute response properties in RPA as discussed in sec. 3. It can then be shown that the RPA expressions for those quantities at zero frequency are identical to those obtained in the fully coupled HF (CHF) method (12) which is the same as the zero-field limit of the widely used finite field method (33). Also these methods are therefore identical to a first order polarization propagator approach.

Another way of estimating the amount of correlation in RPA is to compare with a configuration interaction (CI) calculation. It follows from Eq. (35) that

$$A_{m\alpha,n\beta} = -\delta_{nm}\gamma_{\alpha\beta}\langle HF|H|HF\rangle + \langle HF|a_{\alpha}^{\dagger}a_{m}Ha_{n}^{\dagger}a_{\beta}|HF\rangle \quad (49)$$

$$B_{m\alpha,n\beta} = -\langle HF|a_{\alpha}^{\dagger}a_{m}a_{\beta}^{\dagger}a_{n}H|HF\rangle \quad (50)$$

and the matrix A is the singly-excited CI matrix (minus the ground-state total energy) while B contains matrix elements between $|HF\rangle$ and doubly excited states. When B = 0, Eq. (48) thus yields excitation energies in the singly excited CI approximation, i.e. only excited state correlation is included. In RPA ($B \neq 0$) both ground and excited states are correlated.

An RPA calculation requires a set of SCF molecular orbitals and orbital energies to form the A and B matrices (see Eqs. (43)-(44)). Computationally, this is a simple matter and in most cases an RPA calculation is a very inexpensive relative to an integral transformation. Only when the A and B matrices are very large (of the order several hundred excitations, $m\alpha$) does it become necessary to use special methods (34) for finding only the lowest eigenvalues of Eq. (48).

The literature on RPA calculations is abundant and the reader is referred to some of the recent reviews (13,14,35) for details. Here, we will only give some of the characteristics of RPA calculations. On the positive side, RPA represents an inexpensive way of generating many quite accurate singlet excitation energies, that is, energy differences between states of the same spin symmetry. Typically, errors of the order 0.5 eV, or about 5-10%, occur in RPA singlet excitation energies (15). The errors in the transition moments are generally greater, often about 2o-25%. Singlet response properties, like dynamical polarizabilities (28) and rotatory intensities (36) are also of a satisfactory quality in RPA calculations. Furthermore, RPA has some invarious properties which are useful for practical calculations and which many other, and often more advanced methods, do not possess. The random phase approximation preserve equivalence between oscillator strengths calculated in the dipole length and the dipole velocity formulation (37) provided we use complete basis sets. In other words, the actual disagreement between dipole length and dipole velocity oscillator strengths can be used as a measure of the incompleteness of the basis set. The RPA is also invariant to separate unitary transformations of the occupied and/or virtual SCF molecular orbitals (38) and it is, as mentioned earlier, a consistent first order polarization propagator approximation.

One of the main problems with RPA is that it gives poor triplet $[\Delta S(|0\rangle \rightarrow |n\rangle) = 1]$ excitation energies. It even gives imaginary triplet excitation energies in cases where $|HF\rangle$ is instable with respect to triplet excitation (39), that is, when there exists another state $|\psi\rangle$ with a lower energy than $|0\rangle$ and with a different spin than $|0\rangle$. This occurs for instance in the Be atom and in the CH^+ molecule (40). Also triplet response properties like indirect nuclear spin-spin coupling constant (20) are very poorly described in RPA (=CHF). Theoretically, RPA has one serious deficiency. We use $|HF\rangle$ to compute the A,B etc. matrices but the two-electron density matrix which is computed from the RPA eigenvectors (Z, Y in Eq. (48)) is not the HF two-matrix. This ambiguity indicates (15) that the RPA two-matrix is not N-representable (41), that is, there is no wavefunction that corresponds to the RPA two-electron density matrix. One remedy for this problem would be to find the consistent RPA ground-state, which could be used both to compute the A and B matrices and also would give the RPA two-matrix. Linderberg and Öhrn (42) have suggested that the consistent ground-state can be represented as an antisymmetrized geminal power (AGP) and we will discuss a selfconsistent determination of that state in sec. 7. However, before doing that we will discuss the most obvious extension of RPA, namely a RPA method which uses a multiconfigurational reference state.

6. MULTICONFIGURATIONAL RPA

The use of a single-configuration reference state, $|HF\rangle$, leads to difficulties when trying to describe molecular dissociation or some highly correlated, often open shell, states. If we instead use $|0\rangle = |MCSCF\rangle$, the multiconfigurational SCF state, and choose $h = \{q^\dagger, R^\dagger, q, R\}$ we obtain the multiconfigurational random phase approximation (MCRPA) (43,44) which does not have these problems. The excitation manifold includes in addition to $h_2 = \{q^\dagger, q\}$ the operator

$$R^\dagger = \{|n\rangle\langle o|\} \qquad (51)$$

and its adjoint R. Both R and $R^\dagger$ are row vectors. The set $\{|n\rangle\}$ in Eq. (51) contains the states ($\neq |0\rangle$) that are included in the optimization of the MCSCF state $|0\rangle$. When operating on $|0\rangle$ $R^\dagger$ thus generates all other states that are included in the MC optimization procedure. Actually the unitary operator that performs the configurational variation is given as exp (iS) where (45) S is an hermitian operator

$$S = iP(\tilde{R}^{\dagger}-\tilde{R}) \tag{52}$$

and P is a row vector containing the variational parameters. Analogously, the unitary operator exp (iλ) where

$$\lambda = iK(\tilde{q}^{\dagger}-\tilde{q}) \tag{53}$$

describes the orbital variations with K being a row vector of orbital variational parameters.

The states that originate from $R^{\dagger}|0\rangle$ are mixtures of some of the $q^{\dagger}|0\rangle$, $q^{\dagger}q^{\dagger}|0\rangle$, $q^{\dagger}q^{\dagger}q^{\dagger}|0\rangle$ etc. states and we have thus included certain parts of the higher manifolds $h_4|0\rangle, h_6|0\rangle$... (see Eqs. (30)-(32) in MCRPA. Some of the $q^{\dagger}|0\rangle$states are generated both by $R^{\dagger}|0\rangle$ and by $h_2|0\rangle$ and in order to avoid linear dependencies we must hence exclude the so-called redundant operators from h_2 in exactly the same way as is done in any MCSCF calculation (12). With this choice of excitation manifold the polarization propagator becomes [see Eq. (29)]

$$\langle\langle P;Q\rangle\rangle_{E} = \{(P^{\dagger}|q^{\dagger}),(P^{\dagger}|R^{\dagger}),(P^{\dagger}|q),(P|R)\} \times \left\{E\begin{pmatrix} S & \Delta \\ -\Delta & -S\end{pmatrix} - \begin{pmatrix} A & B \\ B^{*} & A^{*}\end{pmatrix}\right\}^{-1} \begin{Bmatrix} (\tilde{q}^{\dagger}|Q) \\ (\tilde{R}^{\dagger}|Q) \\ (\tilde{q}|Q) \\ (\tilde{R}|Q) \end{Bmatrix} \tag{54}$$

where

$$A = \begin{Bmatrix} (\tilde{q}^{\dagger}|\hat{H}|q^{\dagger}) & (\tilde{q}^{\dagger}|\hat{H}|R^{\dagger}) \\ (\tilde{R}^{\dagger}|\hat{H}|q^{\dagger}) & (\tilde{R}^{\dagger}|\hat{H}|R^{\dagger}) \end{Bmatrix} \tag{55}$$

$$B = \begin{Bmatrix} (\tilde{q}^{\dagger}|\hat{H}|q) & (\tilde{q}^{\dagger}|\hat{H}|R) \\ (\tilde{R}^{\dagger}|\hat{H}|q) & (\tilde{R}^{\dagger}|\hat{H}|R) \end{Bmatrix} \tag{56}$$

and S is defined as A with $\hat{H}$ replaced by $\hat{I}$ in Eq. (55). The other matrix Δ is obtained by performing the same replacement in Eq. (56). The upper left blocks of the A and B matrices are (except that $|HF\rangle \rightarrow |MCSCF\rangle$) the same as the RPA A and B matrices. The off-diagonal blocks of Eqs. (55) and (56) describe the interaction between the orbital and configurational space and the lower right blocks only refer to configuration mixing. The A-B matrix is the Hessian matrix of the second order MCSCF procedure, that is, it is the second order variation of the energy functional (45). The A and B matrices are therefore already computed in the preceeding MCSCF calculation and the MCRPA solutions [the eigenvectors and eigenvalues for the inverse matrix in Eq. (54)] are obtained at little extra cost.

The computational experience with the MCRPA method is at present rather limited (43,44,46,47). However, it is clear that both excitation energies, oscillator strengths and dynamic polarizabilities are significantly better than the corresponding RPA results. This is especially noticeable for triplet excitation energies and for excitations to highly correlated excited states such as e.g. $^3P(2p^2)$, $^1D(3d)$ and $^1D(2p^2)$ of Be (43). The MCRPA also, like RPA, preserves the equivalence between dipole oscillator strengths computed in the length and velocity formulation, again provided we use a complete basis set expansion (44). It can also be shown (12) that MCRPA is identical to the multiconfigurational coupled HF method, just like RPA and CHF were the same approximation.

7. SELF-CONSISTENT APPROXIMATIONS

The observation that |HF> is not the consistent RPA ground-state even though it is used in the construction of the RPA matrices (see sec. 5) has spured a considerable interest in formulating selfconsistent RPA methods. Historically, approximations of this kind represent the first attempts at formulating methods which go beyond RPA (6,48,49). These methods may be divided into two categories: (i) an ansatz is made for the form of the consistent RPA ground-state and some parameters in the trial solution are varied until a self-consistent RPA-type solution is obtained and (ii) from RPA one calculates a two-matrix which is then used to construct new A, B etc. matrices giving a new two-matrix and so on until self-consistency is achieved.

Procedures of the latter kind (ii) encompass the self-consistent polarization propagator approximation (48,50) in which the two-matrix is recovered from the polarization propagator by suitable contour integrations. It turns out that the two-matrix is given as simple products of the Y and Z eigenvectors [see Eq. (48)]. Such methods suffer from an obvious deficiency, namely that we only work with the two-electron density matrix and never with the wavefunction itself and thus do not know if the density matrix is N-representable (41). In fact, we have found (15) that some of the necessary conditions for N-representability are not always fulfilled. Since it is very difficult to formulate a simple sufficient condition for N-representability (51) this means that methods that iterate on the two-particle density matrix will possess inherent difficulties and self-consistent polarization propagator methods of this kind have therefore been abandoned.

Methods that iterate on an RPA-type reference state (i) represent a more viable alternative to simple RPA. The first method of this kind was the higher RPA (HRPA) method of Shibuya and McKoy (6). Their ansatz for the RPA ground-state was

$$|0\rangle = C(|HF\rangle + \sum_{\substack{ab\\ \pi\delta}} \kappa^{ab}_{\pi\delta} q^{\dagger}_{a\pi} q^{\dagger}_{b\pi} |HF\rangle \tag{57}$$

where C is a normalization constant and $\kappa^{ab}_{\pi\delta}$ are the parameters (correlation coefficients) that are varied until a self-consistent RPA solution is obtained. Experience shows (52) that the self-consistent solutions for $\kappa^{ab}_{\pi\delta}$ deviate very little from the first order Rayleigh-Schrödinger correlation coefficients and the last term in Eq. (57) is therefore almost the same as $|0\rangle^{(1)}$ in Eq. (47). Excitation energies calculated in HRPA are always too high and are in worse agreement with experiment than RPA excitation energies (14,15). It is necessary to augment HRPA with part of the h_4 excitation manifold, namely the so-called two-particle, two-hole (2p-2h) corrections (53), to bring the excitation energies down close to the experimental values (see sec. 8 for a discussion of the 2p-2h corrections). Augmented with the 2p-2h corrections HRPA becomes the equation of motion (EOM) method which has been widely used in molecular electronic structure calculations during the last decade (14).

There is no special reason to believe that the consistent RPA ground-state is in the form given in Eq. (57). In fact, Linderberg and Öhrn (42,54) showed that under certain conditions (55) the consistent ground-state of RPA has the form of an antisymmetrized geminal power (AGP)

$$|0\rangle = |AGP\rangle = B^{N/2}\ |VAC\rangle \tag{58}$$

where |VAC> is the vacuum state, N the (even) number of electrons and

$$B = \sum_{i=1}^{s} \xi_i a^{\dagger}_i a^{\dagger}_{\overline{i+s}} \tag{59}$$

The creation operators refer to the orbital basis and s is M/2 where M is the rank of the orbital basis. For systems with an odd number of electrons one uses the generalized AGP of Weiner and

Goscinski (56). The consistent, RPA ground-state must fulfill the 'killer condition' (42) which states that the adjoint of the exact RPA excitation operator, $Q_\lambda^\dagger$, annihilates the RPA ground-state, i.e.

$$Q_\lambda |RPA\rangle = 0 \tag{60}$$

where $Q_\lambda^\dagger$ is the excitation operator referring to state $|\lambda\rangle = Q_\lambda^\dagger |RPA\rangle$. In the self-consistent polarization propagator approximation based on the AGP ansatz for $|0\rangle$ one varies the expansion coefficients ξ_i and the orbitals, i.e. $a_i^\dagger$ in Eq. (59), until conditions are fulfilled (57). The AGP method is rather new and the numerical experience is so far limited to calculations of excitation energies and ground-state correlation energies (58-60). The latter is possible in this case in contrast to most other polarization propagator methods where one does not know the state function.

8. PERTURBATIVE PROPAGATOR METHODS

It was shown in sec. 5 that RPA is consistent through first order in the electronic repulsion. Extension of RPA would thus naturally be methods which are consistent through successively higher orders in perturbation theory. We must thus expand $|0\rangle$ as indicated in Eq. (47) and h as indicated in Eq. (30) such that Eq. (29) is consistent through a particular order in perturbation theory. This is a somewhat tedious but straightforward job and through order two in perturbation theory we find (61) that we need to include $|0\rangle^{(1)}$ and a small fraction of $|0\rangle^{(2)}$, namely the singly excited correlation coefficients, plus the h_4 manifold in order to be able to compute both transition moments and excitation energies through second order in V [see Eq. (40)]. Assuming real SCF orbitals the second order polarization propagator takes the form

$$\langle\langle P;Q\rangle\rangle_E = \{T(P,E),T(P,-E)\}\ F(E) \begin{Bmatrix} \tilde{T}(Q,E) \\ \tilde{T}(Q,-E) \end{Bmatrix} + W_4(E) \tag{61}$$

where the transition moment matrix T is defined as

$$T(P,E) = [(P^\dagger|q^\dagger) - (P^\dagger|q^\dagger q^\dagger)(E1-D)^{-1}C]S^{-\frac{1}{2}} \tag{62}$$

and

$$W_4(E) = (P^\dagger|q^\dagger q^\dagger)[(E1-D)^{-1}+(-E1-D)^{-1}](\tilde{q}^\dagger\tilde{q}^\dagger|Q) \tag{63}$$

The principal propagator, the poles of which determine the particle-hole dominated (principal) excitation energies, is given as (61)

$$F(E) = \begin{Bmatrix} E-S^{-\frac{1}{2}}A(E)S^{-\frac{1}{2}} & -S^{-\frac{1}{2}}BS^{-\frac{1}{2}} \\ -S^{-\frac{1}{2}}BS^{-\frac{1}{2}} & -E-S^{-\frac{1}{2}}A(E)S^{-\frac{1}{2}} \end{Bmatrix}^{-1} \tag{64}$$

where

$$A(E) = A + A' + \tilde{C}(E1-D)^{-1}C \tag{65}$$

and

$$A' = \tfrac{1}{2}[A(0)S(2) + S(2)A(0)] \tag{66}$$

The matrices A,B,C,D and S that appear in Eqs. (62)-(66) are similar to those encountered in RPA except that they are not always computed with $|0\rangle = |HF\rangle$. Sometimes we must instead use $|0\rangle^{(1)}$. This means that they may carry all orders in perturbation theory from zero to two. Also, the inclusion of h_4 gives rise to two new matrices C and D. The A and B matrices are defined in Eqs. (35) and (36) with $|HF\rangle$ replaced by $|HF\rangle + |0\rangle^{(1)}$ and this means that $A = A(0,1,2)$ and $B = B(1,2)$ where the numbers in parentheses indicate the orders of perturbation theory. B(1) and A(0,1) are the RPA matrices (see sec. 5). The matrices which don't appear in RPA are defined as

$$\begin{aligned} C = C(1) &= (\tilde{q}^{\dagger}\tilde{q}^{\dagger}|\hat{V}|q^{\dagger}) \\ D = D(0) &= (\tilde{q}^{\dagger}\tilde{q}^{\dagger}|\hat{F}|q^{\dagger}q^{\dagger}) \end{aligned} \tag{67}$$

and

$$S = S(0,2) = (\tilde{q}^{\dagger}|q^{\dagger}) \tag{68}$$

By comparing Eq. (64) with Eq. (34) we see that the eigenvalue problems in a second order polarization propagator approximation (SOPPA) resemble that of RPA. The main difference is that the A matrix now is energy dependent. The energy dependent term, $\tilde{C}(E1-D)^{-1}C$, originates from the h_4 manifold and is the 2p-2h correction (see sec. 7). The SOPPA excitation energies are thus obtained from an iterative solution of the eigenvalue problem for $F^{-1}(E)$ in Eq. (64): from a initial guess on E in the 2p-2h term one solves an RPA-like eigenvalue problem, obtains a new value for E which is then inserted in $\tilde{C}(E1-D)^{-1}C$ and so on. However, this is not the most time comsuming step in a SOPPA calculation. Most of the computational effort is spent in the construction of the A(2), B(2) and C(1) matrices, with the construction of B(2) alone often accounting for 70-80% of all the computer time (15,62). The A(2) and B(2) matrix elements are sums of terms where each term is a product of two two-electron matrix elements (they are of second order in V). We must therefore do a lot of integral handling in order to construct these matrices. For large basis sets it is not possible to keep all two-electron integrals in fast core which further increase the computational difficulties. We (62) have recently constructed a new polarization propagator program which is part of the MUNICH program system and which does not assume that the two-electron integrals can be kept in the fast core.

The performance of the second order polarization propagator approximation has been studied for a number of years and next to RPA it is the propagator approximation that we know most about. It has been applied to the calculation of a variety of excitation properties, like excitation energies, potential energy surfaces, oscillator strengths, radiative lifetimes, generalized oscillator strengths (63), dynamic polarizabilities, Raman intensities (28), Verdet constants, dispersion forces, and indirect nuclear spin-spin coupling constants. Most of these calculations are of a satisfactory quality when compared with experiments and other calculations. (For a discussion of the individual properties the reader is referred to an earlier review (15).) Compared with RPA we see improvements for all the calculated properties and, like in MCRPA, this improvement is especially marked for triplet excitation energies and response properties. Errors in excitation energies, both singlet and triplet, are in most cases less than 2-4%.

The equivalence between oscillator strengths computed in the dipole length and the dipole velocity formulation is also preserved in an order by order calculation of the polarization propagator (64). Conversely, it can be shown (64) that if we require equivalence between the length and velocity form through a particular order in perturbation theory this implies that excitation energies and transition moments must be obtained from the equation of motion in Eq. (29). It should be pointed out that the equivalence in RPA and MCRPA is exact whereas the equivalence in SOPPA only holds through second order in V and there are thus error terms of order three in V. This means that the agreement in actual finite basis set calculation is slightly better in RPA than in SOPPA (64).

It was stated in sec. 5 that RPA is invariant to separate unitary transformations of the occupied and/or virtual SCF orbitals (33). This invariance does not hold for SOPPA, just like Rayleigh-Schrödinger perturbation calculations of total energies are not invariant to orbital transformations (65). However, we have shown (33) that by using energy-shifted denominators in the 2p-2h corrections and in the correlation coefficients we obtain that the SOPPA is almost invariant to orbital transformations in the virtual orbital subspace.

Extension of perturbative polarization propagator methods through third, fourth etc. orders in perturbation theory is going to be a rather complicated matter. A third order theory was formulated for the principal propagator by Oddershede and Jørgensen (32) and was later derived also within the EOM formalism by Yeager and Freed (66). However, there exists no third order theory for the full polarization propagator and it is thus possible to compute excitation energies [from the poles of F(E)] but not transition moments and response properties throgh third order. Theories which go beyond third order have not yet been formulated.

REFERENCES

(1) McLachlan, A.D., and Ball, M.A.: 1964, Revs. Mod. Phys. 36, pp. 844-55.

(2) Öhrn, Y., and Linderberg, J.: 1965, Phys. Rev. 139, pp. A1063-68.

(3) Thouless, D.J.: 1961, "Quantum Mechanics of Many Body Systems" (Academic Press, New York).

(4) Lindhard, J.: 1954, Kgl. Dan. Vid. Selsk. Mat. Phys. Medd. 28, no 8, pp. 1-57.

(5) Dalgarno, A., and Victor, G.A.: 1966, Proc. Roy. Soc. (London) Ser. A 291, pp. 291-5.

(6) Shibuya, T.-I., and McKoy, V.: 1970, Phys. Rev. A 2, pp. 2208-18.

(7) Linderberg, J., and Öhrn, Y.: 1973, "Propagators in Quantum Chemistry" (Academic Press, London).

(8) Zwicker, J.O., and Yaris, R.: 1975, J. Chem. Phys. 62, pp. 1276-84.

(9) Huo, W.M.: 1975, J. Chem. Phys. 62, pp. 2072-86.

(10) Banerjee, A., Mukherjee, D., and Simons, J.: 1982, J. Chem. Phys. 76, pp. 1972-2002.

(11) Liegener, C.-M.: 1981, Mol. Phys. 43, pp. 1-11.

(12) Jørgensen, P., and Simons, J.: 1981, "Second Quantization-Based Methods in Quantum Chemistry" (Academic Press, New York)

(13) Jørgensen, P.: 1975, Annu. Rev. Phys. Chem. 26, 359-80.

(14) McCurdy, W.C., Rescigno, T.N., Yeager, D.L., and McKoy, V.: 1977 in "Modern Theoretical Chemistry", Vol. 3, pp. 339-86, editor: H.F. Schaeffer (Plenum Press, New York).

(15) Oddershede, J.: 1978, Adv. Quantum Chem. 11, pp. 275-352.

(16) Rowe, D.J.: 1968, Revs. Mod. Phys. 40, pp. 153-66.

(17) Zubarev, D.N.: 1960, Usp. Fiz. Nauk. 71, pp. 71-116 (English translation Sov. Phys. Usp. 3, pp. 320-45).

(18) Öhrn, Y., and Born, G.: 1981, Adv. Quantum Chem. 13, pp. 1-88.

(19) Cederbaum, L.S., and Domcke, W.: 1977, Adv. Chem. Phys. 36, pp. 205-344.

(20) Oddershede, J., Jørgensen, P., and Beebe, N.H.F.: 1977, Chem. Phys. 25, pp. 451-8.

(21) Oddershede, J., Jørgensen, P., and Beebe, N.H.F.: 1975, J. Chem. Phys. 63, pp. 2996-3007.

(22) Linderberg, J., and Öhrn, Y.: 1965, Proc. Roy. Soc. Ser. A (London) 285, pp. 445-56.

(23) Raimes, S.: 1972, "Many-electron Theory" (North-Holland, Amsterdam), Chap. 5.

(24) Ondrechen, M.J., and Ratner, M.A.: 1977, J. Chem. Phys. 66, pp. 938-46.

(25) Zubarev, D.N.: 1974, "Nonequilibrium Statistical Thermodynamics" (Consultants Bureau, New York).

(26) Oddershede, J.: 1982, Int. J. Quantum Chem. (Symposium), in press.

(27) Jørgensen, P., Oddershede, J., and Beebe, N.H.F.: 1978, J. Chem. Phys. 68, pp. 2527-32.

(28) Oddershede, J., and Svendsen, E.N.: 1982, Chem. Phys. 64, pp. 359-69.

(29) Goscinski, O., and Lukman, B.: 1970, Chem. Phys. Lett. 7, pp. 573-6.

(30) Dalgaard, E.: 1979, Int. J. Quantum Chem. 15, pp. 169-80.

(31) Simons, J.: 1976, J. Chem. Phys. 64, pp. 4541-3.

(32) Oddershede, J., and Jørgensen, P.: 1977, J. Chem. Phys. 66, pp. 1541-56.

(33) Cohen, H.D., and Roothaan, C.C.J.: 1965, J. Chem. Phys. 43, pp. S34-9.

(34) Rettrup, S.: 1982, J. Comput. Phys. 45, pp. 100-7.

(35) Hansen, Aa.E., and Bouman, T.D.: 1980, Adv. Chem. Phys. 44, pp. 545-644.

(36) Bouman, T.D., Hansen, Aa.E., Voigt, B., and Rettrup, S.: 1983, Int. J. Quantum Chem. (Proc. 4th Congress in Quantum Chem.), in press.

(37) Harris, R.A.: 1969, J. Chem. Phys. 50, pp. 3947-51.

(38) Oddershede, J., and Sabin, J.R.: 1983, to be published.

(39) Čižek, J., and Paldus, J.: 1967, J. Chem. Phys. 47, pp. 3976-85.

(40) Oddershede, J., Jørgensen, P., and Beebe, N.H.F.: 1977, Int. J. Quantum Chem. 12, pp. 655-70.

(41) Coleman, A.J.: 1963, Revs. Mod. Phys. 35, pp. 668-89.

(42) Linderberg, J. and Öhrn, Y.: 1977, Int. J. Quantum Chem. 12, pp. 161-91.

(43) Yeager, D.L., and Jørgensen, P.: 1979, Chem. Phys. Lett. 65, pp. 77-80.

(44) Dalgaard, E.: 1980, J. Chem. Phys. 72, pp. 816-23.

(45) Dalgaard, E., and Jørgensen, P.: 1978, J. Chem. Phys. 69, pp. 3833-44.

(46) Lynch, D., Herman, M.F., and Yeager, D.L.: 1982, Chem. Phys. 64, pp. 69-81.

(47) Yeager, D.L., Olsen, J., and Jørgensen, P.: 1981, Int. J. Quantum Chem. (Symposium) 15, pp. 151-62.

(48) Linderberg, J. and Ratner, M.: 1970, Chem. Phys. Lett. 6, pp. 37-40.

(49) Linderberg, J., Jørgensen, P., Oddershede, J., and Ratner, M.: 1972, J. Chem. Phys. 56, pp. 6213-9.

(50) Jørgensen, P., Oddershede, J., and Ratner, M.: 1974, J. Chem. Phys. 61, pp. 710-8.

(51) Coleman, A.J.: 1972 in "Energy, Structure and Reactivity" (D.W. Smith and W.B. McRae, Eds.) Wiley and Sons, New York, pp.231-6.

(52) Shibuya, T.-I., and McKoy, V.: 1971, J. Chem. Phys. 54, pp. 1738-51.

(53) Shibuya, T.-I., Rose, J., and McKoy, V.: 1973, J. Chem. Phys. 58, pp. 500-7.

(54) Öhrn, Y., and Linderberg, J.: 1979, Int. J. Quantum Chem. 15, pp. 343-53.

(55) Linderberg, J.: 1980, Physica Scripta 21, pp. 373-7.

(56) Weiner, B., and Goscinski, O.: 1980, Phys. Rev. A 22, pp. 2374-91.

(57) Goscinski, O., and Weiner, B.: Physica Scripta 21, pp. 385-93.

(58) Ortiz, J.V., Weiner, B., and Öhrn, Y.: 1981, Int. J. Quantum Chem. (Symposium) 15, pp. 113-28.

(59) Kurtz, H.A., Elander, N., Goscinski, O., and Sangfelt, E.: 1981, Int. J. Quantum Chem. (Symposium) 15, pp. 143-9.

(60) Jensen, H.J.Aa., thesis, Aarhus University, 1980.

(61) Nielsen, E.S., Jørgensen, P., and Oddershede, J.: 1980, J. Chem. Phys. 73, pp. 6238-46.

(62) Diercksen, G.H.F., Grüner, N., and Oddershede, J.: 1983, Comp. Phys. Comm., to be published.

(63) Allouche, A., Marinelli, F., and Arrighini, G.P.: 1980, Chem. Phys. Lett. 76, pp. 333-6.

(64) Jørgensen, P., and Oddershede, J.: 1983, J. Chem. Phys., to be published.

(65) Silver, D.M., and Bartlett, R.J.: 1976, Phys. Rev. A 13, pp.1-12.

(66) Yeager, D.L. and Freed, K.F.: 1977, Chem. Phys. 22, pp.415-33.

DIAGRAMMATIC MANY-BODY PERTURBATION THEORY

S. Wilson

Theoretical Chemistry Department,
University of Oxford, 1 South Parks Road, Oxford

The application of the diagrammatic many-body perturbation theory to the problem of molecular electronic structure is described. An overview of the basic formalism and properties of the method is given together with a discussion of various aspects of the truncation of the perturbation expansion, emphasizing particularly, the systematic nature of the perturbation theoretic approach. The use of diagrammatic perturbation theory in the analysis of many techniques currently employed to handle correlation in molecules is briefly considered. The main emphasis of this review is on the computational aspects of diagrammatic perturbation theory. Programs written for serial computers are described and it is demonstrated that algorithms based on perturbation theory are both accurate and efficient. The importance of Wigner's (2n+1) rule is noted. The impact of parallel processing computers on perturbative calculations is assessed. An outline of some current research on the application of diagrammatic perturbation theory to the correlation problem in atoms and molecules is presented.

1. INTRODUCTION

Over the past twenty years, the diagrammatic many-body perturbation theory has been developed into an attractive method for the calculation of correlation effects in atoms and molecules, which is both theoretically well founded and computationally efficient. The linked diagram expansion ensures that the method yields energies which are directly proportional to the number of electrons being considered (see, for example, refs 1-11), a property which allows meaningful comparisons of systems of different sizes and treatment of dissociative and reactive chemical processes. The perturbative approach is computationally efficient (see, for

G. H. F. Diercksen and S. Wilson (eds.), Methods in Computational Molecular Physics, 273–298.

example, refs. 12-14) being non-iterative and well suited to implementation on parallel processing computers (15,16).

The first applications of diagrammatic many-body perturbation theory to atoms were reported in 1963, by Kelly (17), who took advantage of the spherical symmetry and employed numerical techniques. Many other calculations for atomic systems were subsequently reported (see, for example, refs. 2, 18-19). The first molecular applications of diagrammatic perturbation theory used a single centre expansion and were, therefore, limited to hydrides containing one non-hydrogenic atom, which was used as an expansion centre by treating the hydrogen nuclei as additional perturbations (see, for example, 20,21). The theory can be applied to arbitrary molecular systems by invoking the algebraic approximation; that is, by parameterization of the orbitals in terms of some finite basis set (see, for example, ref. 9 and references therein; 22).

During the past ten years, the diagrammatic many-body perturbation theory has been shown to afford an accuracy in molecular calculations which is as good as, and in many cases better than, that obtained by other techniques (see, for example, 23,24). Furthermore, the diagrammatic perturbation theory provides a powerful method for the analysis of various techniques for the study of electron correlation in molecules and yields valuable insight into the relations between them (11).

In this article, the application of diagrammatic many-body perturbation theory to molecules is reviewed paying particular attention to the computational aspects of such studies. In section 2, the basic theoretical apparatus of the method is outlined. Various approximations to the full perturbation expansion are described in section 3 and, in section 4, perturbation theory is employed to provide a brief overview of the inter-relation of many contempory theories used in the calculation of correlation energies. In section 5, we turn to the main topic of this article - the computational aspects of atomic and molecular perturbation theory studies which employ finite basis sets. Finally, a brief account of some aspects of current research in this area is given in section 6.

2. DIAGRAMMATIC MANY BODY PERTURBATION THEORY

Only a brief introduction to the formalism of diagrammatic perturbation theory is given here - more detailed descriptions can be found elsewhere (see, for example, refs. 1-11)

A perturbative treatment of correlation effects - the corrections to independent electron, or orbital, models of molecular electronic structure - is developed by writing the total hamiltonian

$$\hat{\mathscr{H}} = \sum_{i=1}^{N} h(i) + \sum_{i>j}^{N} g(i,j) \quad (1)$$

in the form

$$\hat{\mathscr{H}} = (\hat{\mathscr{H}}_o + \lambda \hat{\mathscr{H}}_1)_{\lambda=1} \quad (2)$$

with

$$\hat{\mathscr{H}}_o = \sum_{i=1}^{N} (h(i) + V^N(i)) \quad (3)$$

and

$$\hat{\mathscr{H}}_1 = \sum_{i>j}^{N} g(i,j) - \sum_{i=1}^{N} V^N(i) \quad (4)$$

where h(i) and g(i,j) are one-electron and two-electron operators, respectively. $V^N(i)$ is some effective one-electron potential, such as the Hartree-Fock potential, defining the zero-order hamiltonian and thus the orbital model with respect to which correlation effects are determined. For simplicity, we shall restrict our attention in this section to perturbation series for systems described by a closed-shell Hartree- Fock function in zero-order.

The many-body perturbation theory may be derived from the Rayleigh-Schrödinger expansion corresponding to the reference hamiltonian (3) and the perturbing operator (4) (see, for example, the review of Sadlej in this volume (25)). The zero-order energy is then the sum of the occupied orbital energies, ε, that is

$$E_o = \sum_{i\,=\,\text{occupied}} \varepsilon_i \quad (5)$$

The sum of E_o and the first-order energy, E_1, is the Hartree-Fock energy (or, in a finite basis set, the self-consistent-field energy). The total energy, E, is given by

$$E = E_{scf} + E_2 + E_3 + \dots \quad (6)$$

and, therefore, the correlation energy is given by the sum of the second-order and higher-order energy coefficients:-

$$E_{correlation} = E_2 + E_3 + \dots \quad (7)$$

The Rayleigh-Schrödinger perturbation expansion for the correlation energy, with respect to a reference determinant Φ_o, may be written as

$$E_{correlation} = \sum_{n=1}^{\infty} \langle \Phi_o | \hat{\mathscr{H}}_1 \{ \hat{R}_o (\hat{\mathscr{H}}_1 - E_1 - E_{correlation}) \}^n | \Phi_o \rangle \quad (8)$$

in which $\hat{R}_o$ is the reduced resolvant

$$\hat{R}_o = \frac{1 - |\Phi_o\rangle\langle\Phi_o|}{E_o - \hat{\mathscr{H}}_o} \quad (9)$$

Rayleigh-Schrödinger perturbation theory leads to an expansion for the correlation energy (and indeed other expectation values) which is directly proportional to the number of electrons in the system, N. Beyond second-order, terms arise which have a non-linear dependence on N but such terms can be shown to mutually cancel in each order (26-27). In diagrammatic many-body perturbation theory it is recognized that components of the correlation energy which are directly proportional to N may be associated with linked diagrams, whilst terms which have a non-linear dependence on N may be described by unlinked diagrams. The diagrammatic perturbation theory, therefore, leads to the following expression for the correlation energy

$$E_{correlation} = \sum_{n=1}^{\infty} \langle \Phi_o | \hat{\mathscr{H}}_1 (\hat{R}_o \hat{\mathscr{H}}_1)^n | \Phi_o \rangle_L \quad (10)$$

where the subscript L denotes that only terms corresponding to linked diagrams are included in the summation. This is the well-known linked diagram theorem of Brueckner (26) and Goldstone (27).

The linked diagram theorem is of central importance to the application of perturbation theory methods to problems of chemical interest. Often the quantum chemist is concerned not with the properties of a single molecule but with the comparison of a number of similar species. It is then important that the theoretical apparatus which we use to describe such systems yields energies and other expectation values which are directly proportional to the number of electrons being considered. This property enables meaningful comparisons of systems of different sizes and also facilitates the accurate study of various dissociative and reactive chemical processes, which involve small changes in the total energy.

In many-body perturbation theory, the cancellation of terms corresponding to unlinked diagrams is performed explicitly. Because this cancellation may involve terms arising from different degrees of excitation, the many-body perturbation theory implicitly takes account of some higher-order excitations. For example, in fourth-order there is a cancellation between terms involving doubly excited configurations and quadruply excited configurations.

The diagrammatic perturbation expansion through fourth-order for the correlation energy of a closed-shell molecule, which is described in zero-order by the Hartree-Fock model, is shown in Figures 1 and 2. It should be noted that the use of diagrams in many-body perturbation theory is not obligatory; however, the diagrams are both more physical and easier to handle than the corresponding algebraic expressions and it is well worth the effort required to familiarize oneself with the diagrammatic rules and conventions (see, for example, 1,8,9,11). The detailed description of the diagrammatic rules and conventions which we shall follow in the present work have been presented elsewhere (9). Here we shall give a few examples.

The algebraic expressions corresponding to the diagrams are written in terms of antisymmetrized two-electron integrals

$$g_{pqrs}=\int d\underline{r}_1\int d\underline{r}_2\ \phi_p^*(\underline{r}_1)\phi_q^*(\underline{r}_2)\ \frac{1-(12)}{r_{12}}\ \phi_r(\underline{r}_1)\phi_s(\underline{r}_2) \qquad (11)$$

where ϕ is a one-electron function, r_{12} is the inter-electronic distance and (12) is the permutation operator which interchanges the coordinates of electrons 1 and 2; and in terms of

$$D_{ijk\ldots abc\ldots}=\varepsilon_i+\varepsilon_j+\varepsilon_k+\ldots-\varepsilon_a-\varepsilon_b-\varepsilon_c-\ldots \qquad (12)$$

where ε denotes an orbital energy; i,j,k,... are occupied orbital, or hole, indices and a,b,c... are virtual orbital, or particle, indices. Each hole (particle) is represented by a downward (upward) directed line. Each integral of the type given in (11) is represented by a horizontal dashed line, of which there are n in an nth order diagram. The first diagram displayed in Figure 1 is the only one in second-order; it gives rise to the algebraic expression

$$E_2=\frac{1}{4}\sum_{ij}\sum_{ab}\frac{g_{ijab}g_{abij}}{D_{ijab}} \qquad (14)$$

The remaining three diagrams in Figure 1 are of third-order and lead to the expressions

$$E_3(p\text{-}p)=\frac{1}{8}\sum_{ij}\sum_{abcd}\frac{g_{ijab}g_{abcd}g_{cdij}}{D_{ijab}D_{ijcd}} \qquad (15)$$

$$E_3(h\text{-}p)=\sum_{ijk}\sum_{abc}\frac{g_{ijab}g_{akic}g_{bcik}}{D_{ijab}D_{ikbc}} \qquad (16)$$

and

$$E_3(h\text{-}h)=\frac{1}{8}\sum_{ijke}\sum_{ab}\frac{g_{ijab}g_{klij}g_{abkl}}{D_{ijab}D_{klab}} \tag{17}$$

The denominators in the summands are products of (n-1) factors, each arising between adjacent interactions in the diagram. There is a factor of ½ for each pair of equivalent lines - lines beginning at one interaction and ending at another and both going in the same direction - and a factor of $(-1)^p$ where p is the sum of the number of unique hole lines and the number of fermion loops. Typical expressions arising from the fourth-order correlation energy diagrams, which are shown in Figure 2, are :-

(1) for diagram A_S, which involves singly excited intermediate states

$$E_4(A_S)=\frac{1}{4}\sum_{ijk}\sum_{abcde}\frac{g_{ijab}g_{abcj}g_{ckde}g_{deik}}{D_{ijab}D_{ic}D_{ikde}} \tag{18}$$

(2) for diagram A_D, which only involves doubly excited intermediate states

$$E_4(A_D)=\frac{1}{16}\sum_{ij}\sum_{abcdef}\frac{g_{ijab}g_{abcd}g_{cdef}g_{cfij}}{D_{ijab}D_{ijcd}D_{ijef}} \tag{19}$$

(3) for diagram A_T, which involves triply excited intermediate states

$$E_4(A_T)=-\tfrac{1}{2}\sum_{ijk}\sum_{abcde}\frac{g_{ijab}g_{akcd}g_{cbek}g_{edij}}{D_{ijab}D_{ijkbcd}D_{ijde}} \tag{20}$$

(4) for diagram A_Q, which involves quadruply excited intermediate states

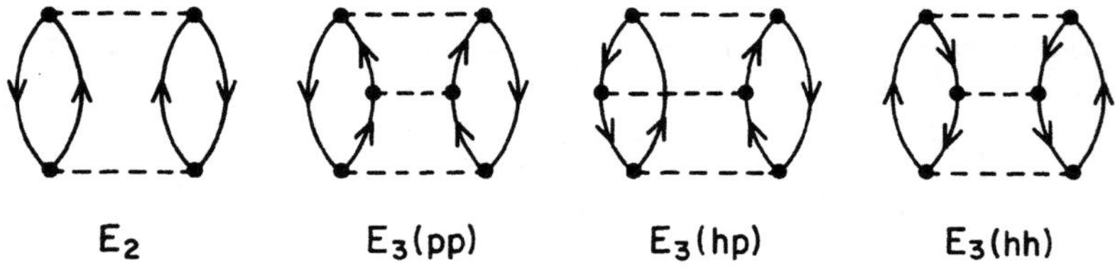

Figure 1. Second-order and third-order diagrams in the diagrammatic perturbation theory series for the correlation energy of a closed-shell system with respect to a Hartree-Fock reference function.

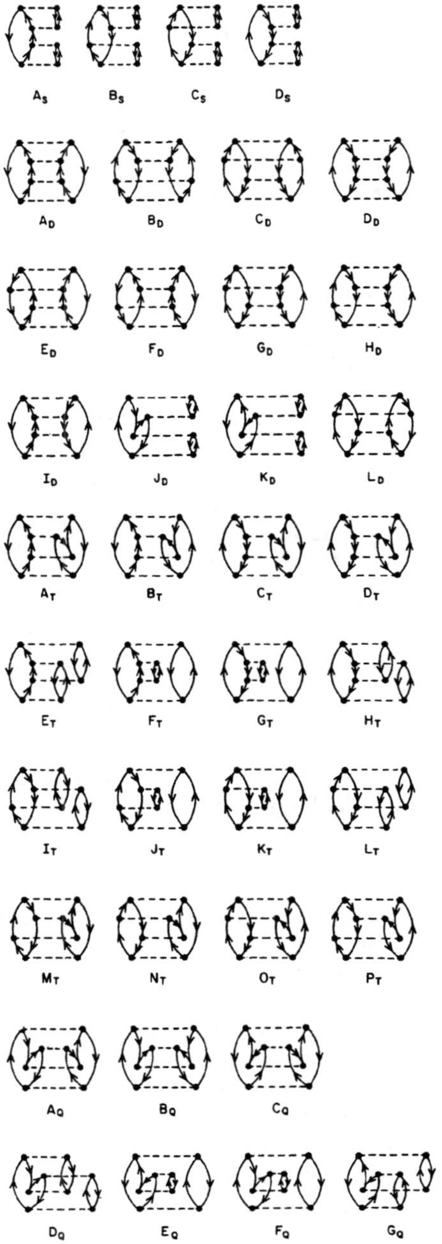

Figure 2. Fourth-order diagrams in the diagrammatic perturbation theory series for the correlation energy of a closed-shell system with respect to a Hartree-Fock reference function.

$$E_A(A_Q)=\frac{1}{2}\sum_{ijk\ell}\sum_{abcd}\frac{g_{ijab}\,g_{k\ell cd}\,g_{cbi\ell}\,g_{adkj}}{D_{ijab}\,D_{i\ell bc}\,D_{jkad}} \tag{21}$$

Some typical higher-order diagrams are shown in Figure 3.

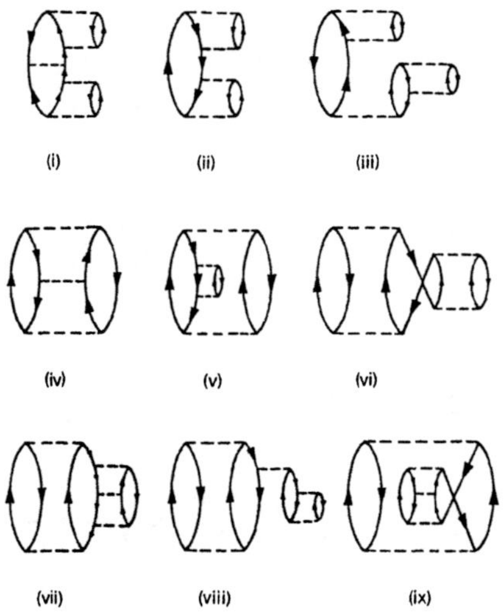

Figure 3. Examples of linked energy diagrams associated with the interaction of intermediate states involving different degrees of excitation: interaction of (i) single- and single-excitations, (ii) single- and double-excitations, (iii) single- and triple-excitations, (vi) double- and quadruple-excitations, (vii) triple- and triple-excitations, (viii) triple- and quadruple-excitations, (ix) quadruple- and quadruple-excitations.

3. APPROXIMATIONS

Diagrammatic many-body perturbation theory forms the basis of a most systematic technique for the evaluation of correlation corrections to independent electron models of atomic and molecular structure in that, once the total hamiltonian has been partitioned, according to equation (2) into a zero-order operator and a perturbation, there is a clearly defined order parameter, λ, in the expansion which affords an objective indication of the relative importance of the various components of the correlation energy expansion. Second-order energies often yield rather accurate estimates of the correlation energy. Through third-order the correlation energy of a closed shell system is represented by just four diagrams, as shown in Figure 1, if antisymmetrized vertices and a Hartree-Fock reference function are employed. Some typical second-order and third-order correlation energies are displayed in Table 1 (28). It has been shown (29-31) that, provided a 'balanced' procedure is followed and all terms through third-order are evaluated, third-order calculations are accurate and very competitive

Table 1

Second-order and third-order energy coefficients for some small atoms and molecules obtained using the Hartree-Fock model perturbation series†

System	E_{scf}	E_2	E_3	E[2/1]
Ne	-128.54045	-210.784	+0.753	-210.034
Ar	-526.80194	-149.200	-11.657	-161.846
BH	-25.12890	-63.700	-16.915	-86.731
FH	-100.06007	-223.633	+ 1.212	-222.427
AlH	-242.45553	-52.514	-14.605	-72.745
ClH	-460.09456	-153.719	-15.311	-170.724
BeH_2	-15.77024	-51.998	-12.597	-68.623
OH_2	-76.05558	-220.304	-3.108	-223.457
MgH_2	-200.72028	-45.934	-12.498	-63.105
SH_2	-398.70077	-150.402	-19.784	-173.182
BH_3	-26.39834	-97.988	-18.943	-121.471
CH_3^+	-39.24533	-112.112	-18.635	-134.462
NH_3 (D_{3h})	-56.20957	-197.500	-10.641	-208.747
NH_3 (C_{3v})	-56.21635	-198.60	-11.003	-210.345
OH_3^+ (D_{3h})	-76.33276	-215.249	-6.161	-221.592
OH_3^+ (C_{3v})	-76.33475	-216.907	-6.224	-223.315
AlH_3	-243.63770	-79.302	-19.460	-105.090
PH_3 (D_{3h})	-342.41887	-141.017	-23.001	-168.516
PH_3 (C_{3v})	-342.47717	-138.965	-23.789	-167.680
N_2	-108.97684	-326.887	+9.201	-317.938
CO	-112.77551	-300.509	+5.083	-295.510
BF	-124.15642	-252.307	-2.390	-254.719
SiO	-363.82790	-284.326	+9.308	-275.312
SiS	-686.48488	-206.813	-16.878	-225.190

†The self-consistent-field energies are in hartree and the correlation energies are in millihartree. The basis sets and nuclear geometries used are given by Guest and Wilson (28).

with other approaches to the correlation problem. The dominant error in such third-order calculations is almost invariably attributable to the truncation of the basis set rather than the neglect of higher-order terms in the perturbation series. The basis set truncation error in electron correlation energy calculations has been discussed elsewhere in this volume (32) and we shall, therefore, not consider it in any detail in this article.

There are two possible ways of improving calculations based on third-order perturbation theory developed with respect to a Hartree-Fock wave function and hamiltonian:- (i) by employing an alternative reference wave function and/or hamiltonian in order to obtain an improved approximation to the correlation energy in third-order; (ii) by extending the perturbation series to fourth-

and higher-orders. We consider these two options in turn.

In the expressions for the components of the correlation energy given in section 2, the N-electron Hartree-Fock model hamiltonian, $\hat{H}_o$, was used as a zero-order operator. This leads to the perturbation series of the type first discussed through second-order, by Møller and Plesset (33,34). However, it is clear that any operator $\hat{X}$ satisfying the commutation relation

$$[\hat{H}_o, X] = 0 \qquad (22)$$

may be used to develop a perturbation series using the same reference wave function. One operator which satisfies (22) is

$$\hat{H}_{shifted} = \sum_k |k\rangle\langle k|\hat{H}|k\rangle\langle k| \qquad (23)$$

in which $|k\rangle$ is an eigenfunction of $\hat{H}_o$, and this gives rise to the shifted denominator, or Epstein-Nesbet (35,36,37), perturbation series. The resulting perturbation expansion has the same diagrammatic representation as that given in section 2. The corresponding algebraic expressions are also as given in section 2 except that (2,9): (i) the denominator factors are each shifted by an amount

$$d_{ijab} = g_{ijij} + g_{abab} + g_{iaai} + g_{ibbi} + g_{jaaj} + g_{jbbj} \qquad (24)$$

(ii) diagonal scattering terms are omitted in third (and higher) order. The use of shifted denominators may also be interpreted as the summation to all orders of certain diagonal terms in the perturbation series based on the Hartree-Fock model hamiltonian.

Following Feenberg (38), let us consider two modifications of a given zero-order operator, $\hat{H}_o$; namely, a uniform displacement of the reference spectrum and a uniform change of scale. This defines the new zero-order hamiltonian as follows:-

$$\hat{H}_o^{\mu,\nu} = \mu \hat{H}_o + \nu\hat{I} \qquad (25)$$

The perturbation is then

$$\hat{H}_1^{\mu,\nu} = \hat{H}_1 + (1-\mu)\hat{H}_o - \nu\hat{I} \qquad (26)$$

so that the full molecular hamiltonian is recovered on adding (25) and (26), that is

$$\hat{H} = (H_o^{\mu,\nu} + \lambda H_1^{\mu,\nu})_{\lambda = 1} \qquad (27)$$

The energy coefficients, $E_i^{\mu,\nu}$, where i denotes the order of perturbation, may be obtained from those of the original series by the expressions

$$E_o^{\mu,\nu} = \mu E_o^{1,0} + \nu \tag{28}$$

$$E_1^{\mu,\nu} = E_1^{1,0} + (1 - \mu)\, E_1^{1,0} - \nu \tag{29}$$

$$E_n^{\mu,\nu} = \left(\frac{1}{\mu}\right)^{n-1} \sum_{k=2} \binom{n-2}{k-2} (\mu-1)^{n-k}\, E_k^{1,0} \quad (n > 1) \tag{30}$$

It can be shown (39-41) that the [M+1/M] Padé approximants, alone among all Padé approximants of order 2M+1 (including the [2M+1/0] Padé approximant), are invariant to arbitrary choices of μ and ν, when λ is set to unity.

The sum of the perturbation expansion, with respect to given reference wave function, through infinite order is, of course, independent of the choice of zero-order operator. The degree of agreement between two perturbation expansions based on different $\hat{H}_o$'s is, therefore, a qualitative measure of the convergence of the series. In Table 2, third-order results are presented for a number of small atoms and molecules obtained using both the Hartree-Fock model perturbation series and the shifted denominator series. For each of these expansions, [2/1] Padé approximants are formed. It can be seen, from this Table, that for systems which are well described by a single determinant, such as Ne, FH or NH_3, the third-order model perturbation series E[3/0] is in close agreement with the [2/1] Padé approximant formed from it, whereas the shifted denominator series leads to a significant change on forming the [2/1] Padé approximant; although the [2/1] Padé approximant is in close agreement with that constructed for the model series. For systems such as the BH, BeH_2 and BH_3 molecules, which involve some degree of quasi-degeneracy, the results obtained from the model perturbation series change significantly on forming the [2/1] Padé approximant and, furthermore, are thereby brought into closer agreement with the shifted denominator results E[3/0] and E[2/1] which for these cases are in close agreement with each other.

The modifications of the perturbation expansion considered above have not departed from the use of the canonical molecular orbitals, both in the reference function and in the excited configuration functions. Let us briefly consider the use of orbitals other than the Hartree-Fock canonical orbitals. If electron-electron interactions are completely neglected, i.e. the bare-nucleus model, in zero-order, we then have to consider in the corresponding diagrammatic perturbation series diagrams which contain 'bubbles'. The second-order diagrams which then arise are displayed in Figure 4 and the full set of third-order diagrams are shown in

Table 2

Third-order diagrammatic perturbation theory calculations for some small atoms and molecules obtained using the Hartree-Fock model perturbation series and the shifted denominator perturbation series[†]

System	Model perturbation expansion		Shift denominator perturbation expansion	
	E[3/0]	E[2/1]	E[3/0]	E[2/1]
Ne	-210.031	-210.034	-201.253	-209.472
Ar	-160.858	-161.846	-152.283	-161.454
BH	-80.615	-86.731	-89.893	-89.893
FH	-222.421	-222.427	-210.262	-221.208
AlH	-67.119	-72.745	-74.334	-74.349
ClH	-169.030	-170.724	-162.270	-169.975
BeH_2	-64.596	-68.623	-69.181	-69.187
OH_2	-223.413	-223.457	-210.657	-221.857
MgH_2	-58.433	-63.105	-63.506	-63.693
SH_2	-170.186	-173.182	-167.182	-172.549
BH_3	-116.931	-121.471	-122.542	-122.593
CH_3^+	-130.748	-134.462	-135.462	-135.451
NH_3 (D_{3h})	-208.141	-208.746	-200.974	-208.077
NH_3 (C_{3v})	-209.600	-210.245	-202.576	-209.543
OH_3^+ (D_{3h})	-221.411	-221.592	-213.513	-220.229
OH_3^+ (C_{3v})	-223.131	-223.315	-216.247	-222.409
AlH_3	-98.762	-105.091	-106.862	-107.067
PH_3 (D_{3h})	-164.029	-168.516	-165.865	-168.229
PH_3 (C_{3v})	-162.763	-167.680	-166.193	-167.891
N_2	-317.686	-317.938	-257.443	-310.457
CO	-295.697	-295.510	-269.302	-293.641
BF	-254.697	-254.719	-247.456	-256.778
SiO	-275.017	-275.312	-244.702	-273.423
SiS	-223.691	-225.190	-205.403	-222.582

[†] All correlation energies are in millihartree. The basis sets and nuclear geometries used are given by Guest and Wilson (28).

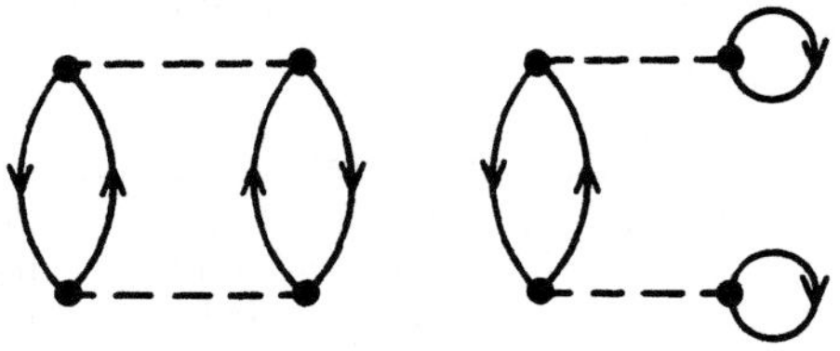

A B

Figure 4. Second-order diagrams in the diagrammatic perturbation theory series for the correlation energy of a closed-shell system with respect to a bare-nucleus reference function.

Figure 5. If instead of using the canonical Hartree-Fock virtual orbitals, we use the orbitals corresponding to the so-called V^{N-1} potential (2,42,43), then the diagram shown in Figure 5N has to be considered together with the diagram obtained from this by replacing the 'bubble' by a 'cross' representing the V^{N-1} potential. It has been demonstrated (43) that results obtained by using the V^{N-1} potential are in fairly close agreement with those corresponding to the Hartree-Fock, V^N, potential if all terms through third-order are included. Similarly, if localized Hartree-Fock orbitals are employed in the reference configurations then Figure 5 M has to be considered together with the corresponding diagram containing a 'cross'.

As an alternative to modification of the reference wave function and/or zero-order hamiltonian in order to obtain an improved approximation for the correlation energy in third-order, the perturbation series can be extended to fourth-order. The fourth-order terms, which correspond to the diagrams shown in Figure 2, can involve singly, doubly, triply and quadruply excited intermediate states (5,44). Diagrammatic perturbation theory provides a systematic scheme for including triply and quadruply, and indeed higher-order, excited intermediate states. The results shown in Table 3 illustrate the importance or fourth-order components of the correlation energy for a number of small molecules (45,46). Provided that all terms are included through fourth-order in the perturbation expansion, it has been demonstrated that chemical accuracy (1 millihartree) can be achieved, within the basis set employed. A comparison of second-order, third-order and fourth-order perturbation theory with full configuration interaction within the same basis set is presented in Figure 6 for the H_2 dimer in various nuclear geometries (47,48). Good agreement between the two calculations is observed for geometries where quasi-degeneracy effects are not significant.

Table 3

Calculations through fourth-order for some small molecules[†]

Molecule	E_2	E_3	E_{4SD}	E_{4T}	E_{4Q}	E[3/0]	E[4/0]
FH	-0.2206	-0.0001	-0.0041	-0.0039	+0.0011	-0.2207	-0.2276
OH_3	-0.2176	-0.0045	-0.0050	-0.0050	+0.0020	-0.2221	-0.2300
NH_3	-0.1989	-0.0118	-0.0003	-0.0050	+0.0027	-0.2107	-0.2133
CH_4	-0.1704	-0.0191	-0.0056	-0.0040	+0.0028	-0.1895	-0.1963
N_2	-0.3291	+0.0098	-0.0178	-0.0180	+0.0065	-0.3193	-0.3486
CO	-0.3076	+0.0048	-0.0172	-0.0161	+0.0049	-0.3028	-0.3312
HCN	-0.3030	+0.0016	-0.0152	-0.0167	+0.0067	-0.3014	-0.3267

[†]Atomic units are used throughout.

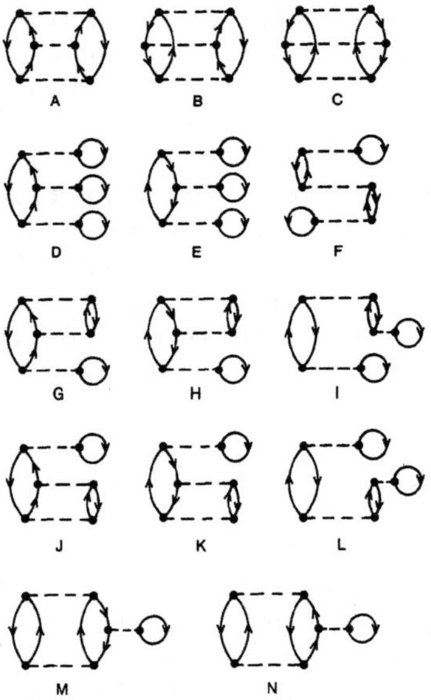

Figure 5: Third-order diagrams in the diagrammatic perturbation theory series for the correlation energy of a closed-shell system with respect to a bare-nucleus reference function

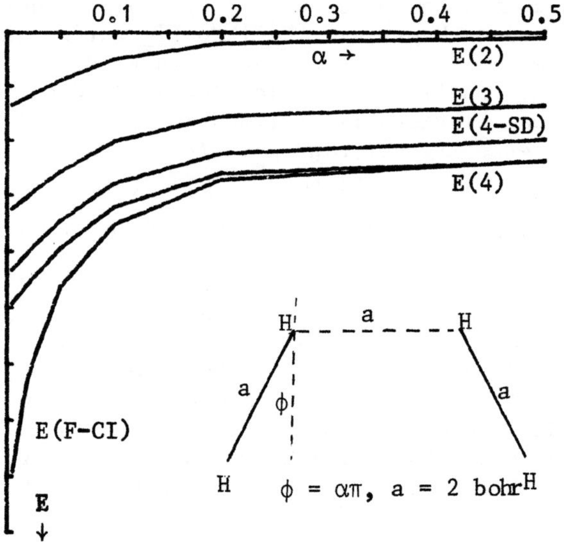

Figure 6: Comparison of second-order, third-order and fourth-order diagrammatic perturbation theory with full configuration interaction calculations for the H_2 dimer.

4. ANALYSIS OF APPROACHES TO THE CORRELATION PROBLEM

All methods for performing accurate calculations of the electronic structure of atoms and molecules and which include a description of correlation effects, involve some finite or infinite expansion for the wave function and corresponding expectation values. This is the case in, for example, the method of configuration interaction, the many-body perturbation theory, the valence bond theory, the group function method and various cluster expansions. Each method leads to the exact wave function and expectation values if all of the terms in the expansion are included. In practice, of course, such expansions have to be truncated, in order to keep the computation tractable, and the various methods differ only in the matter in which this truncation is carried out. However, the method of truncation can significantly affect not only the theoretical properties of a particular approach but also, to some extent, its computational feasibility. The fact that one method may include more terms than another does not necessarily make it superior - we have already indicated that single- and double-excitation configuration interaction, with respect to a single determinantal reference wave function, for example, includes many terms, corresponding to unlinked diagrams in the perturbation theoretic analysis of the method, which are exactly cancelled by other terms involving higher order excitations. An estimate of these unphysical terms is given by the Davidson correction (49-52).

$$\left(\frac{1 - C_o^2}{C_o^2}\right) \quad E_{CI,SD} \tag{31}$$

which is correct through fifth-order. Third-order diagrammatic perturbation theory calculations can also provide an estimate of the unphysical terms present in single and double excitation configuration interaction by derivation of an upper bound from the first order wave function, $\psi_o+\gamma\psi_1$, treating γ as a variational parameter, giving (53)

$$\gamma^3 E_2\Delta_{11}, \quad \Delta_{11} = \langle\psi_1|\psi_1\rangle \tag{32}$$

(Note that this estimate varies as $\sqrt{N}$ in contrast to the fourth-order unlinked diagram terms, $E_2\Delta_{11}$, (54), which vary as N^2 (55,56)). Typical values of the estimate (32) are displayed in Table 4.

Perturbation theory not only provides a very efficient scheme for the computation of correlation effects in atoms and molecules but also affords a highly systematic technique for the analysis of the various approaches employed in many contemporary calculations. In Table 5, an overview of the perturbation theoretic analysis of some of the techniques currently being employed in correlation energy calculations is presented. It should be noted that many of

Table 4

Values of $\gamma^3 E_2 \Delta_{11}$ for some small atoms and molecules†

System	$\gamma^3 E_2 \Delta_{11}$	System	$\gamma^3 E_2 \Delta_{11}$	System	$\gamma^3 E_2 \Delta_{11}$	System	$\gamma^3 E_2 \Delta_{11}$
Ne	-5.605	BeH_2	-2.510	$NH_3(D_{2h})$	-10.914	$PH_3(C_{3h})$	-11.620
AR	-7.144	OH_2	-20.798	$NH_3(C_{3v})$	-11.228	N_2	-25.211
BH	-4.415	MgH_2	-2.460	$OH_3^+(D_{3h})$	-10.099	CO	-21.314
FH	-7.763	SH_2	-11.255	$OH_3^+(C_{3v})$	-10.340	BF	-14.862
AℓH	-4.084	BH_3	-5.803	$A\ell H_3$	-5.805	SiO	-21.775
CℓH	-9.461	CH_4	-5.809	$PH_3(D_{3h})$	-11.804	SiS	-21.628

† All energies are in millihartree. The basis sets and nuclear geometries are given by Guest and Wilson (28).

Table 5

Perturbation theoretic analysis of some approaches to the electron correlation problem in molecules†

Method	Order	Comments	Reference
SD CI	3	includes unlinked diagram components in fourth-order	(57)
SDTQ CI	5	includes unlinked diagram components in sixth-order	(57)
CEPA	3	CEPA(0) includes linked diagram double-excitation components through all orders Other CEPAs include approximations to fourth- and higher-order linked diagram quadruple excitation components CEPA(0) ≡ L-CPMET ≡ DE MBPT	(54),(58) (59),(60)
CPMET	3	neglects triple excitations in fourth-order; these are included in ECPMET	(61)
CC SD	3	neglects triple excitations in fourth-order	(10)

† Acronyms used:- SDCI: single- and double-excitation configuration interaction; SDTQ: single-, double-, triple-and quadruple-excitation configuration interaction; CEPA: coupled electron pair approximation; CPMET: coupled pair many electron theory; CCSD coupled cluster-single- and double-excitation; L-CPMET linear CPMET; De MBPT: double-excitation many-body perturbation theory.

these techniques are third-order theories, in that they neglect or approximate fourth and higher order terms. Perturbation theory can indicate the dominant corrections to these theories.

5. COMPUTATIONAL ASPECTS

In a recent review of molecular quantum mechanics, McWeeny and Pickup (8) summarized the outlook for accurate ab initio calculations as follows:- "In looking to the future, one thing is clear beyond all doubt: ab initio calculations, of 'chemical accuracy', are going to be dominated more and more by the development of computers and highly efficient algorithms. New theories will still be required,..., but formal theory will not be enough; the feasibility of the computational implementation will be of paramount importance". The theoretical properties of the diagrammatic perturbation theory, which were described in section 2, ensure that it leads to a computational scheme of high efficiency, which can treat correlation effects with little more effort than a conventional self-consistent-field calculation. The particular strengths of the diagrammatic perturbation theory from a computational point of view are (12-14,15,62,63):- (a) it leads to a non-iterative algorithm, (b) all possible excited states are included through a given order; configuration selection schemes, such as those widely used in the practical implementation of the method of configuration interaction are avoided, (c) the diagrammatic representation of the correlation energy components can be very easily translated into computer code, (d) it leads to algorithms which are extremely well suited to implementation on parallel processing computers. After integrating over spin coordinates, the various expressions in the perturbation series can be written in terms of integrals over spatial orbitals. Here we use lower case indices to represent spin orbitals and upper case indices to represent spatial orbitals; I,J,K,... and A,B,C,... are used to denote occupied and unoccupied orbitals, respectively. In order to facilitate the evaluation of the algebraic expressions corresponding to each of the diagrams, it is convenient to separate the list of two-electron integrals over molecular orbitals into six separate lists. Each of these lists corresponds to the classes of diagrammatic components shown in Figure 7.

Efficient algorithms for the evaluation of the correlation energy components corresponding to the various terms in the diagrammatic perturbation expansion may be devised by:- (i) taking account of spin orthogonalities, (ii) recognizing certain permutational symmetry properties of various intermediates which arise in the calculation, (iii) using the diagrammatic form of Wigner's (2n+1) rule, (iv) exploiting spatial symmetry properties of the system being studied. The importance of Wigner's (2n+1) rule has been emphasized recently by Kvasnicka and his coworkers (64,65), although it had been employed implicitly in earlier algorithms (12-14).

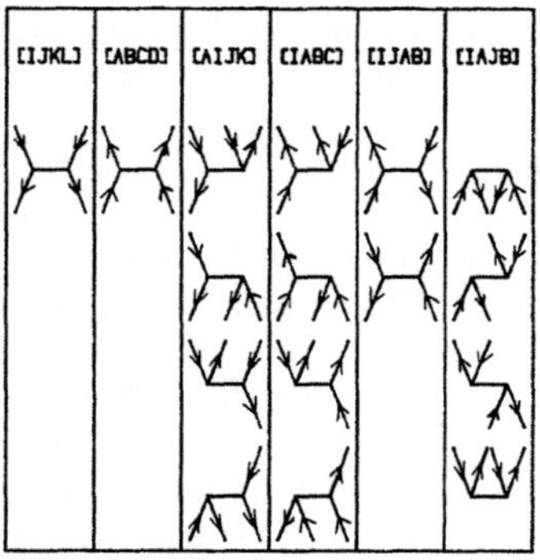

Figure 7: Components of diagrammatic expression corresponding to two-electron interactions. (The "charge-cloud" notation - [PQRS = g_{PRQS} - is used for the integrals over spatial orbitals).

In its diagrammatic form the Wigner (2n+1) rule may be written

$$E_{2n} = \langle\psi_{n-1}|\hat{\mathcal{H}}_1|\psi_n\rangle_L \tag{33}$$

$$E_{2n+1} = \langle\psi_n|\mathcal{H}_1|\psi_n\rangle_L \tag{34}$$

Thus, the determination of the energy through to order 2n+1 only requires a knowledge of the wave function through order n. Algorithms based on Wigner's (2n+1) rule are clearly more efficient than the recursive schemes described in refs. (66).

Schemes for evaluating the diagrammatic perturbation theory energy components have been described in detail elsewhere (12-14, 62-65). Here we shall illustrate these algorithms by considering the computation of the energy corresponding to the fourth-order triple-excitation diagrams. These are the most demanding of the fourth-order diagrams for a closed shell system leading to an algorithm which depends on m^7. The energy components corresponding to the sixteen diagrams, labelled X_T in Figure 2, may all be written in terms of the intermediate quantities (63)

$$f_{ijk;abc} = \sum_d \frac{g_{ijdb}\, g_{dkac}}{D_{ijdb}} \tag{35}$$

and

$$g_{ijk;abc} = \sum_\ell \frac{g_{i\ell ba}\, g_{ik\ell c}}{D_{j\ell ab}} \tag{36}$$

Each diagram leads to an expression of the form

$$K \sum_{ijk} \sum_{abc} \phi_{ijk;abc} P_{ijk} P_{abc} \gamma_{ijk;abc} / D_{ijkabc} \quad (37)$$

where ϕ and γ denote either an f or a g intermediate, K is a constant and $P_{ijk}P_{abc}$ is a permutation operator. The following expressions are typical:-

$$E_4(A_T) = -\frac{1}{2} \sum_{ijk} \sum_{abc} f_{ijk;abc} f_{ijk;acb} / D_{ijkabc} \quad (38)$$

$$E_4(C_T) = -\frac{1}{2} \sum_{ijk} \sum_{abc} g_{ijk;abc} g_{ijk;acb} / D_{ijkabc} \quad (39)$$

$$E_4(I_T) = -\frac{1}{4} \sum_{ijk} \sum_{abc} f_{ijk;abc} g_{ikj;abc} / D_{ijkabc} \quad (40)$$

We note that diagrams L_T K_T N_T P_T are complex conjugates of I_T J_T M_T O_T, respectively and thus for real orbitals are numerically equal. The 9 spin types which arise in the evaluation of the $f_{ijk;abc}$ intermediates are summarized in Table 6. Definition of these intermediates enables a spin-free formulation of the problem to be achieved. If we define the secondary intermediate quantities

$$T_1 = \sum_D \langle IJ|V|DB\rangle\langle DK|V|AC\rangle / D_{IJBD} \quad (41) \qquad T_2 = \sum_D \langle IJ|V|DB\rangle\langle DK|V|CA\rangle / D_{IJBD} \quad (42)$$

$$T_3 = \sum_D \langle IJ|V|BD\rangle\langle DK|V|AC\rangle / D_{IJBD} \quad (43) \qquad T_4 = \sum_D \langle IJ|V|BD\rangle\langle DK|V|CA\rangle / D_{IJBD} \quad (44)$$

where $V = r_{12}^{-1}$, the various spin free intermediates $F_{IJK,ABC}$, are given by

$$F^1_{IJK;ABC} = T_1 - T_2 - T_3 + T_4 \quad (45)$$

$$F^2_{IJK;ABC} = T_1 - T_4 \quad (46)$$

$$F^3_{IJK;ABC} = -T_2 + T_4 \quad (47)$$

$$F^4_{IJK;ABC} = T_1 - T_2 \quad (48)$$

$$F^5_{IJK;ABC} = T_1 \quad (49)$$

$$F^6_{IJK;ABC} = -T_2 \quad (50)$$

$$F^7_{IJK;ABC} = -T_3 + T_4 \quad (51)$$

$$F^8_{IJK;ABC} = -T_3 \quad (52)$$

$$F^9_{IJK;ABC} = T_4 \quad (53)$$

Similar spin-free intermediates may be obtained for the $g_{ijk;abc}$. The energy corresponding to diagram A_T, for example, may then be written as

$$E_4(A_T) = -\sum_{IJK} \sum_{ABC} F^1_{IJK;ABC} F^1_{IJK;ACB} + F^3_{IJK;ABC} F^3_{IJK;ACB} + F^4_{IJK;ABC} F^9_{IJK;ACB} + F^5_{IJK;ABC} F^5_{IJK;ACB}$$

$$+ F^{6}_{IJK;ABC}\, F^{7}_{IJK;ACB} + F^{7}_{IJK;ABC}\, F^{6}_{IJK;ACB}$$
$$+ F^{8}_{IJK;ABC}\, F^{8}_{IJK;ACB} + F^{9}_{IJK;ABC}\, F^{5}_{IJK;ACB}\}/D_{IJKABC} \quad (54)$$

It should be noted that the intermediates $F_{IJK;ABC}$ obey the permutational symmetry properties

$$F_{IJK;ABC} = - F_{JIK;ABC} = -F_{IJK;CBA} = F_{JIK;CBA} \quad (55)$$

and thus only 9 unique intermediates have to be considered. Similar relations exist for the $G_{IJK;ABC}$ intermediates. The implementation of this algorithm for evaluation of the triple-excitation components of the correlation energy on a CRAY 1 vector processing computer has been described elsewhere (15,63). A number of calculations have demonstrated the importance of the triple-excitation energy component, particularly in multiply bonded systems (28,46, 67,68).

Similar algorithms to that described above for the fourth-order triple-excitation energy can be devised for other components of the diagrammatic perturbation expansion of the correlation energy. Pople et al. (69) report an algorithm for the evaluation of E_3(p-p) in which the quantity is determined directly from the integrals over atomic orbitals. This avoids the most time-consuming part of the integral transformation step of the calculation.

The linked diagram theorem of many-body perturbation theory allows the energy of any system to be written as a sum of the energies of its component parts no matter how these components are defined. The linked diagram theorem is, therefore, of considerable computational importance since it provides an additive decomposition

Table 6

Spin types $F^{p}_{IJK;ABC}$

p	g_{IJDB}	g_{DKAC}
1	$\alpha\alpha\alpha\alpha$	$\alpha\alpha\alpha\alpha$
2	$\alpha\alpha\alpha\alpha$	$\alpha\beta\alpha\beta$
3	$\alpha\alpha\alpha\alpha$	$\alpha\beta\beta\alpha$
4	$\alpha\beta\alpha\beta$	$\alpha\alpha\alpha\alpha$
5	$\alpha\beta\alpha\beta$	$\alpha\beta\alpha\beta$
6	$\alpha\beta\alpha\beta$	$\alpha\beta\beta\alpha$
7	$\beta\alpha\alpha\beta$	$\alpha\alpha\alpha\alpha$
8	$\beta\alpha\alpha\beta$	$\alpha\beta\alpha\beta$
9	$\beta\alpha\alpha\beta$	$\alpha\beta\beta\alpha$

of expectation values which aids the development of algorithms for parallel computation by allowing calculations on arbitrarily defined subsystems to be performed concurrently (16). It is clear that a central problem of atomic and molecular physics over the next few years is to develop efficient techniques for the ab initio treatment of electron correlation effects which allow the organization of concurrent computation on a very large scale. Since the linked diagram theorem ensures that the diagrammatic perturbation theory satisfies certain additive separability conditions of the type first discussed by Primas (70), it forms the basis of algorithms affording the maximum degree of parallelism. The paracomputer model, recently introduced by Schwartz (71), plays a useful theoretical role in the determination of the limits of parallel computation in a given method. The paracomputer consists of a very large number of identical processors, each with a conventional order code set, that share a common memory which they can read simultaneously in a single cycle. Such machines cannot be realised physically, since any computing element can have not more than some fixed number of external connections; however, the model is theoretically valuable. Consider, for example, the evaluation of the second-order energy on a paracomputer. The number of terms in the summation is

$$M = \frac{1}{4} N_O N_V (N_O+1) (N_V+1) \tag{56}$$

where N_O and N_V are the numbers occupied and virtual orbitals respectively. On a serial computer the evaluation of the second-order energy will take a time T_{serial} given by

$$T_{serial} = M\tau_1 + (M-1)\tau_2 \tag{57}$$

where τ_1 is the time required to compute the contribution to the sum by a given set of spatial orbitals IJAB and τ_2 is the time required to perform a floating point addition. On a paracomputer with P processors, the computation of the second-order energy will take a time $T_{parallel}$ given by

$$T_{parallel} = \frac{M}{P}\tau_1 + (\frac{M}{P} + P - 2)\tau_2. \tag{58}$$

If the number of processors is considerably less than the number of terms in the summation, M>>P, then

$$T_{parallel} \approx \frac{M}{P}(\tau_1 + \tau_2) \tag{59}$$

and the paracomputer performs the calculation P times faster than a serial machine. The above considerations can be generalized to other components of the correlation energy in the perturbation expansion. Any contribution can be evaluated approximately P times faster on a P-processor paracomputer than on a serial machine.

Considerations, using the paracomputer model, such as those outlined above, clearly demonstrate the suitability of diagrammatic perturbation theory for concurrent computation (16). In a recent monograph, Hockney and Jesshope (72) express the view that "What has changed with the advent of the parallel computer is the ratio between the performance of a good and a bad computer program"; in molecular physics, the quality of a computer program is, to a large extent, dependent on the theory upon which it is based.

6. SOME CURRENT RESEARCH

In this final section, we shall give an outline of some current research on the application of diagrammatic perturbation theory to molecules. The discussion will necessarily be selective and will be restricted to work on correlation energies, since the application of the method to other properties is described elsewhere in this volume by A.J. Sadlej. It was noted in section 3, that the dominant error in most molecular calculations employing diagrammatic perturbation theory,and indeed in most contemporary calculations, is attributable to basis set truncation. A number of perturbation calculations have been reported using universal basis sets and systematic sequences of basis sets (32 and references therein). A scaling procedure has been devised (73) which enables low-order perturbation calculations, using a large basis set, to approximate calculations taken to higher-order in the same large basis set.

Diagrammatic perturbation theory calculations through third-order in the energy have been reported (74-76) for a number of open-shell systems using a restricted Hartree-Fock reference function. The theory has been extended to fourth-order (77).

The use of multideterminantal reference functions in perturbation theory studies often leads to the so-called 'intruder state problem,' i.e. the presence of excited configurations having an energy considerably below that of configurations included in the model space. The presence of intruder states can impair or even destroy the convergence of quasi-degenerate many-body perturbation theory. Two solutions can be advanced: (i) the formation of matrix Padé approximants to the effective interactive matrix - an approach which has proved useful in nuclear physics (28); (ii) the use of an incomplete model space (79-86). The second approach has the disadvantage that the definition of the model space is rather arbitrary.

Finally, we mention the use of valence bond reference functions in perturbation studies (11,87-91). Such functions would appear to offer the most attractive description of dissociative processes since a qualitatively correct picture is obtained in zero-order.

The author is an S.E.R.C. Advanced Fellow.

References

1. March,N.H., Young, W.H. and Sampanthar, S.: 1967, *The Many-Body Problem in Quantum Mechanics*, Cambridge University Press.
2. Kelly, H.P.: 1969, Adv. Chem. Phys. 14, 129.
3. Robb, M.A.: 1974, *Computational Techniques in Quantum Chemistry and Molecular Physics*, edited by G.H.F. Diercksen, B.T. Sutcliffe and A. Veillard, D. Reidel.
4. Paldus, J. and Čížek, J.: 1975, Adv. Quantum Chem. 9, 105.
5. Wilson, S.: 1978, Proc. Daresbury Study Weekend, Dec. 1977, edited by V.R. Saunders, S.R.C.; 1980, Proc. Daresbury Study Weekend, Nov. 1979, edited by M.F. Guest and S. Wilson, S.R.C.
6. Hubač, I. and Čársky, P.: 1978, Top. Curr. Chem. 75, 97.
7. Čarsky, P. and Urban, M.: 1980, *Ab initio Calculations:* Methods and Applications in Chemistry, Springer, and references therein.
8. McWeeny, R. and Pickup, B.T.: 1980, Rep. Progr. Phys. 43, 1065 and references therein.
9. Wilson, S.: 1981, Specialist Periodical Reports: Theoretical Chemistry, 4, 1.
10. Bartlett, R.J.: 1981, Ann. Rev. Phys. Chem. 32, 359 and references therein.
11. Wilson, S.: 1983, *Electron correlation in molecules*, Clarendon Press, Oxford.
12. Silver, D.M.: 1978, Comput. Phys. Commun. 14, 71.
13. Silver, D.M.: 1978, Comput. Phys. Commun. 14, 81.
14. Wilson, S.: 1978, Comput. Phys. Commun. 14, 91.
15. Guest, M.F. and Wilson, S.: 1981, in *Supercomputers in Chemistry*, edited by P. Lykos and I. Shavitt, American Chemical Society, Washington D.C.
16. Wilson, S., 1981, in Proceedings of 5th Seminar on Computational Methods in Quantum Chemistry, edited by P.Th. van Duijnen and W.C. Nieuwpoort, Groningen.
17. Kelly, H.P.: 1963, Phys. Rev. 131, 684.
18. Kelly, H.P.: 2964, Phys. Rev. 136, 896; 1966, Phys. Rev. 144, 39; 1968, Adv. Theoret Phys. 2, 75; 1970, Int. J. Quantum Chem. Symp. 3, 349.
19. Chang, E.S., Pu, R.T. and Das, T.P.: 1968, Phys. Rev. 174, 1, Dutta, N.C., Matsubara, C., Pu, R.T. and Das, T.P., 1969, Phys. Rev. 177, 33; Pu, T.R. and Chang, E.S., 1966, Phys. Rev. 151, 31; Lee, T, Datta, N.C. and Das, T.P.: 1970, Phys. Rev. A 1 995; 1971, Phys. Rev. A. 4, 1410.
20. Kelly, H.P.: 1969, Phys. Rev. Lett. 23, 455; Miller, J.H. and Kelly, H.P.: 1971, Phys. Rev. Lett. 26, 679.
21. Lee, T, Dutta, N.C. and Das, T.P.: 1970, Phys. Rev. Lett. 25, 204, Lee, T. and Das, T.P., 1972, Phys. Rev. A 6, 678; Dutta, C.M., Dutta, N.C. and Das, T.P.: 1970, Phys. Rev. Lett. 25, 1695.
22. Wilson, S. and Silver, D.M.: 1976, Phys. Rev. A14, 1949.
23. Wilson, S. and Silver, D.M.: 1980, J. Chem. Phys. 72, 2159.
24. Wilson, S. and Silver, D.M.: 1982, J. Chem. Phys. 77, (1st Oct).
25. Sadlej, A.J., this volume.

26. Brueckner, K.A.: 1955, Phys. Rev. 100, 36.
27. Goldstone, J.: 1957, Proc. R. Soc. A. 239, 267.
28. Guest, M.F. and Wilson, S.: 1980, Chem. Phys. Lett. 72, 49.
29. Wilson, S. and Silver, D.M.: 1977, J. Chem. Phys. 66, 5400.
30. Wilson, S., Silver, D.M. and Bartlett, R.J.: 1977, Molec. Phys.
31. Bartlett, R.J., Wilson, S. and Silver, D.M.: 1977, Int. J. Quantum Chem. 12, 737; see also ibid 12, 609.
32. Wilson, S.: This volume.
33. Møller, Chr., and Plesset, M.S.: 1934, Phys. Rev. 46, 618.
34. Binkley, J.-J, and Pople, J.A.: 1975, Int. J. Quantum Chem. 9 229.
35. Epstein, P.S.: 1926, Phys. Rev. 28, 695.
36. Nesbet, R.K.: 1955, Proc. R. Soc. London, A. 230, 312, 322.
37. Claverie, P., Diner, S. and Malrieu, J: 1967, Int. J. Quantum Chem. 1, 751.
38. Feenberg, E.: 1958, Ann. Phys. (NY) 3, 292.
39. Amos, A.T.: 1972, Int. J. Quantum Chem, 6, 125; 1970 J. Chem. Phys. 52, 603; 1978, J. Phys. B: At. Mol. Phys. 11, 2053.
40. Wilson, S., Silver, D.M. and Farrell, R.A.:1977, Proc. R. Soc. London, A 356,363.
41. Silver, D.M. and Wilson, S.: 1982, Proc. R. Soc. London (in press)
42. Silver, D.M. and Bartlett, R.J.: 1976, Phys. Rev. A. 13, 1
43. Silver, D.M., Wilson, S. and Bartlett, R.J.: 1977, Phys. Rev. A, 16, 477.
44. Wilson, S. and Silver, D.M.: 1979, Int. J. Quantum Chem. 15, 683.
45. Krishnan, R., Binkley, J.S., Seeger, R. and Pople, J.A.: 1980, J. Chem Phys. 72, 650.
46. Wilson, S. and Guest, M.F.: 1981, J. Phys. B:At. Mol. Phys.
47. Jankowski, K. and Paldus, J: 1980, Int. J. Quantum Chem.
48. Wilson, S., Jankowski, K. and Paldus, J.: 1982, Int. J. Quantum Chem. (submitted).
49. Davidson, E.R.: 1974, The World of Quantum Chemistry, Proc. of First Int. Congr. on Quantum Chem. ed. R. Daudel and B. Pullman. D. Reidel.
50. Langhoff, S.R. and Davidson, E.R.: 1974, Int. J. Quantum Chem. 8, 61.
51. Davidson, E.R. and Silver D.W.: 1977, Chem. Phys. Lett. 52. 403.
52. Siegbahn, P.E.M.: 1978 Chem. Phys. Lett. 55, 386.
53. Wilson, S. and Silver, D.M.: 1979, Theoret. Chim. Acta, 54, 83.
54. Bartlett, R.J. and Shavitt, I.: 1977, Int. J. Quantum Chem, Symp. 11, 165.
55. Meunier, A., Levy, B. and Berthier, G.: 1976, Int. J. Quantum Chem. 10, 1061.
56. Kutzelnigg, W., Meunier, A., Levy, B. and Berthier, G.: 1977, Int. J. Quantum Chem. 12, 77.
57. Shavitt, I.: 1977, in *Modern Theoretical Chemistry, Vol. 3. Methods of Electronic Structure Theory*, ed. H.F. Schaefer III, Plenum.
58. Ahlrichs, R.: 1979, Comput. Phys. Comm. 17, 31.

59. Hurley, A.C.: 1976, *Electron correlation in small molecules*, Academic Press.
60. Blomberg, M.R.A. and Siegbahn, P.E.M.: 1978, Int. J. Quantum Chem. 14, 583.
61. Čížek, J. and Paldus, J.: 1980, Physica Scripta 21, 251 and references therein.
62. Wilson, S. and Silver, D.M.; 1979, Comput. Phys. Comm. 17, 47.
63. Wilson, S. and Saunders, V.R.: 1980, Comput. Phys. Comm. 19 293.
64. Kvasnička, V., Laurinc, V. and Biskupič, S.: 1980, Molec. Phys. 39, 143.
65. Kvasnička, V., Laurinc, V. and Biskupič, S.: 1981, Czeck, J. Phys. B. 31, 41.
66. Bartlett, R.J. and Shavitt, I.: 1977, Chem. Phys. Lett. 50, 190. see also Bartlett, R.J. and Silver, D.M.: 1976, Quantum Science, edited by J.L. Calais et al. Plenum.
67. Wilson, S. and Saunders, V.R.: 1979, J. Phys. B:At. Mol. Phys. 12, L403; 1980 ibid. 13, 5055. Wilson, S. 1979, J. Phys. B: At. Mol. Phys. 12 L657; 1980, ibid 13, 1505; Wilson, S, and Guest, M.F.: 1980, Chem. Phys. Lett. 73, 607; 1981, Molec Phys. 43, 1331; J.P.
68. Krishnan, R., Frisch, M.J. and Pople, J.A.: 1980, J. Chem. Phys. 72, 4244. Frisch, M.J., Krishnan, R. and Pople, J.A.: 1980, Chem. Phys. Letts. 75 66.
69. Pople, J.A., Binkley, J.S. and Seeger, R.: 1976, Int. J. Quantum Chem. Symp. 10, 1, 1965.
70. Primas, H.:1965, Modern Quantum Chemistry II, ed. by. O. Sinanoglu, Academic Press.
71. Schwartz, J.T.:1980, ACM Trans. Prog. Lang. & Systems, 2 484.
72. Hockney, R.W. and Jesshope C.R.:1981, Parallel Computers, Adam Hilger,
73. Wilson, S.: 1981, Theoret. chim. Acta. 59, 71.
74. Hubač, I. and Carsky, P.: 1980, Phys. Rev. A22, 2392.
75. Čarsky, P., Zahradnik, R., Hubac, I., Urban, M. and Kello, V.: 1980, Theoret. chim. Acta. 56, 315.
76. Čarsky, P., Hubač, I. and Staemmler, V.: 1982, Theoret. chim Acta. 60, 445.
77. Wilson, S.: 1982, Theoret. chim. Acta. (in press).
78. Krenciglowa, E.M. and Kuo, T.T.S.: 1974, Nuclear Phys. A235, 171. Hoffmann, H.M., Lee, S.Y., Richert, J. and Weidenmuller.: 1974. Ann. Phys. 85, 410. Leinaas, J.M. and.Kuo, T.T.S.: 1978, Ann. Phys. 111, 19. Schucan, T.H. and Weidenmuller, H.A.: 1972,Ann. Phys. 73, 108. Ruchert, J., Schucan, T.H., Scruber, M.H. and Weidenmuller.: 1976, Ann. Phys. 96, 139. Schucan, T.H. and Weidenmuller, H.A.: 1973, Ann. Phys. 76. 483.
79. Hose, G. and Kaldor, U.: 1979, J. Phys. B:At. Mol. Phys. 12 3827.
80. Hose, G. and Kaldor, U.: 1980, Physica Scripta, 21, 357.
81. Hose, G. and Kaldor, U.: 1982, J. Phys. Chem. 86, 2133.
82. Hegarty, D. and Robb, M.A.: 1979, Molec. Phys. 37, 1455.
83. Hegarty, D. and Robb, M.A.: 1979, Molec. Phys. 38, 1795.
84. Baker, H., Hegarty,D. and Robb, M.A.: 1980, Molec. Phys. 41, 653.

85. Sheppard, M.G. and Freed,K.F.: 1981, J. Chem. Phys. 75, 4507, 4525 and references therein
86. Shavitt, I. and Redman, L.T.: 1980, J. Chem. Phys. 73, 5711 Redmann, L.T. and Bartlett, R.J.: 1982, J. Chem. Phys. 76, 1938.
87. Newman, D.J.: 1969, J. Phys. Chem. Solids, 30, 1709; 1970 ibid. 31, 1143.
88. Moshinsky, M. and Seligman, T.H.: 1971, Ann. Phys. (N.Y.) 66, 311.
89. Gouyet, J.F.: 1973, J. Chem. Phys. 59, 4637.
90. Cantu, A.A., Klein, D.J., Matsen, F.A. and Seligman, T.H.: 1975, Theoret. chim. Acta. 38, 341.
91. Kirtman, B. and Cole, S.: 1978, J. Chem. Phys. 69, 5055.

SCHRÖDINGER SPECTRA

P. W. Langhoff

Department of Chemistry
Indiana University
Bloomington, Indiana 47405

Max Planck Institut für Astrophysik
Karl-Schwarzschild-Strasse 1
8046 Garching bei München
Federal Republic of Germany

INTRODUCTION

Ritz variational methods [1] have long provided a basis for quantitative studies of the eigenstates of atoms and molecules [2]. The wave functions obtained in this way employing appropriate one-electron basis sets and corresponding many-electron configurational state functions generally furnish useful approximations to the ground and low-lying bound excited states of the atom or molecule of interest [3,4]. Closely related computational methods are also employed in construction of approximations to the low-lying poles and residues of molecular many-body Green's functions [5,6]. In the present article, it is shown that useful approximations to all the bound and scattering states of complex molecules are obtained under appropriate conditions from this familiar computational methodology [7]. As a consequence, it proves possible to study molecular excitation and ionization spectra employing a common unified computational procedure.

There are distinct advantages in treating both the discrete and continuum electronic states of a molecule on a common basis. In addition to providing approximations of uniform quality to the entire spectrum, some of the difficulties that have hampered conventional scattering calculations of molecular continuum states are avoid entirely through use of explicit L^2 or Hilbert-space

G. H. F. Diercksen and S. Wilson (eds.), Methods in Computational Molecular Physics, 299–333.

methods. Consequently, a greater range of computational approximations can be employed in this way than has proved possible in conventional scattering approaches [8]. Moreover, the L^2 electronic scattering state functions obtained can be represented in conventional body-frame point-group symmetry, and so provide useful diagnostics for the presence of intravalence or Rydberg transitions in both bound and continuum regions [9]. Finally, much of the computational methodology developed for bound-state studies indicated above can be adopted with little modification for purposes of spectral calculations. As might be expected, however, it is necessary to sacrifice a certain degree of spectral resolution in adopting a global approach to studies of discrete and continuum Schrödinger states.

The theory of the present development is given in Section II, begining with indication of the Schrödinger problem of interest and normalization conventions for the associated bound and scattering states [10]. So-called Stieltjes-Tchebycheff L^2 approximations to discrete and continuum Schrödinger states are defined [7], and certain of their properties indicated. Connections are described between these functions and infinite-order partitioning or optical-potential theory [11], and with aspects of the theory of moments [12-14]. In Section III, a number of computational applications of the development are presented. Results for the bound and scattering states of atomic hydrogen illustrate the overall nature and rate of convergence of the method, whereas calculations on molecular hydrogen clarify the additional computational requirements that arise for molecules [15]. An overview is given of recent applications of the approach to the dipole excitation and ionization spectra of diatomic and polyatomic molecules, and selected results are described. Concluding remarks are made in Section IV.

II. THEORY

A self contained account is presented in this section of the Stieltjes-Tchebycheff methods for explicit Hilbert-space representation of the discrete and continuum Schrödinger states of atoms and molecules. Stieltjes-Tchebycheff functions are defined, certain of their properties are indicated, and connections with previously described moment-theory approximations to spectral densities are clarified.

a) Schrödinger Problem

The Schrödinger problem for an N-electron atom or molecule

$$(H-\epsilon)\Psi_{\epsilon} = 0, \tag{1}$$

where H is the appropriate Coulomb Hamiltonian, generally gives rise to a denumerably infinite number of bound states ($\epsilon = \epsilon_i < 0$) that satisfy the orthonormality condition

$$\langle \Psi_i | \Psi_j \rangle = \delta_{ij}, \tag{2a}$$

as well as an essential spectrum ($\epsilon \geq 0$) of scattering or ionization states that satisfy improper (infinite) or Dirac delta-function orthonormality

$$\langle \Psi_\epsilon | \Psi_{\epsilon'} \rangle = \delta(\epsilon - \epsilon'), \tag{2b}$$

in this case in the energy variable [10]. The bound-states of Eqs. (1) and (2a) carry the usual spin and spatial symmetry labels, in addition to the energy indicated explicitly, whereas the scattering states of Eqs. (1) and (2b) also require appropriate channel labels when two or more ionization processes are degenerate at a given total energy. Of course, rotational and vibrational quantum numbers are required for the bound states, and for the ionic cores of scattering states when the conventional Born-Oppenheimer approximation is employed.

As a consequence of the different normalization conventions employed [Eqs. (2)], methods for constructing the electronic portions of molecular scattering states generally differ considerably from those used for bound states. In the present development, approximations to both types of states are constructed from a single computational procedure. It proves useful in this connection to consider the generalized Schrödinger equation

$$(A(H)-a)\Psi_\epsilon = 0, \quad a = A(\epsilon), \tag{3}$$

with A(H) an explicit operator-valued function of H, as an alternative to Eq. (1). Particular choices of the operator A(H) are indicated below.

b) Lanczos Basis

Introduction of a finite-dimensional L^2 basis of many-electron configurational state functions casts Eqs. (1) and (3) into the form of linear homogeneous matrix equations the solutions of which are generally regarded as providing approximations to the first few bound states of the system [3,4]. With appropriate choice of one-electron basis, however, the configurational Hilbert space obtained can span local portions of scattering states, as well as a significant number of bound states of the system. Moreover, the one-electron basis set incorporates necessary boundary conditions, and choice of configurations specifies the scattering channels of interest, as well as the natures of the bound excited states studied.

To construct the necessary states at any energy ϵ, and to insure that they are appropriately normalized, it is convenient to introduce an orthonormal many-electron basis that brings the Hamiltonian or other appropriate operator A(H) to tridiagonal Jacobi form [16]. This so-called Lanczos basis is defined by the recursive equations [17]

$$\beta_j v_{j+1} = (A(H)-\alpha_j)v_j - \beta_{j-1}v_{j-1}, j = 1,2, \ldots . \quad (4a)$$

where ($v_0 = 0$)

$$v_1 = \Phi \quad (4b)$$

$$\alpha_j = \langle v_j | A(H) | v_j \rangle \quad (4c)$$

$$\beta_j = \langle v_{j+1} | A(H) | v_j \rangle \quad (4d)$$

$$\langle v_j | v_k \rangle = \delta_{jk} \quad (4e)$$

with Φ a unity normalized test function chosen to select the specific subspace of the Schrödinger spectrum of interest. Once the function Φ is selected, Eqs. (4) provide a complete recursive definition of the many-electron Lanczos basis and matrix elements α_j, β_j of the operator A(H) [16,17].

Use of the Lanczos basis in the present development differs considerably from more conventional applications of the procedure [18]. Specifically, the test function Φ in the present case has physical significance, as indicated below, and is not an initial guess at an eigenvector of A(H). Moreover, the operator A(H) is chosen to satisfy particular conditions indicated below that the Hamiltonian H itself does not satisfy. Finally, the functions v_j are expressed in computational applications in a large M-term configurational-state basis, and only the first n ($\ll$ M) of them are employed. These points are further clarified in the computational applications reported in Section III below.

c) Schrödinger States

When the Lanczos functions of Eqs. (4) are employed as a basis for expansion of the Schrödinger states of Eq. (3) [19],

$$\Phi_\epsilon = \sum_{j=1}^{\infty} q_j(\epsilon) v_j, \quad (5)$$

the vector $(\underset{\sim}{q}_\epsilon)_j = q_j(\epsilon)$ is found to satisfy the homogeneous equation

$$(\underset{\sim}{A} - a\underset{\sim}{I})\cdot \underset{\sim}{q}_{\epsilon} = 0 \qquad (6)$$

where

$$(\underset{\sim}{A})_{j,j} = \alpha_j \qquad (7a)$$

$$(\underset{\sim}{A})_{j,j+1} = (\underset{\sim}{A})_{j+1,j} = \beta_j \qquad (7b)$$

is the infinite-order tridiagonal Jacobi matrix representative of the operator A(H). As a consequence of the definition of the Lanczos functions of Eqs. (4), the $q_j(\epsilon)$ are found to satisfy the similar recurrence relation [19]

$$\beta_j q_{j+1}(\epsilon) = (A(\epsilon) - \alpha_j)v_j - \beta_{j-1}v_{j-1}, \; j = 1,2, \ldots, \qquad (8)$$

with $q_0 = 0$ and $q_1 = 1$. As indicated in further detail below, $q_{n+1}(\epsilon)$ is the characteristic or secular polynomial of the n-term Lanczos basis-set problem [20], and, consequently, can be written

$$q_{n+1}(\epsilon) = (-1)^n \det|\underset{\sim}{A}^{(n)} - a\underset{\sim}{I}^{(n)}|/(\beta_1\beta_2 \ldots \beta_n). \qquad (9)$$

Furthermore, these polynomials satisfy the orthonormality condition [21]

$$\int_{-\infty}^{+\infty} q_j(\epsilon)q_k(\epsilon)d\Gamma(\epsilon) = \delta_{jk} \qquad (10)$$

with respect to a distribution $\Gamma(\epsilon)$ defined by the test function Φ according to

$$d\Gamma(\epsilon) = |\langle\Psi_\epsilon|\Phi\rangle|^2 d\epsilon \qquad (11)$$

with $\Gamma(\epsilon \to -\infty) \to 0$. Equation (11) indicates the sense in which the test function Φ has physical significance, since the associated density function

$$g(\epsilon) = |\langle\Psi_\epsilon|\Phi\rangle|^2 \qquad (12)$$

can correspond to a cross section for physical processes [12]. Interest centers here particularly on the case in which $g(\epsilon)$ is proportional to the photo cross section, which includes both discrete and continuum parts [13].

It should be noted that the Lanczos representation of Schrödinger states [Eq. (5)] is not necessarily normalized in the sense of Eqs. (2), since [19]

$$\langle\Phi_\epsilon|\Phi_\epsilon\rangle = \sum_{j=1}^{\infty} q_j^2(\epsilon) \qquad (13a)$$

and

$$\int_{-\infty}^{+\infty} \langle \Phi_\epsilon | \Phi_{\epsilon'} \rangle d\Gamma(\epsilon') = 1. \tag{13b}$$

As indicated in greater detail below, the sum on the right-hand side of Eq. (13a) converges for $\epsilon = \epsilon_i < 0$, whereas it diverges for $\epsilon \geq 0$ [19], suggesting that the states of Eq. (5) can be appropriately renormalized. Of course, it is necessary to consider nth-order ($n < \infty$) approximations to Schrödinger states for practical considerations prior to renormalization.

d) Stieltjes-Tchebycheff Functions

The nth-order approximations to Eq. (5),

$$\Phi_\epsilon^{(n)} = \sum_{j=1}^{n} q_j(\epsilon) v_j \tag{14}$$

are designed Stieltjes-Tchebycheff functions as a consequence of their connection with previously described corresponding moment-theory approximations to spectral densities and distributions [12-14]. To clarify the nature of these functions it is helpful to write the full Schrödinger problem in the Lanczos basis [Eq. (6)] in the partitioned form [11]

$$\begin{bmatrix} \underset{\sim}{A}^{(n)} & \underset{\sim}{\beta} \\ \underset{\sim}{\beta}^{\dagger} & \underset{\sim}{R} \end{bmatrix} \cdot \begin{bmatrix} \underset{\sim}{q}_\epsilon^{(n)} \\ \underset{\sim}{Q}_\epsilon \end{bmatrix} = a \begin{bmatrix} \underset{\sim}{q}_\epsilon^{(n)} \\ \underset{\sim}{Q}_\epsilon \end{bmatrix} . \tag{15}$$

Here, $A^{(n)}$ is n x n tridiagonal [Eqs. (7)], $\underset{\sim}{\beta}$ has only one non-zero element (β_n) in the lower left-hand corner, $\underset{\sim}{\beta}^{\dagger}$ is its adjoint, and $\underset{\sim}{R}$ is an infinite-order tridiagonal remainder matrix. Solving Eq. (15) for the finite vector $\underset{\sim}{q}_\epsilon^{(n)}$ in the standard way gives the equation [11]

$$(\underset{\sim}{A}^{(n)} + \underset{\sim}{V}^{(n)}(\epsilon) - a\underset{\sim}{I}^{(n)}) \cdot \underset{\sim}{q}_\epsilon^{(n)} = 0 \tag{16}$$

where the optical potential matrix $\underset{\sim}{V}^{(n)}(\epsilon)$ is given by the formal expression

$$\underset{\sim}{V}^{(n)}(\epsilon) = \underset{\sim}{\beta} \cdot [\underset{\sim}{I}(a + io) - \underset{\sim}{R}]^{-1} \cdot \underset{\sim}{\beta}^{\dagger}. \tag{17}$$

It is shown by inspection of Eq. (15) and explicit construction that $V^{(n)}(\epsilon)$ has only one nonzero element (n,n) given by the continued fraction [22],

$$[\underset{\sim}{V}^{(n)}(\epsilon)]_{n,n} = - \frac{\beta_n^2}{/a - \alpha_{n+1}}\bigg/ - \frac{\beta_{n+1}^2}{/a - \alpha_{n+2}}\bigg/ - \dots \quad . \tag{18}$$

Making use of the recursive definition of the orthonormal polynomials [Eq. (8)], the infinite-order continued fraction of Eq. (18) can be represented by the rational function

$$[\underset{\sim}{V}^{(n)}(\epsilon)]_{n,n} = \beta_n q_{n+1}(\epsilon)/q_n(\epsilon) \tag{19}$$

for any energy ϵ in the Schrödinger spectrum. As a consequence, Eq. (16) differs from the nxn truncation of Eq. (6) only in the n,n element. Specifically

$$\begin{aligned}[\underset{\sim}{A}^{(n)} + \underset{\sim}{V}^{(n)}(\epsilon)]_{n,n} &= \alpha_n + \beta_n q_{n+1}(\epsilon)/q_n(\epsilon) \\ &= \alpha_n(\epsilon),\end{aligned} \tag{20}$$

where $\alpha_n(\epsilon)$ is a so-called modified recurrence coefficient familiar from the theory of moments [21]. The development of Eqs. (15) to (20) indicates the Stieltjes-Tchebycheff functions are exact optical-potential solutions of the full Schrödinger problem in the Lanczos basis. Consequently, they provide local portions, in the sense of the test function Φ, of the correct bound and continuum Schrödinger states.

The functions of Eq. (14) for all ϵ do not generally provide an orthogonal set, but, rather, have the metric or overlap

$$\langle \Phi_\epsilon^{(n)} | \Phi_{\epsilon'}^{(n)} \rangle = \sum_{j=1}^{n} q_j(\epsilon) q_j(\epsilon') \equiv K^{(n)}(\epsilon,\epsilon'), \tag{21a}$$

and satisfy the normalization condition

$$\int_{-\infty}^{+\infty} \langle \Phi_\epsilon^{(n)} | \Phi_{\epsilon'}^{(n)} \rangle d\Gamma(\epsilon') = 1. \tag{21b}$$

Here, $K^{(n)}(\epsilon,\epsilon')$ is the so-called kernel polynomial of order n having properties in the $n \to \infty$ limit [19]

$$K^{(\infty)}(\epsilon,\epsilon') \to < \infty, \epsilon' = \epsilon = \epsilon_i \tag{22a}$$

$$\rightarrow \infty, \quad \epsilon' = \epsilon \neq \epsilon_i \tag{22b}$$

$$\rightarrow 0, \quad \epsilon' \neq \epsilon \tag{22c}$$

for ϵ', $\epsilon < 0$, and

$$K^{(\infty)}(\epsilon,\epsilon') \rightarrow \infty, \quad \epsilon = \epsilon', \tag{22d}$$

$$\rightarrow 0, \quad \epsilon \neq \epsilon', \tag{22e}$$

in the scattering spectrum ϵ', $\epsilon \geq 0$, that are in accordance with the norms of bound and continuum Schrödinger states [Eqs. (2)]. Consequently, appropriately normalized nth-order approximations to the local portions of discrete and continuum states are obtained from the expression

$$\Psi_\epsilon^{(n)} = (g(\epsilon))^{1/2} \sum_{j=1}^{n} q_j(\epsilon) v_j \tag{23}$$

where $g(\epsilon)$ is the spectral density of Eq. (12). Procedures for construction of $g(\epsilon)$, which have been described previously in considerable detail [12-14], are reviewed briefly in the following subsection (II e). It is further clarified there that the Stieltjes-Tchebycheff functions of Eq. (14) are, in fact, sufficient for determination of $g(\epsilon)$, and so provide their own normalization.

Finally, it is of interest to note that the spectral composition of the correctly normalized Stieltjes-Tchebycheff function of Eq. (23) can be written in the simple form

$$\langle \Psi_{\epsilon'} | \Psi_\epsilon^{(n)} \rangle = (g(\epsilon')g(\epsilon))^{1/2} \sum_{j=1}^{n} q_j(\epsilon') q_j(\epsilon)$$

$$= (g(\epsilon')g(\epsilon))^{1/2} K^{(n)}(\epsilon',\epsilon), \tag{24}$$

making use of the orthonormality of the Lanczos functions. Equation (24) indicates that the nth-order kernel polynomial and the appropriate densities provide the spectral composition of the Stieltjes-Tchebycheff functions, without explicit reference to the correct Schrödinger states. This is largely a consequence of the specific form of the spectral content of individual Lanczos functions, which can be written

$$\langle \Psi_{\epsilon'} | v_j \rangle = g(\epsilon')^{1/2} q_j(\epsilon), \tag{25}$$

emphasizing the spatial-spectral correspondence between Lanczos functions and orthogonal polynomials.

e) Connections with Moment Theory

The test function Φ of Eqs. (4) and (12) provides a set of spectral moments

$$\begin{aligned} \mu_k &= \langle \Phi | A(H)^k | \Phi \rangle \\ &= \int_{-\infty}^{+\infty} A(\epsilon)^k |\langle \Psi_\epsilon | \Phi \rangle|^2 d\epsilon \\ &= \int_{-\infty}^{+\infty} A(\epsilon)^k d\Gamma(\epsilon), \quad k = 0,1,\ldots , \end{aligned} \tag{26}$$

that help to clarify connections between the Stieltjes-Tchebycheff functions and previously described moment-theory approximations to spectral densities and distributions [12-14]. To obtain convergent moments it is evident that the operator A(H) must be chosen appropriately. In the case of dipole excitation spectra [23], for example, the test function and operator A(H) are given by [15]

$$\Phi = \mu\Psi_0 / \langle \mu\Psi_0 | \mu\Psi_0 \rangle \tag{27a}$$

$$A(H) = (H - \epsilon_0)^{-1}, \tag{27b}$$

where μ is a particular polarization component of the dipole moment operator, and ϵ_0, Ψ_0 are the target-state energy and eigenfunction, respectively. With these choices an infinite number of the moments of Eq. (26), which take the form

$$\begin{aligned} \mu_k &= \int_{-\infty}^{+\infty} (\epsilon - \epsilon_0)^{-k} |\langle \Psi_\epsilon | \Phi \rangle|^2 d\epsilon \\ &= \int_{-\infty}^{+\infty} (\epsilon - \epsilon_0)^{-k} \, d\Gamma(\epsilon), \quad k = 0,1,\ldots, \end{aligned} \tag{28}$$

are found to converge [24]. The moments of Eq. (28) can be rewritten as the metric

$$\mu_{k+\ell} = \langle \theta_k | \theta_\ell \rangle, \quad k,\ell = 0,1, \ldots \tag{29a}$$

of L^2 linearly-independent ($\det|\mu_{k+\ell}| > 0$) Cauchy functions [24]

$$\begin{aligned} \theta_k &= A(H)^k \Phi \\ &= (H-\epsilon_0)^{-k} \Phi, \quad k = 0,1,\ldots \end{aligned} \tag{29b}$$

that span the same invariant subspaces as do corresponding numbers of the Lanczos functions of Eqs. (4) [25]. Consequently, it is seen that the existence of an infinite number of convergent moments, guaranteed by appropriate choice of A(H), insures that a complete set of L^2 Lanczos functions can be constructed for representation of Schrödinger states [Eq. (5)]. When the moments of Eqs. (26) to (29) diverge, the corresponding Cauchy and Lanczos functions do not form a Hilbert space.

Stieltjes-Tchebycheff approximations to the spectral density $g(\epsilon)$ and distribution $\Gamma(\epsilon)$ of Eqs. (11) and (12) are obtained from the nth-order Radau quadrature points and weights satisfying the moment equations [12]

$$\Gamma^{(n)}(\epsilon)\, a^{-k} + \sum_{j=1}^{n-1} \Gamma_j^{(n)}(\epsilon)\, a_j^{(n)}(\epsilon)^{-k} = \mu_k, \quad k=0,1,\ldots 2n-2. \tag{30}$$

Here, ϵ is an arbitrary point in the spectrum, $\Gamma^{(n)}(\epsilon)$ is the Radau weight at this point, and the n-1 other points and weights $a_j^{(n)}(\epsilon)$, $\Gamma_j^{(n)}(\epsilon)$ are functions of this preselected energy. Convergent approximations to the spectral density are obtained from these points and weights in the form [23]

$$g^{(n)}(\epsilon) = \frac{d}{d\epsilon}\left[\sum_{j=1}^{k} \Gamma_j^{(n)}(\epsilon) + \frac{1}{2}\Gamma^{(n)}(\epsilon) \right] \tag{31}$$

for $\epsilon \geq 0$. The sum on the right-hand side is over those quadrature weights having energies $\epsilon_j^{(n)}(\epsilon)$ less than the preselected point ϵ. Consequently, it is seen that Eq. (31) is the derivative of the quadrature approximation to the spectral distribution $\Gamma(\epsilon)$. In low orders ($n \leq 15$) it is generally preferable to replace Eq. (31) with a finite difference approximation [23], or to first smooth the quadrature distributions of various orders with polynomial fits prior to differentiation [24]. In the discrete region of the spectrum ($\epsilon < 0$), $g^{(n)}(\epsilon)$ is simply the weight $\Gamma^{(n)}(\epsilon)$, which provides the appropriate normalization factor for bound states, as indicated further below.

The Radau points of Eq. (30) are also obtained from the roots of the characteristic or secular polynomial corresponding to Eqs. (16) to (20) [19,20]

$$\tilde{q}_{n+1}(\epsilon',\epsilon) = (-1)^n \det|\underset{\sim}{A}^{(n)}+\underset{\sim}{V}^{(n)}(\epsilon)-a'\underset{\sim}{I}^{(n)}|/(\beta_1\beta_2\ldots\beta_n)$$

$$= q_{n+1}(\epsilon') - q_{n+1}(\epsilon)q_n(\epsilon')/q_n(\epsilon), \qquad (32)$$

which is a so-called quasi-orthogonal polynomial of order $n + 1$ [21]. Moreover, the weight $\Gamma^{(n)}(\epsilon)$ at the point ϵ is given by [19-21]

$$\Gamma^{(n)}(\epsilon) = \left[\sum_{j=1}^{n} q_j(\epsilon)^2\right]^{-1}$$

$$= \langle\Phi_\epsilon^{(n)}|\Phi_\epsilon^{(n)}\rangle^{-1} \qquad (33)$$

with similar expressions obtained for the n-1 other weights $\Gamma_j^{(n)}(\epsilon)$ by replacing ϵ in Eq. (33) with the other Radau points $\epsilon' = \epsilon_j^{(n)}(\epsilon)$. Consequently, it is seen that the eigenvalues of the so-called orthonormal nth-order Tchebycheff eigenfunctions of Eq. (15) provide the Radau quadrature points of Eq. (30), and that their norms give the corresponding quadrature weights. It is particularly important that the so-called Tchebycheff eigenfunctions of Eq. (16) provide appropriate information for normalization in the continuous spectrum, since the lack of an asymptotic part prevents their delta-function normalization in the standard way [10].

When the preselected point ϵ is set equal to a root of $q_{n+1}(\epsilon)$ of Eq. (9), the optical potential of Eqs. (16) to (20) vanishes identically and Eq. (16) becomes an (n x n) Lanczos problem, the solutions of which give generalized Gaussian quadrature points and weights. The corresponding functions of Eq. (14) at these energies are designated Stieltjes eigenfunctions [7], in accordance with previous usage [23]. Since the Stieltjes energies are the generalized Gaussian quadratures of the density $g(\epsilon)$, they are sure to appear in the dense portions of the spectrum. In this sense the corresponding Stieltjes functions provide an optimal n-term L^2 representation of the spectrum; they can be regarded as natural pseudostates of the system [26]. It should also be noted that the optical potential of Eq. (19) diverges at the (n-1)th-order Stieltjes energies or roots of $q_n(\epsilon)$. At these energies, the solutions of Eq. (16) include one state at infinity having zero weight $\Gamma^{(n)}(\epsilon \to \infty) \to 0$, the others corresponding to the (n-1)th-order Stieltjes functions. Of course, as indicated above, all Stieltjes-Tchebycheff functions of a given order are obtained from Eq. (14) by scanning the energy ϵ over the entire spectral interval. It is nevertheless of interest to identify the special sets of Tchebycheff and Stieltjes eigenfunctions, in view of their close connections with Radau and generalized Gaussian quadratures, respectively.

III. COMPUTATIONAL APPLICATIONS

In this Section, the Stieltjes-Tchebycheff development is applied to construction of the discrete and continuum (dipole) Schrödinger states of hydrogen atoms and molecules. In the former case, the Lanczos functions and polynomial recurrence coefficients required to construct 1s → np/kp spectra are obtained in closed form. The static-exchange approximation is employed for molecular hydrogen in order to illustrate the computational requirements that arise in molecular applications. Large Gaussian basis sets and canonical orthogonalization methods are used in this case to construct the necessary Lanczos functions. The resulting cross sections are reported and Stieltjes-Tchebycheff orbitals are presented in three-dimensional graphical forms to analyze the natures of continuum states in this simple molecule. Finally, a brief review is given of recent applications of the Stieltjes-Tchebycheff development to the dipole excitation and ionization spectra of diatomic and polyatomic molecules, and selected results are described.

a) Atomic Hydrogen

In the case of the regular p-wave spectrum in atomic hydrogen, Eqs. (4) can be solved in closed form using the complete discrete set of Laguerre functions of order $n+\ell+1/2\ell+2$ with constant exponent [27],

$$\alpha_j = (j+1)/j \tag{34a}$$

$$\beta_j = \frac{1}{2}\,[(j+3)/(j+1)]^{1/2}. \tag{34b}$$

These are seen to approach finite values in the limit $j \to \infty$, in accordance with the existence of an infinite set of convergent (reciprocal) power moments [Eqs. (26) to (29)] in this case [23]. By contrast, positive-integer power moments of the dipole density

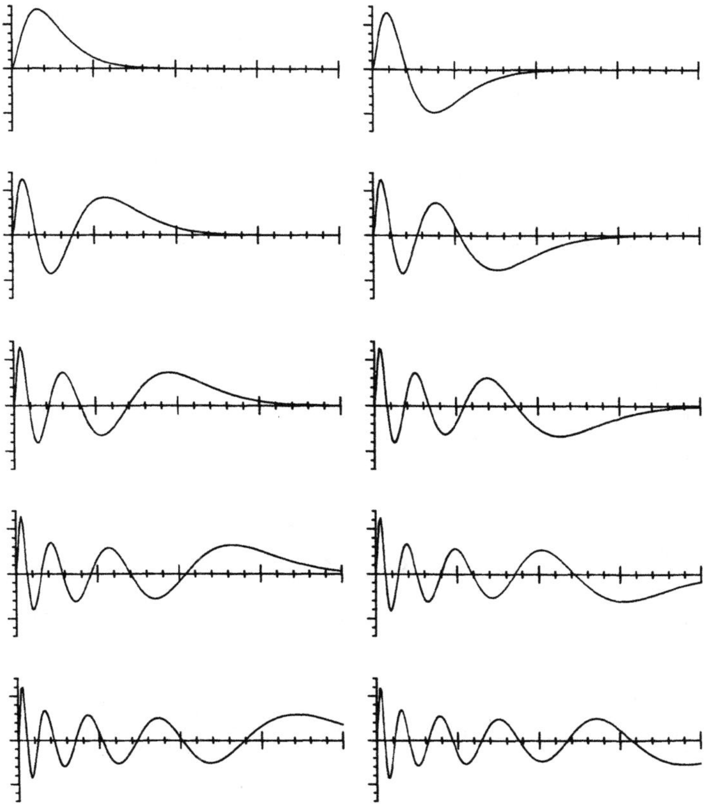

Figure 1. Lanczos functions $rv_j(r)$ of Eqs. (4) for dipole excitation of atomic hydrogen (j = 1 to 10), constructed as discussed in the text. The abscissa spans $24a_0$, the ordinate 0.07 a.u.

are known to diverge for $k \geq 3$ [28]. Consequently, use of the Hamiltonian H in Eqs. (3) and (4), rather than A(H) of Eq. (27b), results in an invalid Lanczos problem, related to the divergence of the norms of the Cauchy functions of Eqs. (29), and the consequent failure of Hilbert space techniques. Such developments, although they can always be implemented numerically in finite basis sets, will generally not converge to the correct Schrödinger states [29].

In Figure 1 are shown the first ten Lanczos functions $rv_j(r)$ of Eqs. (4) and (27) for the p-wave spectrum in atomic hydrogen. Starting with the defining test function $v_1(r) \sim re^{-r}$, these are seen to span increasingly larger portions of the spatial interval

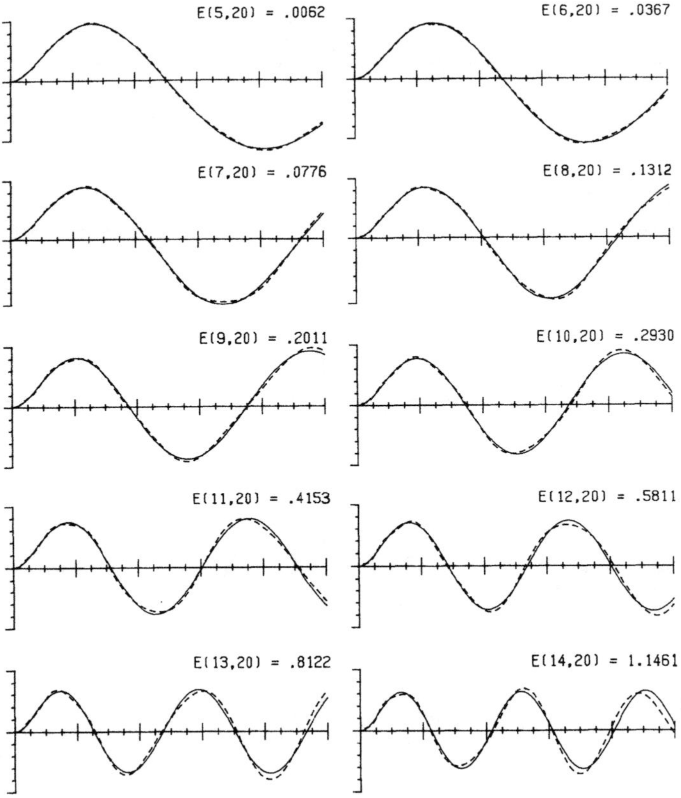

Figure 2. Regular p waves in atomic hydrogen, $rR_{\epsilon\ell}(r)$ ($\ell = 1$); (----), 20th-order Stieltjes eigenfunctions, constructed as discussed in the text; (——), correct values at the indicated Stieltjes energies (a.u.), constructed in accordance with reference [30]. The abscissa spans 10 a_o, the ordinate 1 a.u.

shown (0 to 12 a_o). Moreover, since their nodes continue to change position with increasing j, in contrast to the behaviors of the familiar hydrogenic discrete states [30], they remain linearly independent with increasing order, and achieve completeness in finite orders over local portions of the spatial interval. Although the first ten corresponding Cauchy functions [not shown] of Eqs. (29) are linearly independent, they do not exhibit the nodal structures of the Lanczos functions of Figure 1, since they are not an orthonormal set. As a consequence, the higher-order Cauchy functions include information already provided by the lower-order functions, and the associated higher-order spectral moments provide new information regarding the spectrum only in very high significant figures. This circumstance results in a numerically unstable eigenvalue problem when the Cauchy functions are employed directly as a basis set [24], and in a correspondingly unstable power moment problem [Eq. (30)]. These instabilities are avoided entirely through use of the Lanczos basis, corresponding recurrence coefficients α_j, β_j, and recursive solution of the moment problem [23].

The Lanczos functions of Figure 1 and related recurrence coefficients [Eq. (34)] are used to construct 20th-order Stieltjes eigenfunctions in atomic hydrogen. These are the Stieltjes-Tchebycheff functions of 20th-order [Eq. (14)] having energies at which the corresponding optical potential of Eqs. (16) to (20) vanishes identically. Consequently, their energies are the 20th-order generalized Gaussian quadrature points of the photoabsorption density, and their norms provide the corresponding quadrature weights [Eq. (33)]. In Figure 2 are shown the first ten of the correctly normalized 20th-order Stieltjes eigenfunctions that appear in the continuous spectrum [Eq. (23)]. Their energies are seen to span the lower portion of the ionization spectrum in a sensible manner, in accordance with their Gaussian quadrature nature. That is, they are most dense where the dipole matrix element of Eqs. (12) and (27) for atomic hydrogen [30],

$$g(\varepsilon) = \frac{8}{3}\,\frac{\exp[-4\varepsilon^{-1/2}\,\tan^{-1}(\varepsilon^{1/2})]}{(\varepsilon+\frac{1}{2})^3\,\{1-\exp[-2\pi\varepsilon^{-1/2}]\}}\,,\varepsilon(\text{a.u.}) \geq 0, \qquad (35)$$

is largest.

Figure 2 indicates the correctly normalized 20th-order Stieltjes functions [Eq. (23)] are converged to the correct corresponding regular Coulomb p-waves [30], also shown in the figure, over the spatial interval indicated. Of course, for sufficiently large r the Stieltjes functions decay, and do reproduce the Coulomb p waves. As a consequence, the 20th-order Stieltjes functions are not eigenstates of A(H) and H, but are spectrally broad. This point is illustrated in Figure 3, where the spectral content (squared) [Eq. (24)] of each of the ten functions of Figure 2 is shown. Evidently, the lower-lying Stieltjes functions are spectrally narrower than are the higher-lying ones, in general accord with the observation that higher frequencies are required to reproduce correctly the high-energy Coulomb p waves. The spectral interval shown is seen to be spanned fully by the ten Stieltjes functions of Figure 2, with the spectral distribution of each state in Figure 3 having a zero at the energies of the nine other states, in accordance with their

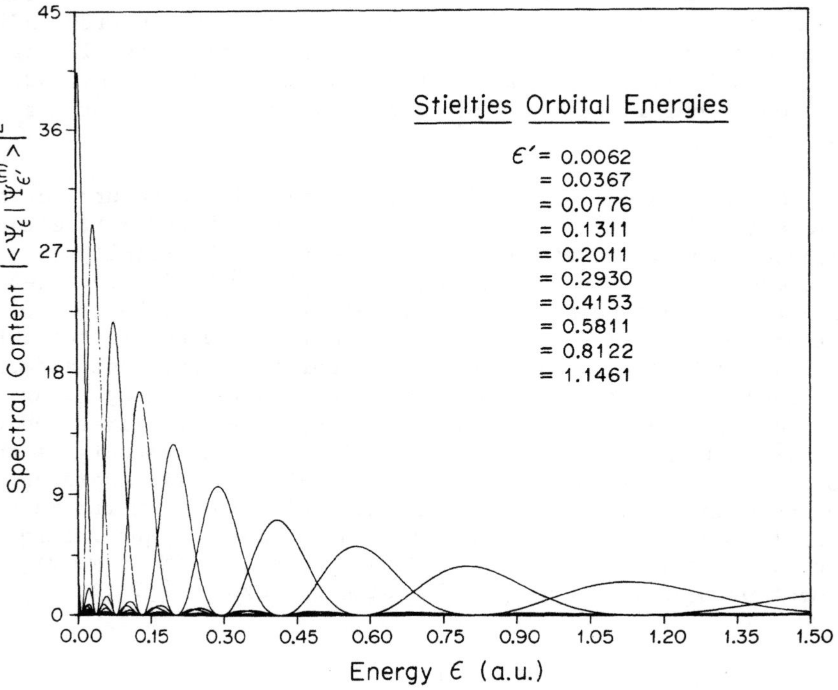

Figure 3. Spectral composition of Eq. (24) squared for the 20th-order Stieltjes eigenfunctions of Figure 2. Hartree atomic units are employed.

orthogonality. Moreover, the (four) lower-lying 20th-order Stieltjes functions (not shown) are found to reproduce the 2p, 3p, and 4p states accurately, and to provide a fourth bound state corresponding to a spectral sum of all the other higher-lying discrete np hydrogenic states. Finally, the remaining (six) Stieltjes functions appear at higher energies in the continuous portion of the spectrum (not shown). Of these, the first four provide excellent approximations to local portions of the corresponding regular p waves, in spite of the fact that they are spectrally broad.

Stieltjes-Tchebycheff functions of various orders (10 to 100) and energies ($\epsilon \leq 5$ a.u.) are evaluated for the p-wave spectrum of atomic hydrogen. An overview of the results obtained appears in Figure 4, where the weight functions $\Gamma^{(n)}(\epsilon)$ or reciprocal norms of Eq. (33) are shown for n = 10 to 100 in the neighborhood of the

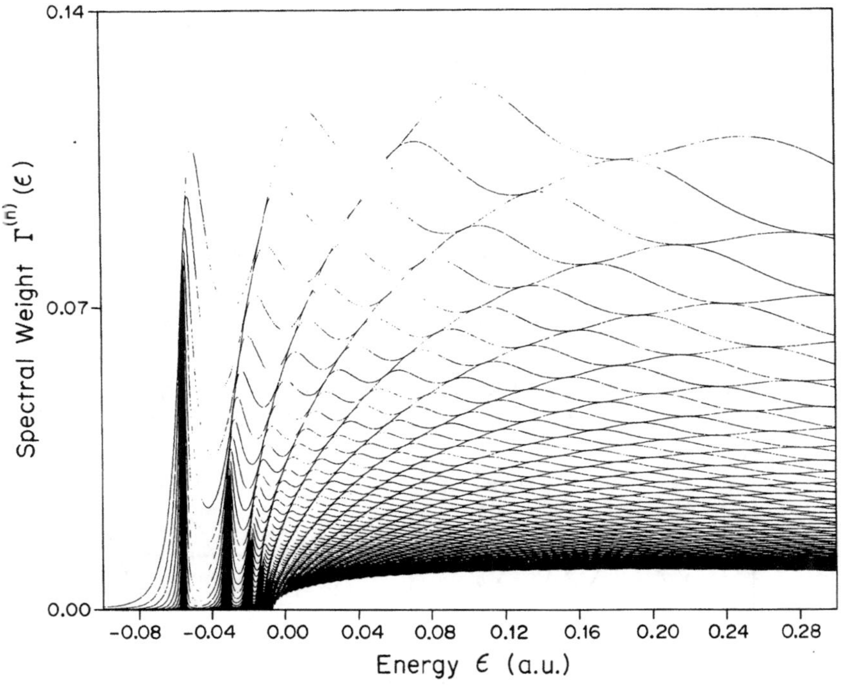

Figure 4. Spectral weights $\Gamma^{(n)}(\varepsilon)$ or reciprocal norms of Tchebycheff functions [Eq. (33)] for n = 10 to 100 for the regular p-wave spectrum in atomic hydrogen. The 1s → 2p resonance appears at lower energy not shown in the figure. All values in Hartree atomic units.

ionization threshold. The weight functions evidently converge monotonically from above with increasing n to a spectrum that includes discrete features (f numbers) at the appropriate Coulomb eigenvalues. Above threshold in the essential spectrum ($\varepsilon \geq 0$) the weights evidently decrease slowly but without limit, in accordance with the improper (infinite) normalization of the associated scattering states [Eq. (2b)]. The points at which adjacent (n and n+1) weight distributions touch give the nth-order Stieltjes energies. These points of coincidence are seen to form apparent rays with increasing n that converge from above to the discrete Coulomb eigenvalues, in accordance with the Hylleraas-Unheim theorem. The n Stieltjes energies of the nth-order provide n + 1 intervals in which there appear one and only

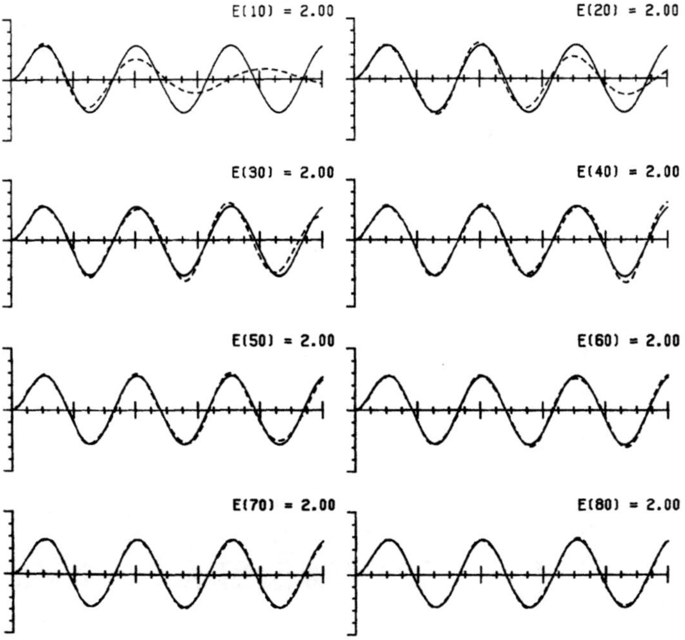

Figure 5. Regular p waves in atomic hydrogen, $rR_{\epsilon\ell}(r)$ ($\ell = 1$, $\epsilon = 2$ a.u.); (----) Stieltjes-Tchebycheff functions of indicated order [Eq. (23)]; (——), correct values constructed in accordance with reference [30]. The abscissa spans 10 a_o, the ordinate 1 a.u.

one of the (n + 1)th-order Tchebycheff eigenvalues [12]. Consequently, as the energy variable ϵ is scanned over one of the Stieltjes intervals, the n other Tchebycheff eigenenergies will correspondingly traverse the n other Stieltjes intervals. In this way a dense set of Stieltjes-Tchebycheff orbitals can be selected from the interval $(-\infty, +\infty)$. Alternatively, the entire interval can be spanned using a fine mesh and Eqs. (14), (23), and (33) employed directly.

The Stieltjes-Tchebycheff orbitals of Eqs. (14) and (23) corresponding to the weights of Figure 4 are found to converge with increasing n to the correct discrete p states and regular p waves in atomic hydrogen, respectively, in the appropriate energy intervals ($\epsilon < 0$ and $\epsilon \geq 0$). For $\epsilon < 0$ but $\neq \epsilon_i$, convergence is obtained to highly oscillatory unphysical states having improper (infinite) norm. Typical continuum results are shown in Figures 5 and 6, depicting Stieltjes-Tchebycheff functions for $\epsilon = 2$ a.u.

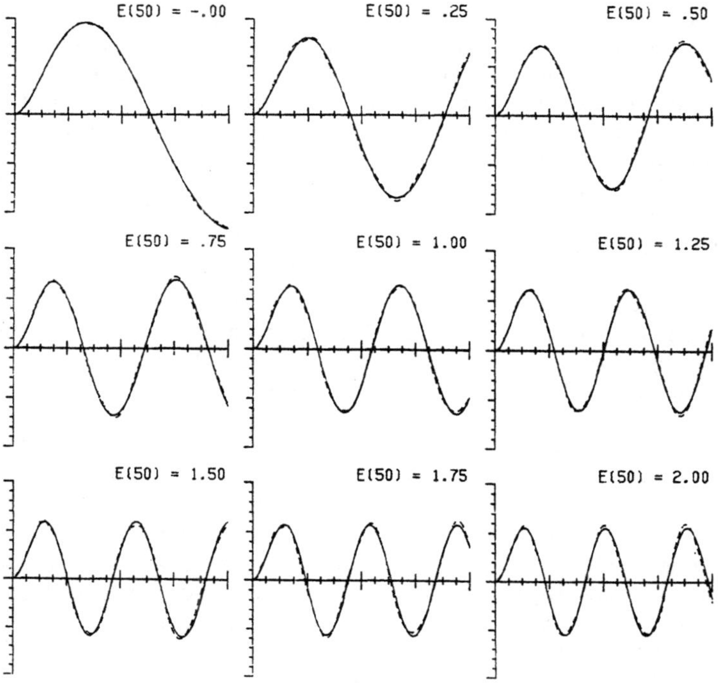

Figure 6. Regular p waves in atomic hydrogen, $rR_{\epsilon\ell}(r)$ ($\ell = 1$), at the indicated energies (a.u.); (-----) Stieltjes-Tchebycheff functions of 50th order; (——), correct values constructed in accordance with reference [30]. The abscissa spans 8 a_0, the ordinant 1 a.u.

for n = 10 to 80 in steps of 10, and n = 50 at 0.25 a.u. energy intervals, respectively. The correctly normalized Stieltjes-Tchebycheff functions of Figure 5 are seen to converge with increasing n to the correct regular Coulomb p wave for $\epsilon = 2$ a.u. also shown in the figure [30]. In accordance with the spatial characteristics of the Lanczos functions [Figure 1], the Stieltjes-Tchebycheff functions are seen to converge first at smaller r values, although for sufficiently large n there is convergence over the full spatial interval shown (0 to 10 a_0). Similarly, the n = 50 results of Figure 6 are seen to be converged to the corresponding regular p waves for the energies shown over the indicated spatial interval. It should be noted that even low-order Stieltjes-Tchebycheff functions provide useful information in the innermost spatial interval, which determines the density of Eq. (12).

In addition to the p wave results presented here, ℓ-wave spectra in atomic hydrogen for $\ell \le 15$ have also been evaluated employing the Stieltjes-Tchebycheff development with appropriate test functions ($\Phi \sim r^{\ell} P_{\ell}(\cos\theta)\Psi_0$). Point-wise convergence to the appropriate discrete and regular scattering states similar to the p-wave results presented here is also obtained in these cases [31]. It should be noted that no particular difficulties are encountered in the present development from the Coulomb potential treated. Incorporation of the appropriate boundary conditions ($\sim r^{\ell}$) at the origin in the Lanczos basis is sufficient to insure that the Stieltjes-Tchebycheff functions have the correct asymptotic behavior at large r, from which the Coulomb phase shifts can be determined, if required. Moreover, a second linearly independent solution can be constructed at any energy by replacing the $q_i(\epsilon)$ in Eq. (14) with so-called (numerator) polynomials $p_j(\epsilon)$ of the second kind that satisfy the recurrence relation of Eq. (8) with initial conditions $p_1 = 0$ and $p_2 = 1/\beta_1$. The associated functions are found to be solutions of an inhomogeneous Schrödinger equation with the test function Φ providing the appropriate source term. Consequently, these second solutions are regular at the origin but are linearly independent of the regular Coulomb functions in the asymptotic region. Finally, it should also be noted that solutions of the Weyl type can be constructed for complex energy in the Lanczos basis employing appropriate linear combinations of complex valued polynomials of the first and second kind [19]. Since solutions of the irregular and Weyl type are not required in photoionization studies they are not considered further here.

b) Molecular Hydrogen

In order to illustrate the computational requirements that arise in applications of the Stieltjes-Tchebycheff development to molecules, detailed results are presented here for molecular hydrogen [15]. To keep the development simple, computations are made in the Born-Oppenheimer, vertical-electronic, static-exchange approximation [8]. In this approximation the photo cross section is obtained from body-fixed $1\sigma_g \to n\sigma_u/k\sigma_u$ and $n\pi_u/k\pi_u$ excitations, and appropriate Franck-Condon and London-Hönl factors [32]. Consequently, attention is focused here on the required electronic excitation/ionization calculations in the body frame. The $k\sigma_u$ and $k\pi_u$ scattering functions obtained from the Stieltjes-Tchebycheff development in these cases correspond to linear combinations of eigenchannel functions as obtained from conventional K-matrix normalized body-frame calculations [8].

Large Cartesian Gaussian basis sets $\{\phi_i;\ i=1,N;\ N \cong 150\}$ are employed to construct the ground-state $1\sigma_g$ orbital at R_e, and to solve the Lanczos equations in σ_u and π_u symmetry in the static-

Table I. Supplemental Gaussian basis functions employed, in addition to the (4s;3p)/[3s;1p] valence basis, in molecular hydrogen calculations.

Location $[R_i$(a.u.)$]$[a]	Type	Exponents
0.0 (cm)	p_x, p_z	0.200 0.100 0.050 0.025 0.0125 0.00625 0.003125 0.0010 0.0005 0.0002
± 0.70 ($R_e/2$)	s	0.01569 0.00663 0.00276 0.00100
	p_x, p_z	0.0933 0.0280 0.0140 0.0070
± 1.3 and ± 2.8	s, p_x, p_z	2.0 1.0 0.5 0.2
± 4.3	s, p_x, p_z	1.0 0.5 0.2 0.1
	p_x, p_z	2.00 0.05 0.02 0.01 0.005 0.002

a. All functions are located at centers on the internuclear line at the indicated positions.

exchange or single-excitation configuration-mixing approximation. In Table I are shown the supplemental functions, in addition to a standard (4s;3p)/[3s,1p] valence basis [33], employed in the calculations. These functions are chosen by experimentation, in accordance with the frequencies of the bound and scattering states expected, and on basis of a previously reported preliminary study of the H_2 ionization spectrum [34]. To avoid the numerical difficulties that can arise from near linear dependence, the overlap matrix in the Gaussian basis of Table I,

$$(\underset{\sim}{S})_{ij} = \langle \phi_i | \phi_j \rangle \qquad (36)$$

is constructed and diagonalized. Those linear combinations of basis functions that correspond to very small diagonal elements ($\leq 10^{-6}$) are discarded, and the remaining M-term ($M < N$) canonical basis set used to construct the ground-state Fock matrix. This matrix is diagonalized and a $1\sigma_g$ Fock orbital with energy of -16.1286 eV obtained, as compared with an experimental vertical ionization potential of ~16.0 eV [35]. Next, the single-excitation configuration-mixing (static-exchange) Hamiltonian matrices are constructed in the σ_u and π_u subspaces of the M-term canonical basis set, and these are diagonalized. The resulting orthonormal static-exchange pseudospectra are then used in solving the Lanczos equations in each case. Since the static-exchange Hamiltonian is diagonal in these basis sets, there is no difficulty in constructing the matrix A(H) of Eqs. (3), (4) and (27). It should be noted, consequently, that the Lanczos basis is not being used to diagonalize a matrix in the present development, since the static-exchange Hamiltonian is already diagonal. Rather, an n-term Lanczos basis, is employed in constructing accurate approximations to the Stieltjes-Tchebycheff functions of Section II, in this case in the static-exchange approximation. In the present development, in which $N \cong 150$, it is found that converged results are obtained for $n \leq 10$, as is clarified further below.

In Table II are shown the calculated values of the recurrence coefficients for both σ_u and π_u symmetry in H_2 obtained in the static-exchange approximation. These coefficients are seen to approach the correct asymptotic values also shown in the table with increasing n [23]. The small oscillations present are a consequence of having both discrete and continuum states in the spectrum. For $n > 10$ [not shown], the coefficients become more irregular, and oscillate about the asymptotic values. These observations indicate the discrete pseudospectrum employed in the solution of the Lanczos equations begins to show through above 10th-order, suggesting that valid results can be expected in the basis set employed for $n \leq 10$.

In Figure 7 are shown the weight functions or reciprocal norms of Eq. (33) ($n = 5$ to 10) for the $1\sigma_g \rightarrow k\sigma_u$ and $k\pi_u$ spectra in H_2.

Table II. Dipole recurrence coefficients for $1\sigma_g \rightarrow k\sigma_u$ and $k\pi_u$ spectra in H_2 in the static-exchange approximation.[a]

	$1\sigma_g \rightarrow k\sigma_u$ spectrum		$1\sigma_g \rightarrow k\pi_u$ spectrum	
	α_n	β_n	α_n	β_n
1	1.6866	0.4653	1.3979	0.5434
2	1.3433	0.4994	1.2332	0.5150
3	1.0615	0.5471	1.1259	0.5022
4	0.9498	0.5038	1.0356	0.4879
5	1.0226	0.4494	1.0094	0.4729
6	1.0289	0.4554	0.9765	0.4732
7	0.9593	0.4781	0.9445	0.4798
8	0.8785	0.4612	0.9330	0.4348
9	1.0524	0.4026	0.9194	0.4893
10	0.8133	0.4958	1.1094	0.2404
∞[b]	0.8436	0.4218	0.8436	0.4218

a. Values, in Hartree atomic units, obtained from the Lanczos development of Eqs. (3), (4), and (27) in the static-exchange approximation [15].

b. Asymptotic values obtained from the expressions [23] $\alpha_\infty = 1/(2\epsilon_t)$, $\beta_\infty = 1/(4\epsilon_t)$, where $\epsilon_t = 0.5927$ a.u. is the Koopmans ionization potential.

These are calculated from the pseudospectral distributions of points and weights of Eqs. (30) to (33) also used in constructing the recurrence coefficients of Table II. Distributions of dimensionality $M \cong 50$ are employed in each case. Although the dimensionality of the primitive basis set of Table I is much larger ($N \cong 150$) than this, not all basis functions contribute to a given (σ_u or π_u) symmetry, reducing the number of pseudostates obtained in each case. Moreover, the canonical orthogonalization procedure indicated above also effectively reduces the number of primitive basis functions employed in the calculation. Nevertheless, the resulting pseudospectra are found to span satisfactorily the discrete and scattering states of interest here.

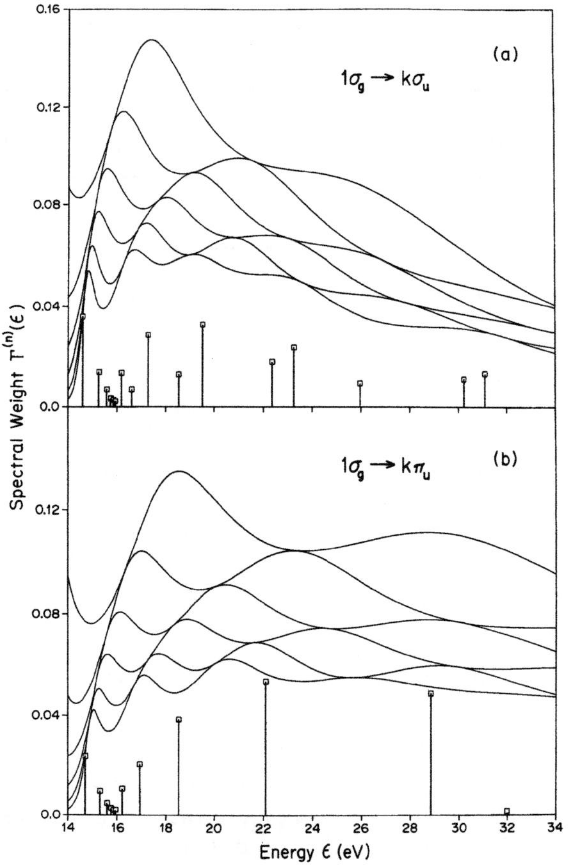

Figure 7. Reciprocal norms $\Gamma^{(n)}(\varepsilon)$ of Tchebycheff functions [Eq. (33)] for n = 5 to 10 for $1\sigma_g \rightarrow k\sigma_u$ and $k\pi_u$ spectra in H_2 [15]. The static-exchange pseudospectra are indicated as vertical lines in the figure. The discrete resonance transitions in each channel appear at lower energy, and are not shown in the figure.

Not shown are the vertical-electronic, static-exchange approximations to the $X^1\Sigma_g^+ \rightarrow (2p\sigma_u)\ B^1\Sigma_u^+$ and $(2p\pi_u)\ C^1\Pi_u$ resonance transitions in each polarization, which appear at lower energies below the ionization threshold. Since the f numbers for these transitions are relatively large (~ 0.3) [15], they dominate the discrete spectral region, and the 10th-order weight distributions are converged to sharp peaks at these energies. The second Rydberg transitions, $1\sigma_g \rightarrow 3p\sigma_u$ and $3p\pi_u$, have relatively weak f numbers

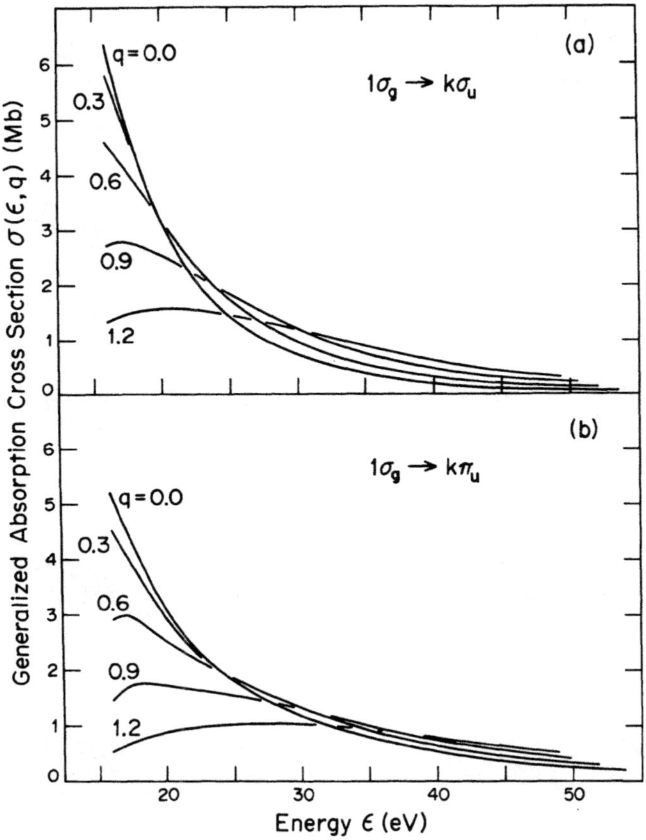

Figure 8. Generalized $1\sigma_g \rightarrow k\sigma_u$ and $k\pi_u$ photoionization cross sections of Eq. (37) in H_2 constructed in the static-exchange approximation as discussed in the text for the indicated q values (a.u.).

($\sim$ 0.05), and are barely discernible as discrete states in the two spectra. Above threshold, there are small waves present in both results of Figure 7, due to the required coincidence of adjacent distributions at the Stieltjes energies. Some of the higher-order ($10 < n$) weight distributions obtained are also satisfactory (not shown); with increasing n, however, the results show more pronounced waves that are unrelated to the coincidence of adjacent distributions. These reveal the presence of the underlying discrete basis set and corresponding pseudospectral weight distributions, also shown in the figures. When

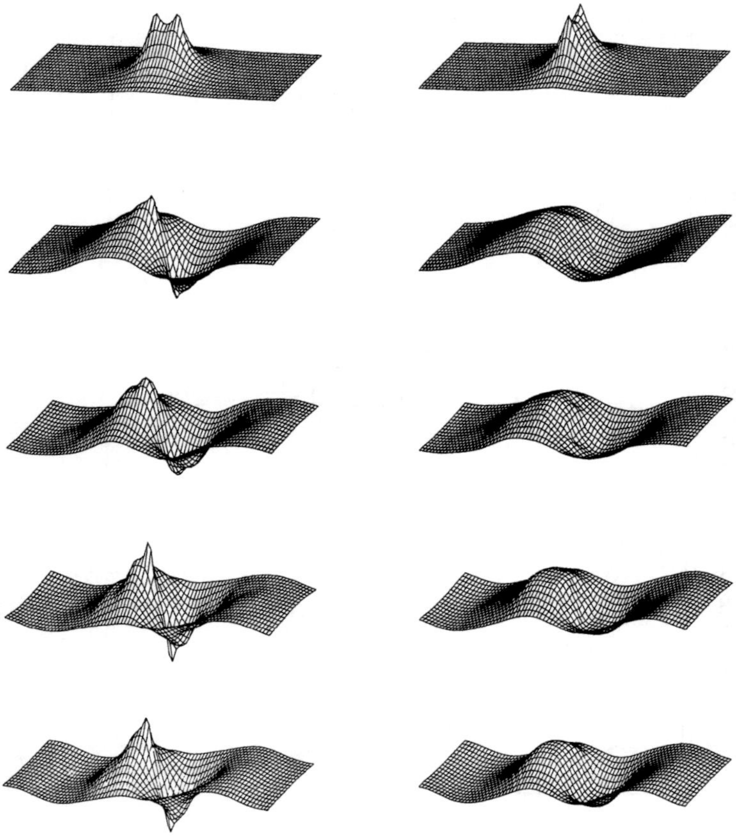

Figure 9. Hartree Fock $1\sigma_g$ orbital and 10th-order $k\sigma_u$, $k\pi_u$ Stieltjes-Tchebycheff functions in H_2 determined at R_e in the static-exchange approximation [15]. The Tchebycheff functions are given in 2 eV intervals starting at 16 eV. Note that the internuclear line runs from left to right for the $k\sigma_u$ functions in the left-hand column, whereas, the out-of-plane line runs from left to right for the $k\pi_u$ functions in the right-hand column.

n = M, the completely discrete pseudospectral weight distribution obtained from the basis set static-exchange calculations is reproduced by the results of Eq. (33). It is clear that the graphs of Figure 7 provide convenient means for accessing the extent of convergence of the Lanczos basis set and corres-

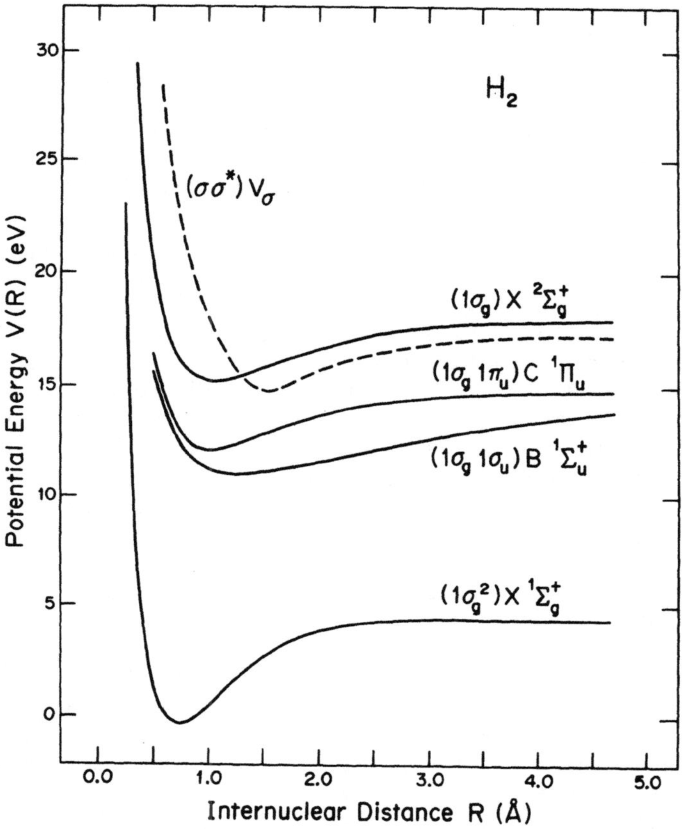

Figure 10. Potential energy curves for H_2 and H_2^+. The $X^1\Sigma_g^+$, $B^1\Sigma_u^+$, $C^1\Pi_u$ and $X^2\Sigma_g^+$ results are adopted from reference [36], whereas the V_σ curve is constructed from $\sigma_g 1s$ and $\sigma_u 1s$ orbitals as discussed in the text.

ponding polynomial recurrence coefficients. In the following, Stieltjes-Tchebycheff functions up to 10th-order are employed in construction of the $1\sigma_g \rightarrow k\sigma_u$ and $k\pi_u$ cross sections and corresponding wave functions in H_2.

In Figure 8 are shown generalized momentum-transfer dependent cross sections

$$\sigma(q,\varepsilon) = \hbar(\varepsilon-\varepsilon_0)\left|\langle k\phi_\varepsilon|\frac{1}{q}\, e^{i\underset{\sim}{q}\cdot\underset{\sim}{r}}|1\sigma_g\rangle\right|^2 \tag{37}$$

for $1\sigma_g \to k\phi_\varepsilon$ (= $k\sigma_u$, $k\pi_u$) ionization in H_2 obtained from 10th-order Stieltjes-Tchebycheff functions in the static-exchange approximations [15]. These correspond to first Born inelastic cross sections when appropriate kinematical factors are included [31]. For present purposes, they constitute somewhat more general and sensitive probes of the scattering functions than do the dipole results, to which they reduce in the limit $q \to 0$. It should be noted that the vector $\underset{\sim}{q}$ has been fixed along the lab frame z axis, and the results of Figure 8 obtained by angle averaging Eq. (37) over all (equally weighted) orientations of the internuclear line. Evidently, the two $q = 0$ results of Figure 8 are highly similar, although the $1\sigma_g \to k\pi_u$ cross section is somewhat smaller than the $1\sigma_g \to k\sigma_u$ result near threshold. This small difference at $q = 0$ becomes fractionally larger with increasing q, as a consequence of the different portions of the continuum functions $k\phi_\varepsilon$ the momentum-transfer dependent operator of Eq. (37) samples. The behaviors of the cross sections of Figure 8 can be clarified by examining the spatial characteristics of the scattering orbitals $k\sigma_u$, $k\pi_u$, for which purpose the Stieltjes-Tchebycheff development is highly suitable.

In Figure 9 are shown three-dimensional graphical representations of the $1\sigma_g$ orbital and 10th-order $k\sigma_u$ and $k\pi_u$ Stieltjes-Tchebycheff orbitals. The functions are evaluated in planes containing the internuclear line and at energies in two eV intervals starting at 16 eV for the scattering orbitals. The internuclear line runs from left to right for $1\sigma_g \to k\sigma_u$ ionization, whereas the out-of-plane axis runs from left to right for $1\sigma_g \to k\pi_u$ ionization, corresponding to the polarization directions in the two cases. When viewed from these perspectives, the $k\sigma_u$ and $k\pi_u$ scattering functions at equal energies are seen to be highly similar, although there are differences in each case. Specifically, the $k\sigma_u$ Stieltjes-Tchebycheff functions of Figure 9 include cusp-like features at the nucleii that are absent from the $k\pi_u$ functions. These features arise from compact σ_u 1s or σ^* contributions to the $k\sigma_u$ spectrum, and are largely responsible for the differences in the $1\sigma_g \to k\sigma_u$ and $1\sigma_g \to k\pi_u$ cross sections of Figure 8. Since the cross sections of Figure 8 sample the inner portions of the scattering orbitals at higher q values, it is appropriate that the differences in the two polarization results become more noticeable in these cases.

An alternative clarification of the presence of σ^* contributions to the $k\sigma_u$ spectrum in H_2 is obtained from Figure 10, which shows potential curves for ground state H_2 and H_2^+, as well as for the $(2p\sigma_u)B^1\Sigma_u^+$ and $(2p\pi_u)C^1\Pi_u$ states in H_2 [36]. Also shown in the figure is a diabatic curve associated with the configurational state function $(\sigma\sigma^*)^1\Sigma_u^+$, constructed using $1\sigma_g 1s$ and $1\sigma_u 1s$ orbitals for $R \leq 3.0$ Å, and made to approach smoothly the H + H^- ground state in the $R \to \infty$ limit [37]. The latter configurational state is seen to cross the ground-state H_2 Franck-Condon zone in the ~ 20-30 eV interval, clarifying the origin of the differences in the cross sections of Figure 8, and confirming the presence of σ^* contributions to the $k\sigma_u$ functions of Figure 9. It is interesting to note that the non-Rydberg behavior of the $(2p\sigma_u)B^1\Sigma_u^+$ state at large R can also be attributed to the $\sigma\sigma^*$ configuration, which contributes to this state for $R \geq 1.2$ Å. By contrast, the $(2p\pi_u)C^1\Pi_u$ state exhibits pure Rydberg behavior, following closely the H_2^+ ground-state potential curve for all R.

c) Molecular Applications

Stieltjes-Tchebycheff studies of dipole excitation and ionization spectra have been reported for a number of gas-phase molecular compounds. The diatomic molecules studied in various approximations include H_2 [15,34,38,39], N_2 [40-44], O_2 [45,46], F_2 [47,48], CH [49], HF [50], CO [51], and NO [52], whereas the polyatomics studied include H_2O [53-55], CO_2 [56], O_3 [57], H_2S [58], C_2H_2 [59, 60], H_2CO [61,62], H_2CS [58], NH_3 [63], and CH_4 [63]. Aspects of the photoexcitation and partial-channel photoionization cross sections of certain of these molecules have been reviewed earlier [64-66]. In the present report, only a few selected applications of the Stieltjes-Tchebycheff development are described, with particular emphasis placed on more recent results.

The $(3\sigma_g^{-1})\ X^2\Sigma_g^+$ channel photoionization cross section in N_2 has been the subject of considerable interest, due in part to the presence of a $\sigma \to \sigma^*$ shape resonance in the $3\sigma_g \to k\sigma_u$ polarization channel [40-44]. Single-excitation configuration mixing (static-exchange) calculations using a very large seven-center Gaussian basis set indicate that the $k\sigma_u$ functions in this compound are largely of Rydberg character in the threshold region, whereas ~ 10-12 eV above threshold they acquire a distinct compact σ^* character [34]. As a consequence, the corresponding photoionization cross section exhibits a pronounced maximum at these energies [41]. The $\sigma \to \sigma^*$ transition in N_2 has gone unassigned for many years, since the σ^* orbital appears localized largely in the $k\sigma_u$ photoionization continuum, and contributes to the discrete spectral region only through $V_\sigma - V_\pi$ mixing in the $b'^1\Sigma_u^+$ state [43]. The latter effect is verified by the results of coupled-channel single-excitation calculations of $3\sigma_g \to k\sigma_u$ and $1\pi_u \to k\pi_g$ ionization in N_2 employing a large three-center Gaussian basis [43]. When coupling between the $3\sigma_g$ and $1\pi_u$ channels is neglected, the $1\pi_u \to k\pi_g$ cross section is found to include a spurious $\pi \to \pi^*$ contribution $\sim$ 3-4 eV above threshold. The effects of $V_\pi - V_\sigma$ mixing lowers this peak into the discrete spectral region, where it correctly corresponds to the strong $^1\Sigma_g^+ \to b'^1\Sigma_g^+$ resonance.

A somewhat similar situation to that in N_2 obtains for the $\sigma \to \sigma^*$ transition in F_2, which has been the subject of recent theoretical and experimental study [47,48]. In this case, the $\sigma \to \sigma^*$ or $N \to V_\sigma$ transition falls below the $3\sigma_g$ threshold, although it mixes so strongly with the $1\pi_g \to np\pi_u$ Rydberg series that the latter takes on significant intravalent character, and the former is largely depleted in intensity. An assignment of the relevant portion of the F_2 spectrum based on electron energy-loss measurements and large-basis-set coupled-channel single-excitation configuration-mixing calculations has been reported recently [48]. As a consequence of the strong $1\pi_g \to np\pi_u/3\sigma_g \to 3\sigma_u$ mixing in $X^1\Sigma_u^+ \to {}^1\Sigma_u^+$ polarization, which is adequately described theoretically only in the very large compact and diffuse basis sets employed in Stieltjes-Tchebycheff studies [47,48], the F_2 spectrum has been generally misassigned experimentally, with the resonance transition at 12.8 eV incorrectly designated as $\sigma \to \sigma^*$ on basis of its intensity. The recent measurement and calculation [48], however, place this latter transition at $\sim$ 17 eV, with its intensity significantly depleted, as indicated above.

A third example of the use of very large basis sets in Stieltjes-Tchebycheff calculations of dipole excitation spectra is given by recent calculations on H_2O in a large Gaussian basis set [55]. In this case there are a great many Rydberg series present, as well as a number of intravalence transitions involving the $4a_1$ (γ^*) virtual orbital [55]. The latter are found to result in broad bands, associated with photodissociative processes, which underlie the sharper Rydberg series. Distinct nsa_1, npa_1, and two types of nda_1 Rydberg series are all obtained from calculations in particular ($^1A_1 \to {}^1A_1$, $^1B_1$, or $^1B_2$) polarization symmetry. When smaller basis sets are employed, as in most earlier studies, this separation of Rydberg states is not achieved, and spuriously mixed series are obtained. Such refined discrete-state calculations are a natural consequence of the use of very large basis sets employed in the Stieltjes-Tchebycheff development.

A final example of discrete-basis-set studies of molecular photoionization continua is provided by recent studies of C_2H_2 in the static-exchange approximation [59,60]. Of particular interest are the $1\pi_u \rightarrow k\sigma_g$, $k\pi_g$, and $k\delta_g$ polarization components of the $(1\pi_u^{-1})X^2\Pi_u$ partial channel. These cross sections are highly similar to atomic $2p \rightarrow ks$, kp, and kd results, respectively, and are typical of $1\pi_u$ cross sections in light diatomic and linear polyatomic molecules. The experimental data obtained for this channel are found to include a dispersion lineshape centered at ~ 15 eV not present in the calculations [59]. This feature can be attributed to autoionization of a strong $2\sigma_u \rightarrow 1\pi_g$ transition predicted by the calculations to appear at this energy [59,60]. Because the Stieltjes-Tchebycheff development provides scattering functions at any energy, these can be used as background continuum states in an appropriate autoionization lineshape study of the $2\sigma_u \rightarrow 1\pi_g$ transition, one of a number of such calculations presently in progress.

IV. CONCLUDING REMARKS

Explicit Hilbert-space methods are described for construction of approximations of uniform reliability to both the discrete and continuum portions of Schrödinger spectra. When sufficiently large one-electron basis sets and appropriate configurational state functions are employed, the familiar computational methodology of bound-state molecular studies furnishes highly accurate photoexcitation and partial-channel photoionization cross sections. Although the corresponding Stieltjes-Tchebycheff functions are not eigenstates of the Hamiltonian operator, they are converged in the spatial region of an appropriate test function. Consequently, three-dimensional graphical representations of these functions, which can be evaluated at any energy in the spectrum, provide useful diagnostics of discrete and continuum Schrödinger eigenstates. Of particular interest and importance in these calculations is the use of conventional molecular body-frame symmetry, allowing for the description of continuum functions in a manner familiar from bound-state studies. The presence of intravalence transitions, common in the photoionization continua of many molecules, can therefore be verified in a direct fashion, and their absence in the corresponding discrete spectrum clarified. In this way, long standing puzzles in molecular electronic spectroscopy have been resolved by recent studies of discrete and continuum states.

Although the theoretical situation is generally satisfactory, and separated-channel static-exchange calculations of photoexcitation and ionization cross sections of many molecules reported to date, further refinement and application of the Stieltjes-Tchebycheff development would be helpful. Of particular interest are applications of the method to autoionization lineshape studies, and to determinations of dissociative recombination rates, in which cases more conventional methods have proved difficult to implement. The Stieltjes-Tchebycheff functions also provide a basis for studies of the angular distributions of photoejected electrons, and studies of both gas phase and surface adsorbate species are in progress. Finally, more elaborate computational approximations, including polarization effects and the incorporation of interchannel coupling, are in order.

REFERENCES

1. S. T. Epstein, The Variational Method in Quantum Chemistry (Academic, NY, 1974).
2. H. F. Schaefer, III, Editor, Methods of Electronic Structure Theory (Plenum, NY, 1977).
3. S. D. Peyerimhoff and R. J. Buenker, in Excited States in Quantum Chemistry, C. Nicolaides and D. Beck, Editors (Reidel, Boston, 1979).
4. H. H. Michels, in The Excited State in Chemical Physics, J. W. McGowan, Editor (Wiley, NY, 1980), Vol. II.
5. J. Linderberg and Y. Öhrn, Propagators in Quantum Chemistry (Academic, London, 1973).
6. L. S. Cederbaum and W. Domcke, Advan. Chem. Phys. 36, 205 (1978).
7. M. R. Hermann and P. W. Langhoff, J. Math. Phys. xx, xxx (1982).
8. N. F. Lane, Rev. Mod. Phys. 52, 29 (1980).
9. M. B. Robin, Higher Excited States of Polyatomic Molecules (Academic, NY, 1974), Vols. I and II.
10. R. G. Newton, Scattering Theory of Waves and Particles (Springer, Berlin, 1982).
11. M. R. Hermann and P. W, Langhoff, Int. J. Quantum Chem. S17, xxx (1982).
12. C. T. Corcoran and P. W. Langhoff, J. Math. Phys. 18, 651 (1977).
13. P. W. Langhoff, in Electron-Molecule and Photon-Molecule Collisions, T. Rescigno, V. McKoy, and B. Schneider, Editors (Plenum, NY, 1979), pp. 183-244.
14. P. W. Langhoff, in Theory and Applications of Moment Methods in Many-Fermion Systems, B. J. Dalton, S. M. Grimes, J. P. Vary, and S. A. Williams, Editors (Plenum, NY, 1980), pp. 191-212.

15. M. R. Hermann and P. W. Langhoff, Phys. Rev. Axx, xxx (1982).
16. C. Lanczos, J. Res. Nat. Bur. Stand. 45, 367 (1950).
17. J. H. Wilkinson, The Algebraic Eigenvalue Problem (Clarendon, Oxford, 1965).
18. R. R. Whitehead, in Theory and Applications of Moment Methods in Many-Fermion Systems, B. J. Dalton, S. M. Grimes, J. P. Vary, and S. A. Williams, Editors (Plenum, NY, 1980), pp. 235-255.
19. N. I. Akhiezer, The Classical Moment Problem (Oliver and Boyd, London, 1955).
20. Yu. V. Vorobyev, Method of Moments in Applied Mathematics, (Gordon and Breach, NY, 1965).
21. J. A. Shohat and J. D. Tamarkin, The Problem of Moments, Mathematical Surveys 1 (American Mathematical Society, Providence, R.I., 1943, revised 1970).
22. H. S. Wall, Analytic Theory of Continued Fractions (Van Nostrand, NY, 1948).
23. P. W. Langhoff, C. T. Corcoran, J. S. Sims, F. Weinhold, and R. M. Glover, Phys. Rev. A14, 1042 (1976).
24. P. W. Langhoff, J. S. Sims, and C. T. Corcoran, Phys. Rev. A10, 829 (1974).
25. R. Haydock, in Computational Methods in Classical and Quantum Physics, M. B. Hooper, Editor (Hemisphere Publishing, Washington, D.C., 1975), pp. 268-283.
26. P. W. Langhoff, Chem. Phys. Letters 22, 60 (1973).
27. H. Shull and P.-O. Löwdin, J. Chem. Phys. 23, 1362 (1955).
28. P. W. Langhoff, J. Chem. Phys. 57, 2604 (1972).
29. J. D. Broad, Phys. Rev. A18, 1012 (1978).
30. H. Bethe and E. Salpeter, Quantum Mechanics of One- and Two-Electron Atoms (Academic, NY, 1957).
31. D. J. Margoliash and P. W. Langhoff, J. Comp. Phys. xx, xxx (1982).
32. G. Herzberg, Spectra of Diatomic Molecules (Van Nostrand, NY, 1950) 2nd ed.
33. T.H. Dunning and P. J. Hays, in Modern Theoretical Chemistry, H. F. Schaefer, III, Editor (Plenum, NY, 1976), Vol. 3, Chap. 1.
34. M. R. Hermann and P. W. Langhoff, Chem. Phys. Letters 82, 242 (1981).
35. J. Berkowitz, Photoabsorption, Photoionization, and Photoelectron Spectroscopy (Academic, NY, 1979).
36. T. E. Sharp, Atomic Data 2, 119 (1971).
37. J. C. Slater, Quantum Theory of Molecules and Solids (Addison-Wesley, Reading, 1963), Vol. 1.
38. P. W. Langhoff and C. T. Corcoran, J. Chem. Phys. 61, 146 (1974).
39. S. V. ONeil and W. P. Reinhardt, J. Chem. Phys. 69, 2126 (1978).

40. T. N. Rescigno and P. W. Langhoff, Chem. Phys. Letters 51, 65 (1977).
41. T. N. Rescigno, C. F. Bender, B. V. McKoy, and P. W. Langhoff, J. Chem. Phys. 68, 970 (1978).
42. T. N. Rescigno, A. Gerwer, B. V. McKoy, and P. W. Langhoff, Chem. Phys. Letters 66, 116 (1979).
43. G. R. J. Williams and P. W. Langhoff, Chem. Phys. Letters 78, 21 (1981).
44. P. W. Langhoff, S. R. Langhoff, T. N. Rescigno, J. Schirmer, L. S. Cederbaum, W. Domcke, and W. von Niessen, Chem. Phys. 58, 71 (1981).
45. P. W. Langhoff, A. Gerwer, C. Asaso, and B. V. McKoy, Int. J. Quantum Chem. S13, 645 (1979); J. Chem. Phys. 72, 713 (1980).
46. J. J. Delaney, I. H. Hillier, and V. R. Saunders, J. Phys. B15, L37 (1982).
47. A. E. Orel, T. N. Rescigno, B. V. McKoy, and P. W. Langhoff, J. Chem. Phys. 72, 1265 (1980).
48. A. P. Hitchcock, C. E. Brion, G. R. J. Williams, and P. W. Langhoff, Chem. Phys. 66, 435 (1982).
49. J. Barsuhn and R. K. Nesbet, J. Chem. Phys. 68, 2783 (1978).
50. K. Fraegri, Jr., and H. P. Kelly, Phys. Rev. A23, 52 (1981).
51. N. Padial, G. Csanak, B. V. McKoy, and P. W. Langhoff, J. Chem. Phys. 69, 2992 (1978).
52. J. J. Delaney, I. H. Hillier, and V. R. Saunders, J. Phys. B15, 1477 (1982).
53. G. R. J. Williams and P. W. Langhoff, Chem. Phys. Letters 60, 201 (1979).
54. J. J. Delaney, V. R. Saunders, and I. H. Hillier, J. Phys. B14, 819 (1981).
55. G. H. F. Diercksen, W. P. Kraemer, T. N. Rescigno, C. F. Bender, B. V. McKoy, S. R. Langhoff, and P. W. Langhoff, J. Chem. Phys. 76, 1043 (1982).
56. N. Padial, G. Csanak, B. V. McKoy, and P. W. Langhoff, Phys. Rev. A23, 218 (1981).
57. N. Padial, G. Csanak, B. V. McKoy, and P. W. Langhoff, J. Chem. Phys. 74, 4581 (1981).
58. G. H. F. Diercksen and P. W. Langhoff, Chem. Phys. (to be published).
59. P. W. Langhoff, B. V. McKoy, R. Unwin, and A. M. Bradshaw, Chem. Phys. Letters 83, 270 (1981).
60. L. E. Machado, E. P. Leal, G. Csanak, B. V. McKoy, and P. W. Langhoff, J. Electron Spectrosc. Relat. Phenom. 25, 1 (1982).
61. P. W. Langhoff, S. R. Langhoff, and C. T. Corcoran, J. Chem. Phys. 67, 1722 (1977).
62. P. W. Langhoff, A. E. Orel, T. N. Rescigno, and B. V. McKoy, J. Chem. Phys. 69, 4689 (1978).
63. G. R. J. Williams and P. W. Langhoff, J. Chem. Phys. (to be published).

64. P. W. Langhoff, Int. J. Quantum Chem. S11, 301 (1977).
65. P. W. Langhoff, N. Padial, G. Csanak, T. N. Rescigno, and B. V. McKoy, Int. J. Quantum Chem. S14, 285 (1980).
66. P. W. Langhoff, N. Padial, G. Csanak, T. N. Rescigno, and B. V. McKoy, J. de Chim. Phys. (Paris) 77, 590 (1980).

Acknowledgments

It is a pleasure to acknowledge the assistance and cooperation of various coworkers in these studies, particularly M. R. Hermann, K. Greenwald, and G. R. J. Williams. Financial support provided by the Chemistry and International Programs Divisions of the National Science Foundation, by the National Aeronautics and Space Administration, and by the Petroleum Research Fund, administrated by the American Chemical Society, is also gratefully acknowledged. Finally, the author thanks G. H. F. Diercksen and S. Wilson for their kind invitation to attend the NATO Institute and to contribute to its Proceedings.

COMPUTERS AND COMPUTATION IN MOLECULAR PHYSICS

G.H.F. Diercksen, N.E. Grüner, and J. Steuerwald

Max-Planck-Institut für Physik und Astrophysik
Institut für Astrophysik
Karl-Schwarzschild-Strasse 1
D-8046 Garching bei München

Computers have become a basic tool in physics over the past two decades. The 4th generation of computers available since the beginning of the eighties has increased the number of problems accessible to investigation considerably and experience with these systems is now available. The 5th generation of computers being discussed at present and expected to be available in the early nineties will certainly have an impact on physics hardly predictable.

In fact, molecular physicists were among the first users of computers and mostly physicists carried out the task of writing and maintaining the necessary programs. Physicists, together with representatives of which later became computer science, proposed or developed solutions to what the market did not offer and created environments in which the programs needed could be developed, tested, and executed.

Today computer science has established itself as a separate subject and methods and tools have been developed to use computers intelligently and economically and a market of commercial software products has been build up.

Still the task of program writing and maintaining is carried out by physicists, who are, as a rule, without training or specific interest in computer science. This lack of communication between physics and computer science has made the process of writing and maintaining programs by physicists and the generated code itself to fall behind acceptable software standards. This is an unacceptable situation as these programs constitute an enormous

G. H. F. Diercksen and S. Wilson (eds.), Methods in Computational Molecular Physics, 335–350.

investment of manpower and financial resources of the physics community. These series of lectures is an attempt to solve the problem of lack of communication arising from history.

It is certainly beyond the scope of these lectures to present an introduction into computer science for physicists. A great number of introductory textbooks are available in computer science. Although none of these textbooks is in particular aimed at physicists, some of the available books will serve this purpose. In the present lectures an introduction into some basic subjects of computer science will be given which are considered to be of particular importance to computation in molecular physics. The lectures will concentrate on the fundamentals of computer organisation and on the principles of software engineering.

1. FUNDAMENTALS OF COMPUTER ARCHITECTURE

History shows that parallelism has been used to improve the effectiveness of computer organisation since the times of the earliest mechanical calculations. The impact of parallelism on computer organisation is uniquely demonstrated by the fact that about half the increase in computer speed between 1950 and 1975 is due to advances in the technology of circuit design and production, especially of very large scale integration techniques (VLSI) while the other half is due to the parallel organisation of suitable tasks. In order to make efficient use of parallel computer systems users must be aware of the overall organisation of the computer, described as its architecture.

Parallelism has been applied on several distinct levels of computer organisation which may be described as follows:

- i, arithmetic level
- ii, instruction level
- iii, program level
- iv, job level

This pattern will be followed to develop stepwise a general model of parallel computer organisation.

The simplest model of a computer consists of the following fundamental units, called resources:

- i, central processing unit (CPU)
- ii, memory
- iii, input/output (I/O) unit
- iv, data lines (data busses)

The resources are controlled by programs which consist of sequences of instructions and of data. Both instructions and data are called collectively data if no distinction between them is necessary. Programs are entered into the computer by the I/O unit which may be thought of as a typewriter. Instructions and data are stored in the memory. The CPU executes serially the instructions specified in the program. For execution the instructions and data must be transfered to the CPU and the resulting data must be returned to the memory by data busses. The execution of each instruction including fetching and storing of data, requires a certain amount of time, usually measured in units of cycles/sec. The cycle time itself is characteristic for each computer and is very much a measure of the hardware technology used.

In this simple computer model, obviously, the processor is idling while the bus is busy with loading and storing data and the bus is idling while the CPU executes instructions. This computer organisation can be considerably improved if the load and store instructions are executed in parallel to all other instructions. This means to provide a separate functional unit in addition to the CPU for executing the data transfer instructions. To make this improved model work efficiently the load operations have to be initiated well ahead before the data are needed by the CPU. The data will be stored until needed in buffers, storage areas local to a functional unit with a short access time. The buffers local to the processor are traditionally called registers. The concept of buffering is complementary to parallel instruction execution and has always to be applied to achieve overlapped processing if information is used non consecutively in particular for data transfer between CPU and memory.

By this simple example the fundamental terms and techniques used in computer organisation have been introduced.

In general, optimum computer organisation is aimed at maximum overlap of mutually independent operations. Parallelism may be introduced into computer organisation in the following, principally different ways:

i, by providing several processors each obeying its own instructions, usually communicating via a common memory.
- multiprocessing

ii, by providing an array of identical processing elements under common control, each performing the same operation but on different data stored in their private memories.
- arrayprocessing

iii, by providing several independent units under common control for performing different different functions operating simultaneously on different data.
- functional parallelism

iv, by providing independent units for subtasks of the same operation working on different data in parallel under common control.
- pipelining

Usually some or all of the different concepts described are

applied in actual computer organisation.

First, the optimum choice of the approaches used in central processor organisation will be discussed. The choice is determined by an analysis of actual programs to be executed. The following elements of programs are easily recognized by inspection:

i, branches
ii, loops
iii, logical expressions
iv, arithmetic expressions

Most arithmetic expressions are either integer arithmetic or floating point arithmetic or are easily reformulated in this form. Arithmetic expressions involving more than one operation can always be reformulated as a sequence of expressions consisting of only one operation in such a way that a maximum logical independence between the simple expressions is achieved. In particular scientific programs contain in addition arithmetic expressions involving operations on long vectors of data.

A central processing unit efficiently processing such programs may be designed by introducing functional parallelism. Such a processor organisation will certainly include the following functional units: instruction unit, fetch/store unit, logical unit, and arithmetic unit. The latter may be subdivided further into integer arithmetic and floating point arithmetic units. Each of these two resources may be further subdivided into separate add-, multiply-, and divide units. Each of these units may have its private buffers and registers, or different units may share certain buffers and registers. The function of most of the resources is clear from their names. The instruction unit is in control of all other functional units. It decodes serially each instruction, determines the operation to be executed, and the operands needed and schedules the operation such that seriality is preserved.

The performance of some functional units may be improved further by applying pipelining techniques. Vector operations are particularly important examples to discuss the principles of pipelining in some detail. The addition of two vectors to obtain a sum vector will be used for demonstration. The addition of any two floating point numbers each represented for simplicity by a mantissa and a decimal exponent, may be divided into the following subtasks:

i, determination of the shift of one mantissa relative to the other to line up the decimal point
ii, shifting of the mantissa to actually line up the decimal points
iii, adding of the mantissa to obtain the mantissa of the result

iv, normalizing the result, that is shifting the mantissa until the leading digit is next to the decimal point and calculation of the exponent

Obviously the four subtasks cannot be executed in parallel for the same two numbers because each subtask depends on the result of one or more previous subtasks. But in case of the addition of two vectors the subtasks may be executed in parallel such, that at any one time subtask 1 works on vector element n+1, subtask 2 on vector element n+2, and so on, where n is increased by 1 every time all subtasks are completed. The pipelined process, after being initiated, works continuously until the specified number of operations has been processed or the number of operand registers associated to the functional unit has been exhausted. It follows immediately, that a pipeline works at its maximum efficiency if the number of operations to be processed is equal to the number of operand registers. Because the start up of a pipelining process requires some extra time, the efficiency of a pipeline process decreases with decreasing number of operations which is equivalent to decreasing vector length.

In general, pipeline techniques will increase the performance of computer organisation for operations that can be subdivided into two or more subtasks and be executed on long sequences of mutually independent operations. The application of pipelining techniques to the organisation of floating point operations has resulted in the design of floating point vector processors which are part of the so called "vector processors", used in all powerful computer systems available today.

Another obvious candidate for pipelined organisation is the instruction processing. Because of the importance of pipelining techniques for computer organisation this example will be discussed as well. Instruction processing may be divided into the following subtasks:

i, instruction decoding
ii, calculation of operand addresses
iii, initiation of operand fetch
iv, sending the command to the functional unit
v, fetching the next instruction

Again, these subtasks cannot be performed overlapped on the same instruction, but only on consecutive instructions. Obviously, the execution of consecutive instructions will usually depend on each other, but not the processing of the instruction itself. Pipeline techniques are used in the actual organisation of most instruction units.

Alternatively, the performance of functional units may be improved by providing arrays of identical processors that perform simultaneously identical operations on different data stored in their privat memory. For array processing the operations must not be separable into subtasks.

Efficient processors require fast access to large amounts of data in order to work at maximum rate. But with increasing storage size the access time increases strongly because of delays introduced by additional decoding logic and by transmission delays. Usually it is not absolutely necessary, although often very convenient, to hold exceedingly large amounts of data in storage with fast access. Often it is sufficient to keep data on external storage with slow access like mass-storage, magnetic disks and tape, and to load them into memory when needed. Thus the storage may be organized in a hierarchy according to increasing access time as follows:

i, buffer storage (cache)
ii, memory
iii, external storage

The memory is the storage directly addressable by the CPU or more correctly by instructions. Buffer storage is an optional feature of storage organization. If buffer storage is included, access to data in memory is only via the buffer storage. Whenever data are needed from memory which are not resident in buffer storage, data not needed at present must be returned from buffer storage and the referenced data must be loaded instead. The data are transfered in units of pages. The pages to be unloaded from buffer storage are selected according to the history of the individual pages. Actually that page is returned that has been used furthest back in time (aging). The loading and unloading of the buffer storage is controlled by special hardware. All external storage media involve mechanical motion and therefore their access time is orders of magnitude slower than that of buffer storage and even of memory. Mass storage media are in general permanently mounted to the unit, some disks may be demountable, but most installation prefer not to change disk-packs, magnetic tapes finally have to be mounted whenever needed. Data transfer between external storage and memory must be requested by the program, that is by so called input/output instructions. The data transfer is controlled by special functional units, called channels, which work parallel to the CPU. Buffering is provided by software (to be discussed later) to avoid unnecessary delays of other functional units waiting for data from external storage.

There are limitations on the access to memory due to hardware logic: only one word of n bits can be transfered from or to memory at a time, but not simultaneously. To increase data transfer between CPU registers and memory parallelism has been introduced into the memory organisation by memory interleaving. This is achieved by subdividing the memory into a number of memory banks each of which is separately accessible. Different words (addresses) are distributed between the different memory banks in such a way that as many words at consecutive addresses can be retrieved in parallel as there are memory banks. Storage conflicts and access delays will occur, if words are required from the same memory bank.

The different computer resources are managed by a special computer program, the operating system. For maximum efficiency all computer resources have to work in parallel. This can hardly be achieved by a single job. Therefore most operating systems allow handling of the different computer resources and simultaneously more than one job. Obviously, access to the different resources is only granted serially. If more than one job is handled at a time in such multiprogramming environment, there is a competition between the jobs for the different resources. Access is granted to the resources by algorithms tailored to optimize the usage of the individual resources. Multiprogramming environments are aimed in particular at an optimum overlap between CPU and I/O processes. This optimum overlap can usually be realized reasonably well if an appropriate "mix" of jobs is handled simultaneously. In such an environment the time the job actually spends in the computer, the elapsed time, is usually larger than the larger of the times for the CPU and for the I/O processes. If only one job is handled at a time by the operating system, special techniques of arranging data have to be used, known as buffering, to achieve as much overlap as possible between the CPU and I/O processes of the job. In such environments, assuming complete overlap between CPU and I/O processes, the elapsed time of a job is determined by the larger of the CPU time and the I/O time. Although it is always good practice to aim for balanced and overlaped CPU and I/O processes, this becomes of primary importance, if only one job can be handled separately by the operating system because for example it requires all available primary storage. Finally some operating systems even allow to manage different virtual computers simultaneously which may be used to execute jobs. The virtual resources of the virtual machines are associated by the operating system to adequate resources of the real system. The virtual machine concept is of special interest for developing programs, in particular operating systems, because it allows essentially to define any system configuration (that is any type and number of virtual resources). Virtual machines are not especially suited for large numerical calculations because of the additional overhead involved in the administration of the different virtual machines.

The number of jobs that can be executed simultaneously in a multi-programming environment is limited by the available memory space. By the virtual storage concept, this limitation is avoided to a large extent. In a virtual storage system the content of the storage defined for the job is actually resident on secondary storage. A real primary storage area is assigned to the job that is usually smaller than the storage area requested for the job. Only data in primary storage can be accessed by the central processor(s) and by the I/O processor(s). If data are addressed by these units which are not resident in primary storage at that time, data must be returned from primary to secondary storage and the referenced data must be loaded instead. The data are transfered in units called pages and the process has been termed paging. For optimum efficiency paging is usually controlled by computer hardware. The pages to be unloaded from primary storage are selected according to the "history" of the individual pages (aging). Therefore exceedingly large virtual storage should not be defined to avoid (programming) conventional I/O operations between memory and external storage, in particular it should not be used to avoid searching and ordering of data. Whenever it is possible to define the order in which data are needed in primary storage by the mathematical algorithm involved it is definitely preferable and more economic to search and order the data by using external storage and conventional I/O operations rather than to store these data in virtual storage. This is in particular true for small ratios of real to virtual storage.

2. PRINCIPLES OF PROGRAMMING LANGUAGES

High-level languages have been developed as a programming tool to express algorithms in a concise and machine independent form. The most commonly used language in scientific computation, FORTRAN, like many other programming languages, has been developed in the early 1950's. It strongly reflects the architecture of the machines at that time performing sequences of operations on individual data items. Programming in these languages requires the decomposition of algorithms into sequences of operations on scalar data objects. In sequential programming languages some ordering of the operations must be specified although the resulting order is often arbitrary. This overspecification required by sequential languages may prevent algorithms to execute efficiently even on conventionally organized machines. In practice, the deficiency of the existing sequential programming languages has resulted in obscure programming practices to generate code efficiently executing on different target machines. Obviously such code is not portable and its use is very much restricted to machines of similar architecture.

New architectural features require suitable algorithms and programming languages to take full advantage of the development. In the case of parallel computers the need for introducing parallelism into algorithms and programming languages is best illustrated by considering the development of code from the algorithm to the execution on a target machine. The following four steps are easily recognized in this process:

i, choosing a suitable algorithm for the problem

ii, expressing the algorithm in a high-level language

iii, compilation of the language into machine readable object code

iv, execution of the code on the target machine.

The degree of parallelism is defined as the number of independent operations that can be executed in parallel. In the ideal situation parallelism should not increase through this development process. After some thought it becomes clear that this principle of conservation of parallelism is desireable: to translate a parallel process into a sequential requires simply an arbitrary ordering of independent operations. But to translate a sequential process into a parallel process requires an analysis called vectorisation. This analysis must check that the ordering is in fact arbitrary and that no sequential dependencies exist in the process. To avoid this principally unnecessary and often complicated analysis it is necessary to preserve the parallelism of a process during development as much as possible.

In the case of parallel computers designed to work most efficiently on one- and two-dimensional list of data, called array objects, parallelism is most naturally achieved by introducing the mathematical concept of matrices and matrix operations directly into the programming language. This has first been done in a consistent form in the programming language APL in order to achieve concise expressions rather than to preserve parallelism. Any parallel language has to include syntax to specify and manipulate data objects and slices and subsets of data objects which may themselves be parallel objects. The more important techniques to define slices and subsets include the selection of rank reduced objects, the selection of ranges of values, the selection (mapping) using integer or boolean arrays and index shifting. Various languages, including some extentions to FORTRAN, have been introduced partly on an experimental level that include syntax for all or part of the specifications and options listed to preserve parallelism. Most of these languages very much reflect the architecture of the target machine.

Vectorizing compilers take a code written in a sequential language and whenever possible generate parallel instructions from sequences of sequential operations. This process can be considered as an optimisation of the sequential source code. It is clear that the most likely place to find vectorizable sequences of operations are codes specifying vector and matrix operations that is within respective calculations, in FORTRAN in DO-loops. Therefore most vectorizing compilers are restricted to an analysis of DO-loops, often only of the innermost, sometimes of more. In sequential languages the order of execution implied in DO-loops is statement by statement for each value of the loop index. For parallel processing the order required is each statement for all given index values. The transformation performed on DO-loops whenever possible and necessary for parallel processing will consist in a reordering of the DO-loops and/or of individual instructions. Some vectorizing compilers actually require the programmer to perform this reordering in the source code.

There are barriers to vectorization. Some constructs are principally not vectorizable while others are difficult for a vectorizing compiler to analyze. The most frequently used constructs that may inhibit the vectorization of repetive calculations are:

- i, conditional branches
- ii, sequential dependencies
- iii, nonlinear and indirect indexing
- iv, subroutine calls.

The portability of code between different computers requires some common standard in a programming language available on the different target machines. For the array processing features of FORTRAN, the most widely accepted scientific programming language, such a standard is not available and will most likely not become available before the late 1980's. Therefore each manufacturer has chosen his own approach most of them reflecting the architecture of the underlying target machine. This has already resulted and will certainly continue to result in a large numbers of nonportable programs which usually have to be completely redeveloped to execute efficiently on target machines with different architectural features. Therefore the problems involved in transporting programs between different parallel computers are severe and most likely will increase during the next years.

There are some approaches to the problem of program portability none of which has any great merrits: obviously standard FORTRAN may be used together with vectorizing compilers. Unless obscure constructions are used, different of course for different compilers, the code may not vectorize properly, or not at all, and the program may execute inefficiently. Alternatively, automatic code translation between different FORTRAN dialects may be used. Experience shows that usually the generated code executes inefficiently, even for similar dialects on target machines with comparable architectural features. The most pragmatic approach consists in separating the most important vectorizable instruction sequences that is vector and matrix operations, in individual subroutines or procedures, respectively, and to provide for each language dialect and/or target machines efficiently coded subroutines or procedures. This approach seems to be an acceptable technique in the absence of a FORTRAN standard although it includes some additional organisation in program development and maintenance.

3. PRINCIPLES OF SOFTWARE ENGINEERING

Vector and parallel processors having become available recently have increased the number of problems accessable to investigations in physics considerably and the impact of computers on physics in future is hardly predictable. Therefore it seems timely and relevant to reconsider the way programs are written and computers are used in physics. Powerful methods and tools have been developed in computer science to write, test and maintain software systematically, to execute programs, and to acquire, administrate, retrieve, and analyze data (results) efficiently. Some fundamental aspects of software engineering and of data processing will be presented in the following.

The software engineering process may be described by a model, the software life cycle, which consists of the following major phases:

i, requirement specification
ii, design
iii, programming
iv, verification
v, maintenance

The software engineering process is an iterative scheme with some interaction between successive development phases. The individual phases will be discussed in some detail below.

i, Requirement specification

Requirement definitions are supplied by the user of the software or are developed jointly by the designer and the user. Requirement specifications are constructed by the designer after the requirement definitions supplied by the user. The requirement specifications must contain a description of the required functions of the software. It may contain as well a specification of the host system environment and of host system components to be used. The requirement specifications must not contain any design or implementation details. In general this phase includes the development of a test plan based on the expected functions, test data and master output.

ii, Design

In the design phase the software logic and structure is developed. There are two approaches to software design called "bottom up" and "top down" design. In the bottom up approach software modules are developed before addressing the integration and interface issues. Most of the helpful techniques for software design are based on the top down approach. The top down design provides a procedure for organizing and developing the control structure of the software in a way which focuses early attention on the critical issues of modularisation, integration, and interface definition. It begins with the top level operation, or a set of it, and proceeds to develop for each operation the more basic imported (called) operations. Experience shows that some general ideas about the bottom level design proofs useful in most software projects. But it is dangerous and in general it should be avoided to mix the bottom up with the top down approach in developing operations.

It has been pointed out repeatedly that the amount of complexity that the human mind can cope with at any instant of time is considerably less than that embodied in much of the software existing and to be build. Thus the key problem in the design and implementation of large software systems is to reduce the amount of complexity or detail that must be considered at any time. One way to do this is via the process of abstraction.

One of the most significant and most widely used aid to abstraction in programming is the subroutine (or procedure, resp.), which performs a specific arbitrary abstract function by means of an unprescribed algorithm. Thus, at the level where it is invoked it separates the relevant detail of "what" from the usually unrelevant detail of "how". Similarly, on the level where it is implemented, it separates the relevant detail of "how" from the usually unrelevant detail of "why". By nesting subroutines a hierarchy of abstractions may be developed.

Unfortunately, subroutines are well suited for the description of abstract events (operations) but are not particularly well suited for the description of abstract objects. This is a serious limitation because in a great many applications the complexity of the data objects contributes substancially to the overall complexity. At most points of a program only the behavioral characteristics of the data objects are of concern. It is only of interest to know what operations can be performed on the data objects and not the way the data are physically represented and how the operations on them are implemented. If at any given level of refinement only the behavioral characteristics of certain data objects are of concern then any attempt to abstract data must be based on the behavioral characteristics and only on those characteristics. The class of objects defined by a representation independent specification has been termed "abstract data type".

There are of course many possible approaches to the specification of an abstract data type. Most of these can be placed in two categories: operational and defintional. Only the operational approach is of any practical interest. In this approach the abstract data type is described by the specification of the operations applicable to it. Operational specifications are usually relatively easy to construct. As the operations to be specified become complex, however, the operational specifications tend to become too long to permit substantial confidence in their aptness. A more serious problem in the operational approach is the over-specification of the abstraction. The extraneous detail constrains the choice of the implementation and, actually, may eliminate the best solution to the problem, and it complicates the problem of proofing the correctness.

A very efficient software modularisation scheme is based on the abstract data type concept. In this scheme operations are constructed of simpler, more fundamental operations step by step. On each level of refinement all operations accessing the same information are grouped together in what is called a module. All operations contained in a module define the abstract data type of the module. In any practical application of this scheme it is important to carry it all the way down to the most fundamental operations accessing the actual variables. Experience shows that in applying this scheme it is extremely dangerous to think in any other terms than in operations accessing information, except on the fundamental level of implementation.

iii, Programming

Various high level languages have been developed to express algorithms in a concise and machine independent form. The most well known of these languages are FORTRAN, ALGOL, PASCAL and ADA. Standards have been defined for these languages, except for PASCAL. Manufacturer supplied extentions have been introduced for most of the languages, in particular for FORTRAN. Each of these languages has its own characteristics. Actually for each software project the most suitable programming language should be carefully selected. There is general agreement that numerical algorithms expressed in FORTRAN can be executed very efficiently. This characteristic has contributed very much to the wide acceptance and use of FORTRAN for numerical applications. For expressing algorithms mostly involving non-numerical operations programming languages like ALGOL and PASCAL are definitely more suitable. There is no easy way to link code written in different high level languages. A key issue of programming is the software portability. To keep programs portable the language standards have to be strictly obeyed. Unfortunately most language standards, in particular of FORTRAN, do not contain any statements to express algorithms in a form executing efficiently on vector and array processors. Therefore sometimes it cannot be avoided to make use of manufacturer supplied language extensions to write code that executes efficiently on such processors. Code involving nonstandard statements must be kept to an absolute minimum and must be carefully isolated in order to minimize the problems involved in transporting the software to different environments. It is safe to assume that most physicist have never used any other programming language than FORTRAN and that most scientific software projects are implemented in this language. In particular FORTRAN provides only limited possiblities to control the program flow. Actually, the conditional GOTO statement is the most widely used statement to express the logic of FORTRAN programs. This lack of suitable control statements has resulted in FORTRAN code with very little structure which is difficult to analyse by inspection. The FORTRAN 77 standard finally includes a number of ALGOL-like control statements which allow to write "structured" code. Compilers for this language standard are available from most manufacturers. Therefore, if programming in FORTRAN, it is strongly suggested to write structured code using FORTRAN 77.

iv, Verification

A careful verification of each individual operation is of vital importance for the correct function of any software. All errors encountered in a later phase are difficult to trace and to correct. Various theoretical models for rigorous program verification have been described. None of these methods is really of any practical interest. Two approaches have proofed most useful for practical software verification: inspection and testing. The analysis of the program logic and the numerical expressions by inspection, preferably by other persons than the programmer, is a very effective method to detect errors in the code. The execution of operations for a selected number of parameters, called testing, is a standard method of software verification. A test plan should be developed for each individual operation based on the expected function. It is very advisable to develop the test plan before and independent of the construction and coding of the operation. This helps to avoid that the test does not check the required function of the operation but instead only the implemented function. Particular attention must be given to the choice of the parameters used in the test. Most errors encountered at a later phase usually result from an incomplete choice of test parameters.

v, Maintenance

Efficient maintenance depends heavily on a well defined modularisation and a careful documentation of the software. Maintenance includes both the correction of errors not detected in the test phase and the extention of the functions provided by the software to satisfy additional requirements that have proofed necessary or convenient. In extending the functions of the software particular attention has to be paid to the issue of integration and interfacing of the required operations. It must strictly be avoided to violate the modularisation concept of the software. Therefore it may become necessary to redesign complete modules to introduce new operations to satisfy the required functions. Experience shows that in physics it is hardly predictable which approaches are needed to model and interpret the studied physical properties and processes before certain results have become available. Therefore most scientific software is subject to continuous modification and extention and the problem of software maintenance is of particular importance.

Various methods and automated tools are available or could be built using existing technology to support the work required in the software engineering process. The set of tools available for software development are collectively called the software engineering environment. The more well known tools include text editors, compilers, assemblers, linkers, method banks and project development libraries.

The documentation plays a central role in software engineering in two ways: as a document of the software engineering process and of the software product and as an automated tool for the software engineering process. To serve as automated tool the documentation must be machine processable and must contain all information necessary for the required services.

A documentation scheme and a machine processable documentation language has been developed as a basic tool of the software engineering environment of the Physics Information Processing Environment (PIPE). Clearly a separate document based on this scheme is necessary for each individual module. This documentation includes the information summarized below.

i, Introduction
- contains a concise description of the motivation for defining the module and of the characteristics of the abstract data type

ii, Requirements
- contains the complete specification of the functions required from the module

iii, Specification
- contains a description of the user commands processed by the module, of the operations exported by the module, and of any restriction on the commands and operations

iv, Construction/Implementation
- contains a description of the construction of the exported operations

v, Verification
- contains a test plan, test input data, and master output data for each user command and exported operation

This short review of basic software engineering principles cannot be expected to prepare the reader for using the presented methods and tools correctly and efficiently. Additional reading about the subject and personal experience with the individual methods and tools will be necessary to explore their full power for software development.

SELECTED LITERATURE

Lorin, H.: "Parallelism in Hardware and Software", Prentice-Hall, Englewood Cliffs, 1972

Hockney, R.W., and Jesshope, C.R.: "Parallel Computers", Adam Hilger, Bristol, 1981

Fox, J.M.: "Software and its Development", Prentice-Hall, Englewood Cliffs, 1982

Sommerville, I.: "Software Engineering", Addison-Wesley, London, 1982

Kernighan, B.W., and Plauger, P.J.: "Software Tools", Addison-Wesley, London, 1976

Kernighan, B.W.: "The Elements of Programming Style", McGraw-Hill, New York, 1974

PARTICIPANTS

H. Abgrall
Observatoire de Meudon
LAM
F-92190 Meudon
France

H. Agren
Physical Chemistry 2
Chemical Center
P.O.B. 740
S-2207 Lund
Sweden

R. Ahlrichs
Lehrstuhl für Theoretische Chemie
Institut für Phys. Chemie und Elektrochemie
Universität Karlsruhe
Kaiserstr. 12
D-7500 Karlsruhe
Germany

K. van Alsenoy
Department of Chemistry
Universitaire Instelling Antwerpen
Universiteitsplein 1
B-2610 Wilrijk
Belgium

P. Archirel
Laboratoire de Chimie
Ecole Normale Superieure de Jeunes Filles
1 Rue Maurice-Arnoux
F-29000 Montrouge
France

R.A. Barrett
Department of Chemistry
University of York
Heslington
York, YO1 5DD
England

G.L. Bendazzoli
Instituto di Chimica Fisica
Via Risorgimento 4
I-40136 Bologna
Italy

M. Blomberg
Institute of Theoretical Physics
University of Stockholm
Vanadigsvaegen 9
S-11346 Stockholm
Sweden

R. Chapron
Centre d'etudes de Limeil
BP 27
F-94190 Villeneuve-St. Georges
France

D.L. Cooper
Harvard-Smithsonian Center for Astrophysics
60 Garden Street
Cambridge, Mass 02138
U.S.A.

N.S. Correia
Quantum Chemistry Group
Uppsala University
P.O.B. 518
S-75120 Uppsala
Sweden

M.T. Costa
Dep. Quimica
Fac. Ciencias
P-4000 Porto
Portugal

C. Daniel-Flageollet
Laboratoire de Chimie Quantique
Universite L. Pasteur
4 Rue B. Pascal
F-6700 Strasbourg
France

E.R. Davidson
Department of Chemistry
University of Washington
Seattle, Wash 98195
U.S.A.

P. Decleva
Instituto di Chimica
Pile Europa-1
I-34127 Trieste
Italy

S.A.H. Demetraki-Paleolog
Department of Chemistry
Queen Elizabeth College, University of London
Atkins Building, Campden Hill Rd., Kensington
London, W8 7AH
England

G.H.F. Diercksen
Institut für Astrophysik
Max-Planck-Institut für Physik und Astrophysik
Karl-Schwarzschild-Str. 1
D-8046 Garching bei München
Germany

S.T. Elbert
218 Office and Laboratory Building
Ames Laboratory - USAOE
Iowa State University
Ames, Iowa 50011
U.S.A.

M.O. Fanning
Chemistry Department
Thomond College
Plassey Campus
Limerick
Ireland

P.C. Ford
Chemistry Department
University of Manchester
Brunswick Street
Manchester, M13 9PL
England

P. Froelich
Quantum Chemistry Institute
Uppsala University
P.O.B. 518
S-75120 Uppsala
Sweden

N.E. Grüner
Institut für Astrophysik
Max-Planck-Institut für Physik und Astrophysik
Karl-Schwarzschild-Str. 1
D-8046 Garching bei München
Germany

T.S. Hansen
Chemistry Department
Odense University
Campusvej 55
DK-5320 Odense M
Denmark

A. Heiberg
Department of Chemistry
University of Oslo
Blindern
N-Oslo 3
Norway

T.U. Helgaker
Department of Chemistry
University of Oslo
Blindern
N-Oslo 3
Norway

J.A. Hunter
Department of Chemistry
Heriot-Watt University
Riccarton, Currie
Edinburgh, EH14 4AS
Scotland

G.J.M. Janssen
Laboratory of Chemical Physics
University of Groningen
Nyenborgh
NL-9747 AG Groningen
The Netherlands

R. Jaquet
Lehrstuhl für Theoretische Chemie
Abteilung Chemie
Ruhr-Universität Bochum
D-4630 Bochum-Querenburg
Germany

J. Kalcher
Institute of Theoretical Chemistry
Mozartgasse 14
A-8010 Graz
Austria

J.J. Kaufman
Dept. of Chemistry
The Johns Hopkins University
Baltimore, Md 21218
U.S.A.

P.J. Knowles
Univ. Chemical Laboratory
Lensfield Rd.
Cambridge, CB2 1EW
England

W.P. Kraemer
Institut für Astrophysik
Max-Planck-Institut für Physik und Astrophysik
Karl-Schwarzschild-Str. 1
D-8046 Garching bei München
Germany

P.W. Langhoff
Department of Chemistry
Indiana University
Chemistry Building
Bloomington, Ind 47405
U.S.A.

P. Lazzeretti
Instituto di Chimica Organica
Universita di Modena
Via Campi 183
I-41100 Modena
Italy

A. Metropoulos
Theor. & Phys. Inst.
Nat. Research Foundation
48 Vassileos Constantinou Ave
GR-Athens 501/1
Greece

D. Moncrieff
Department of Chemistry
Manchester University
Manchester, M13 9PL
England

W. von Niessen
Institut für Physikalische Chemie
Technische Universität Braunschweig
D-3300 Braunschweig
Germany

J.J. Novoa
Dep. Quimica Fisica
Fac. Quimica
Univ. Barcelona
Av. Diagonal 647
E-Barcelona (28)
Spain

J. Oddershede
Department of Chemistry
Odense University
Campusvej 55
DK-5320 Odense M
Denmark

Fernando R. Ornellas
Instituto de Estudos Avancados
Divisao de Fisica Teorica
Centro Tecnico Aerospacial
Sao Jose, Dos Campos
SP, 12200
Brasil

A. Palma
Via D. Alighieri
00040 Monteporzio Catone (Roma)
Italy

L. Pettersson
Institute of Theoretical Physics
University of Stockholm
Vanadisvaegen 9
S-11346 Stockholm
Sweden

J.R.M. Polder
Rijksuniversiteit Utrecht Vakgroep
Theoretische Chemie
Padualaan 8
NL-De Uithof-Utrecht
The Netherlands

M. Ramek
Institut für Physikalische
und Theoretische Chemie
Technische Universität Graz
Rechbauerstraße 12
A-8010 Graz
Austria

A.E. Reed
Theoretical Chemistry Inst.
University of Wisconsin
1101 University Avenue
Madison, Wis 53706
U.S.A.

S. Rettrup
Chemistry Department 4
H.C. Orsted Institut
Copenhagen University
Universitetsparken 5
DK-2100 Copenhagen
Denmark

I. Roeggen
Inst. Math. Phys. Sci.
University of Tromso
P.O. Box 953
N-9022 Tromso
Norway

B. Roos
Chemical Center
Physical Chemistry 2
P.O.B. 740
S-22007 Lund 7
Sweden

J.R. Sabin
Department of Chemistry
Odense University
Campusvej 55
DK-5320 Odense M
Denmark

A.J. Sadlej
Chemical Center
Physical Chemistry 2
P.O.B. 740
S-22007 Lund 7
Sweden

E. Sangfelt
Quantum Chemistry Group
Uppsala University
P.O.B. 518
S-75120 Uppsala
Sweden

V.R. Saunders
Daresbury Laboratory
Science Research Council
Daresbury
Warrington, WA4 4AD
England

W.D. Sepp
Physics Department
University of Kassel
Heinrich-Plett-Str. 40
D-3500 Kassel
Germany

G.J. Sexton
Dept. of Theor. Chem.
Univ. Chemical. Labs.
Lensfield Rd.
Cambridge, CB2 1EW
England

P. Siegbahn
Institute of Theoretical Physics
University of Stockholm
Vanadigsvaegen 9
S-11346 Stockholm
Sweden

E. Otto Steinborn
Institut für physikalische
und theoretische Chemie
Universität Regensburg
Universitätsstr. 31
Postfach
D-8400 Regensburg
Germany

J. Steuerwald
Institut für Astrophysik
Max-Planck-Institut für Physik und Astrophysik
Karl-Schwarzschild-Str. 1
D-8046 Garching bei München
Germany

A. Strich
Laboratoire de Chimie Quantique
Universite L. Pasteur
4 Rue B. Pascal
F-6700 Strasbourg
France

B.T. Sutcliffe
Department of Chemistry
University of York
Heslington
York, YO1 5DD
England

B. Themans
Laboratoire Chimie
Theorique Appliquee
Facultes Universitaires
N.D. de la Paix
61 Rue de Bruxelles
B-5000 Namur
Belgium

P. Tomasello
Instituto di Fisica
Dell'Universita
Corso Italia 57
I-95129 Catania
Italy

F. Visser
Instituut voor Theoretische Chemie
Toernooiveld
NL-Nijmegen
The Netherlands

T.M. Wecker
Inst. of Organic Chemistry
University of Munich
Karlstraße 23
D-8000 München 2
Germany

U. Wedig
Institut für Theoretische Chemie
Pfaffenwuldring 55
D-7000 Stuttgart 80
Germany

E.J. Weniger
Institut für Physikalische
und Theoretische Chemie
Universität Regensburg
D-8400 Regensburg
Germany

K.B. Wenzel
Jathoweg 22
D-1000 Berlin 26
Germany

S. Wilson
Theoretical Chemistry Dept.
University of Oxford
1 South Parks Road
Oxford, OX1 3TG
England

R. Zanasi
Instituto di Chimica Organica
Universita di Modena
Via Campi 183
I-41100 Modena
Italy

J.P. Zwart
Gorlaeus Laboratoria
Wassenaarseweg 76
P.O. Box 9502
NL-2300 RA Leiden
The Netherlands

INDEX

GPSR Compliance

The European Union's (EU) General Product Safety Regulation (GPSR) is a set of rules that requires consumer products to be safe and our obligations to ensure this.

If you have any concerns about our products, you can contact us on ProductSafety@springernature.com

In case Publisher is established outside the EU, the EU authorized representative is:

Springer Nature Customer Service Center GmbH
Europaplatz 3
69115 Heidelberg, Germany

Batch number: 10815974

Printed by Printforce, the Netherlands